The New American Cinema

Edited by **Jon Lewis**

The New American Cinema

Duke University Press Durham & London 1998

Third printing, 1999

© 1998 Duke University Press   All rights reserved

Printed in the United States of America on acid-free paper ∞

Typeset in Dante with Franklin Gothic display by Keystone Typesetting, Inc.

Library of Congress Cataloging-in-Publication Data appear on the last

printed page of this book.

for Q, little a and Big G,

and (in loving memory) Richard deCordova

and Donie Durieu

Contents

Acknowledgments

Big thanks: to Dana Polan, Eric Smoodin, Richard deCordova (whom I miss every day), Eric Zinner, and Ken Wissoker for the encouragement and smart advice; to the contributors one and all for keeping (pretty close) to deadlines and for taking all those comments in the margins in the spirit they were given; and to Martha and the boys, Guy and Adam, for the love and stuff at home.

Introduction

The public exhibition of motion pictures has just entered its second century here in the United States. This celebration of film's centenary is at once a timely and a particularly tenuous moment to lay claim to a new American cinema. But here goes.

I first began using the title "The New American Cinema" for a course I introduced back in 1983. At the time I was interested in the apparent decline of one new American cinema—the so-called auteur renaissance that took shape after the studios adopted the MPAA Ratings System in 1968—and the emergence, around 1980, of another new American cinema that seemed largely a consequence of a new conglomerate, multinational Hollywood.

I have taught a course on this new American cinema almost every year since then, although these days the word "new" in the title gets me into trouble. What was new in 1983, or even in 1993, is hardly so today. Soon the term "cinema" will be just as problematic. Scholars, reviewers, and filmmakers have been hawking "the end of cinema" (as we know it) for years now, and soon enough they may finally get their wish. Someday, fairly soon, I will find myself in front of a class talking nostalgically about movies as some sort of crude artifact screened once upon a time in public theaters nationwide.

In the 1990s, as I persist in teaching and writing about contemporary American film, I find that I spend a lot less time than I used to lamenting the end of the seventies. Instead, these days, in class and in print, I tend to talk and write more expansively, less nostalgically, about the much-lauded golden age of the 1970s and less judgmentally about the new American cinema that took shape after 1980, after the release of *Apocalypse Now* (1979) and *Heaven's Gate* (1981), after the pivotal Kirk Kerkorian antitrust decision, the failed Screen Actors Guild strike, the deal between Sony and the studios over the commercial launch of the VCR. Lately, I have found it necessary to stop focusing on *important* films and filmmakers as if they matter much anymore. Instead, I have begun to look more seriously at the ways in which the movie and money industries overlap and the ways in which the business of making movies now informs, underscores, and underwrites what we see and hear and do with the films in our lives.

The contributors to this book examine a number of new American cinemas—the two I suggest above and a number of other incarnations found within the commercial mainstream and, as well, on the margins of Hollywood. The contributors do not all share my interest in money matters. But for the most part, they share my willingness, my anxiousness, to take a fairly large, interdisciplinary view of filmmaking and filmgoing in the United States as it has evolved over the past few decades.

What follows is a collection of solicited essays by scholars interested in, but drawn to, very different arguments about contemporary film in the United States. The result is panoramic, interdisciplinary, speculative, and, I hope, provocative. No implicit or explicit argument emerges that speaks to the entirety, the enormity, of the subject at hand. Instead, this book contents itself with offering a showcase to a series of different critical and historical perspectives, all of them attending one or another aspect of contemporary American film.

The Essays

The essays that make up this collection are presented here under three section headings: "Movies and Money," "Cinema and Culture," "Independents and Independence." The titles of the three sections and the order of the essays no doubt imply a larger argument about contemporary American film, one that moves from economic issues to more textual concerns, and from the mainstream to the more marginal and marginalized. I have toyed with a number of alternative groupings, chronologies, thematics, all of which still seem reasonable to me. This juggling of possibilities I take as an indication that the book is flexible enough to be read and taught and studied in a variety of ways.

As to the table of contents, I tried to respond to the ways in which the essays seemed to gravitate toward certain key issues: the industrial subculture that supports mainstream production and distribution, the political and cultural significance of (mostly studio-produced) American movies, the nature and function of American films produced and/or exhibited outside the commercial mainstream. Yet I remain aware that these are hardly the only three areas worth discussing; indeed, these are not the only three categories into which these essays might otherwise fit.

The first two essays in the collection, David A. Cook's "Auteur Cinema and the Film Generation in the 1970s" and Timothy Corrigan's "Auteurs and the New Hollywood," contextualize American cinema's golden age with regard to what Corrigan terms "the commerce of auteurism," the notion that late sixties' and seventies' American cinema has at its root an economic imperative. Both essays examine the key critical and industrial debate attending the auteur theory, the notion that the director is, or at least should be, considered the sole author of a film. As both Cook and Corrigan point out, because cinema is an industrial art, the literary-historical concept of authorship and ownership promoted by the auteur theory at once conflates and confuses issues of art and commerce that are essential when one considers studio-produced movies. In the final analysis, both Cook and Corrigan argue, it is impossible to talk about movies, even outstanding, original, auteur movies like *The Godfather* (1972) or *Taxi Driver* (1975), without first accounting for the industrial conditions that enabled or disabled their production and release.

Justin Wyatt's study of film distribution strategies, "From Roadshowing to Saturation Release: Majors, Independents, and Marketing/Distribution Innovations," also focuses on the complex relationship between the aesthetic and the economic in Hollywood. Wyatt claims that significant aesthetic changes—changes in visual style or narrative—more often than not reveal and occur in response to new studio marketing ideas. By focusing on distribution strategies, Wyatt ends up talking about movies in much the same way as studio executives do—as products of an industrial process and only secondarily as works of art or entertainment.

In "Money Matters: Hollywood in the Corporate Era," I take the approach taken in the first three essays to its logical extreme in proposing that, especially today, one can write about film history without ever stopping to talk about specific movies. The principal historical argument posed by the essay is that the new Hollywood has taken shape as a result of a series of critical confrontations between federal regulatory agencies and the Hollywood studios, confrontations that, since 1980, have been settled very much in the studios' favor. The megamergers and acquisitions that now characterize the entertainment business—Time Warner Turner, Disney/Capital Cities/ABC, Viacom/Blockbuster/Paramount, for example—reveal the extent to which antitrust regulation has become irrelevant or at least unenforceable in the new Hollywood. These new entertainment trusts foreshadow and foreground an entertainment industry that may someday be (or maybe already is) controlled by two or three highly diversified companies that—given the extent and range of their holdings—may well have and be able to implement a cultural or political agenda.

The second section of the book features cultural studies essays, the first three of which examine how films depict gender roles and relationships and the way such roles and relationships are understood, perhaps even reproduced, in the larger culture. All three essays focus on mainstream genre pictures—for Tania Modleski, war and Vietnam films; for Fred Pfeil, action-adventure/male rampage films; for Sabrina Barton, the psycho-thriller—and all three ask: what do women mean in these films? what do these films mean to women? what do men mean in these films? what do these films mean to men?

In "A Rose Is a Rose?: Real Women and a Lost War," Tania Modleski focuses on cinema's progressive potential, asking specifically whether or not a film about war—moreover, a visceral, thrilling film about war like *Platoon* (1986), *Born on the Fourth of July* (1989), or *Apocalypse Now*—could ever be convincingly and progressively antiwar. In broaching such a complex question, Modleski offers for consideration Nancy Savoca's *Dogfight* (1991), a film that suggests the potential for some future popular, progressive, even feminist cinema, one that might someday transcend the very genre restrictions that make the male-themed and male-targeted movies today so financially successful and so politically unsatisfying.

Modleski's focus on Vietnam is instructive in that the war is the most common backstory in contemporary American cinema. As Fred Pfeil points out in "From Pillar to Postmodern: Race, Class, and Gender in the Male Rampage Film," the straight, white male hero of so many action-adventure pictures (the *Rambo* and *Lethal Weapon* movies, for example) has his roots in some past and most likely awful experience in Vietnam. Reading through a wide range of contemporary American films—films rooted in but not limited to the action-adventure genre—Pfeil explores how these popular pictures reflect and refract "real" domestic and global political issues and do so in terms that embrace and perpetuate the politics of the Reagan revolution—an anti-big government sentiment, a certain style of machismo (an embattled straight, white masculinity), uncritical notions of right and wrong, idealized notions of America (at Christmastime!).

In "Your Self Storage: Female Investigation and Male Performativity in the Woman's Psychothriller," Sabrina Barton shares Modleski's interest in the possibility of a progressive American cinema—especially one that is progressive with regard to the role of women in contemporary U.S. culture—and she finds such a body of work in the most unlikely of places, the psychothriller. Despite the fact that these films—*Sleeping with the Enemy* (1991), *The Stepfather* (1987), *The Silence of the Lambs* (1991)—often feature protracted scenes of physical and psychical violence against women, Barton argues that they nevertheless provide for female *action*, a small window of hope that may well be for women viewers of mainstream American film as progressive as it gets these days.

The final two essays in the "Cinema and Culture" section deal with movies and history. In "Conspiracy Theory and Political Murder in America: Oliver Stone's *JFK* and the Facts of the Matter," Christopher Sharrett offers an ideological study of a self-consciously ideological motion picture. The complex structure of Stone's film, Sharrett suggests, functions as an apt correlate for the problematics of re-presenting history in postmodern America; so much of our history is on film, yet we cannot, we should not, believe everything that we see. The specific ways in which *JFK* has come to stand in for history—and the ways in which Stone has been vilified in the popular press for daring to tell such a history on-screen—says less about Stone's intent or ability or good citizenship than it does about the complex set of relationships at work and at stake between the American political system and the various media that report on it.

Focusing on the period's most persistent genre, science fiction, Scott Bukatman offers another sort of historical argument and makes another set of historical claims about popular cinema. In "Zooming Out: The End of Offscreen Space," Bukatman examines contemporary cinematic special effects and other spectacular visual media by tracing their roots back to turn-of-the-century amusement parks and early silent pictures. In doing so, he observes changes in these media as they regard America's transition from the Industrial Age to the Information Age, and he implicitly argues that in the process of tracking the history of a single film genre or media phenomenon, one also tracks the trajectory of a larger American cultural history.

The book's final section focuses on films produced primarily outside the studio system, movies at once marginal to and marginalized by Hollywood. Although this alternative cinema could well support a book-length study on its own, its inclusion here reminds us that cinema in the United States is more than just Hollywood, more than just the films that play at the local multiplex.

The roots of this alternative, independent cinema—and of Hollywood's various accommodations and assimilations of popular alternative work—can be found in a variety of places, from a cinematic avant-garde that dates from the turn of the century, to 1950s' and 1960s' B movies. And its practice at present includes everything from esoteric, autobiographical films made by and for participants in specific American subcultures to straight-to-video genre and exploitation pictures.

In "John Cassavetes: Amateur Director," Ivone Margulies examines the work of a fiercely independent American filmmaker who exerted a signifi-cant influence on a range of new American independents as well as on several of the key figures in 1970s' mainstream cinema. As Margulies argues, Cassavetes was *the* key figure in a New York-based alternative to Hollywood, and his films evince a number of important differences be-tween the independent project in general and the sorts of pictures made in and by the studios: theater vs. film, location vs. the studio, improvisation vs. studio / star acting, character vs. narrative, low production values with roots in low-cost realist cinema vs. the high production values that in and of themselves depend on and answer to studio money and technical expertise.

Chuck Kleinhans's essay, "Independent Features: Hopes and Dreams," focuses on the economics of the independent film and video industries, and in so doing he provides something of a primer on the practicalities and impracticalities of indie production. While examining the filmmaking aspiration—the American dream of someday writing and directing one's own films (a dream that for the present generation of high school and college students has supplanted the desire to write the great American novel)—Kleinhans cautions that independent filmmaking is hardly a ro-mantic, let alone a lucrative, enterprise.

In "A Circus of Dreams and Lies: The Black Film Wave at Middle Age," Ed Guerrero explores the complexity of industrial / cultural assimilation. Guerrero concedes that since the mid-1980s a number of African Ameri-can filmmakers have found opportunity in Hollywood. But, he argues, these filmmakers—who are given budgets significantly tighter than white filmmakers and are held to different, often unreasonable box-office expec-tations—continue to occupy a position at the periphery of the mainstream industry. Their assimilation into studio Hollywood has been accompanied by an institutionalized marginality. Additionally, he shares with Fred Pfeil a concern about popular black-white buddy films, like *Lethal Weapon* (1987) or *Pulp Fiction* (1994), that feature African American actors in prominent roles but nonetheless evince the tangled workings of race in Hollywood.

In "Culture as Fiction: The Ethnographic Impulse in the Films of Peggy Ahwesh, Su Friedrich, and Leslie Thornton," Catherine Russell focuses on three women who make movies far off-Hollywood. Financed on arts grants and produced at a fraction of the cost of even other independent features, the films made by these three women provide an alternative to the seamless Hollywood style by systematically alluding to a range of realist and ethnographic films released in the 1960s and 1970s. Not only do these films question the truth claims made in both documentary and fiction films, they ably reveal how the medium can and might someday be more extensively used as a form of personal and political expression.

The collection begins with essays about commercially successful and artistically interesting movies and ends with a look at a persistent alternative to that work that has emerged from outside the studio system. Such a critical / historical trajectory suggests that one cannot discuss alternative cinema without first examining the mainstream tradition. The terms *alternative* and *independent* are not only relative, they are only partially accurate. A lot more work needs to be done on this aspect of the new American cinema—much more than space or time here allows.

The final four essays in the collection remind us—significantly, I think—that American cinema in any of its various forms and traditions is an odd mix of industrial process and artistic inspiration. That we have come to talk about independent film as a genre, that so many of us have agreed more or less to talk about independence as an ideal in the process of making movies in America, evinces just how the duality of art and commerce is weighted and has been weighted in these last days of the twentieth century.

The New American Cinema

As the summary above suggests, this collection is designed to provide not only a survey of an especially interesting period in American film, but also an introduction to—and a showcase for—a range of scholarly projects. A number of new American cinemas have taken shape in the past few decades, and as the contributors to this volume suggest, there are a number of useful, provocative, instructive ways to talk about them.

At the risk of concluding on a self-critical note, the project here is, alas, complicated not only by the range of films, the changes in the industry, the seismic shifts in the American zeitgeist that feeds and feeds off contemporary motion pictures in the United States, but by the difficulty, in any form of art criticism and history, of discussing *the contemporary* with any real accuracy, any real conviction. Such is also the hook here, the very reason why we are so interested in these films, in the industry and mass culture as it is right now, today.

For better or worse, these are the movies in our lives. This book presents a number of ways we might begin to understand them.

Jon Lewis
September 1997

Movies and Money

Auteur Cinema and

the "Film Generation" in 1970s Hollywood

David A. Cook

The concept of authorship in the cinema is nearly as old as the medium itself, but it was first formally articulated by François Truffaut in his 1954 *Cahiers du cinéma* essay, "Une certaine tendance du cinéma français." There, he spoke of the *politique des auteurs* ("the policy of authors") whereby film should ideally be a means of personal artistic expression for its director (or director-writer, as he originally intended it), bearing the signature of his or her personal style, rather than being the work of some corporate collective. The definition naturally privileged Hollywood film-makers like Welles, Hitchcock, Hawks, and Ford, who had worked within the studio system but transcended it to achieve a cinema of personal vision. The idea was not imported into American critical discourse until the 1960s, however, when Andrew Sarris christened it "the auteur theory" in an essay in *Film Culture*[1] and began to construct the pantheon of Ameri-can directors that became the subject of *The American Cinema: Directors and Directions, 1929–1968.*[2] By the time this influential volume had ap-peared, even critics like Pauline Kael, who were initially hostile to the idea (see her "Circles and Squares"),[3] had begun to accept its fundamental premise, if only by inverse corollary (e.g., in her 1971 essay, "Raising Kane,"[4] Kael went to inordinate lengths to try to demonstrate that script-writer Herman Mankiewicz, not Orson Welles, was the principal author of *Citizen Kane*).

In popular terms, authorship became associated with the work of the New American cinema announced in 1967 by Arthur Penn's *Bonnie and Clyde.* This sensational film—heavily influenced by the *Cahiers*-inspired

French New Wave (in fact, it was originally to have been directed by Truffaut, and, after he proved unavailable, by Jean-Luc Godard)—took both critics and industry by surprise in its revolutionary mixing of genres and styles and its unprecedented violence. When it was released in August, the film was universally denounced by old-line critics, but by November its immense popularity had prompted a second look and forced many retractions, including unprecedented ones by both *Newsweek* and *Time,* which ran a cover story on *Bonnie and Clyde* as an avatar of an American New Wave ("The New Cinema: Violence . . . Sex . . . Art . . ."). Ultimately, the film was nominated for ten Academy Awards and received two (best cinematography: Burnett Guffey; best supporting actress: Estelle Parsons) and won the New York Film Critics' Award for best screenplay (David Newman and Robert Benton) and best film of 1967. It also became one of the most profitable films of 1967–68, returning nearly $22.8 million in theatrical rentals on an investment of $2.5 million. Significantly, the highest-grossing film of 1968 was another European-influenced, youth-oriented vehicle—Mike Nichols's *The Graduate,* released in December 1967—which earned $39 million to become the box-office champ of the decade and won Nichols the 1967 Oscar for best director. What the success of both films demonstrated was that a large new American audience existed for the kind of films that the *Cahiers* critics had been writing about (and subsequently making, as independent directors at the margins of the French film industry)—films that were visually arresting, thematically challenging, and stylistically individualized. Demographically, this U.S. audience was younger, better-educated, and more affluent than Hollywood's traditional audience. It had grown up with the medium of television, learning to process the audiovisual language of film on a daily basis.

Bonnie and Clyde was produced by its 29-year-old star Warren Beatty for Warner Bros., and *The Graduate* was directed by the 34-year-old Nichols, inspiring several studios to hire younger, nontraditional producers and directors to appeal to a younger clientele. This strategy met with some success, and 1968 and 1969 witnessed studio productions like Stanley Kubrick's *2001: A Space Odyssey* (Warner Bros., 1968), Arthur Penn's *Alice's Restaurant* (UA, 1969), and Sam Peckinpah's *The Wild Bunch* (Warner Bros., 1969), whose uniqueness contributed to the auteurist stature of their directors and whose popularity confirmed the box-office power of the new audience. But the studios also continued to produce a large number of expensive flops for the old audience, such as Fox's *Star!* (Robert Wise, 1968) and *Hello, Dolly!* (William Wyler, 1969) and Paramount's *Paint Your Wagon* (Josh Logan, 1969), at a time when structural change and financial crisis were taking place. As a direct result of overproduction during the sixties, all of the studios but Twentieth Century-Fox, Columbia, and Disney had been taken over by conglomerates by the end of the decade. Beginning with the purchase of Universal by MCA in 1962, extending through Paramount's sale to Gulf & Western Industries in 1966, the acquisition of United Artists by Transamerica Corporation in 1967, and the sale of Warner Bros. to Kinney National Service Corporation and of MGM to Las Vegas financier Kirk Kerkorian in 1969, the studios were one-by-one absorbed by larger, more diversified companies.

In this context, the runaway success of the generationally savvy road film *Easy Rider* (Dennis Hopper, 1969)—produced by the independent BBS Productions for $375,000 and returning $19 million—convinced producers that inexpensive films could be made specifically for the youth market and that they could become blockbusters overnight. This delusion led to a spate of low-budget "youth culture" movies and the founding of many short-lived independent companies modeled on BBS. But it also drove the studios to actively recruit a new generation of writers, producers, and directors from the ranks of film schools like USC, UCLA, and NYU where the auteur theory had become institutionalized as part of the curriculum. (As Martin Scorsese, one of the most successful new directors would later remark of this era, "Sarris and the 'politique des auteurs' was like some fresh air.")[5] Reaching out to the youth market in the late sixties was not enough to prevent an industry-wide recession from 1969 to 1971, which produced $500 million in losses for the majors, and by 1970 left 38 percent of Hollywood filmmakers unemployed. But it did substantially help to create the "Hollywood Renaissance" of the 1970s in which, as Michael Pye and Lynda Myles have put it, the "film generation took over Hollywood"[6] and attempted to create an American auteur cinema based in large part on the European model.

Whereas the generation of directors who descended from classical Hollywood came largely from Broadway and the theater, and the recruits of the fifties and sixties were trained in television, many new directors of the seventies had studied *film as film* in university graduate programs and professional schools. They had taken film history, aesthetics, and production as formal academic subjects, and they had learned the technical aspects of production, as well as budgeting and marketing, more thoroughly than any generation before them. Many of them also had apprenticed with producer-director Roger Corman at American International Pictures (AIP), where low-budget, youth-oriented genre films were the stock-in-trade. (As a creative exploitation producer in the sixties, Corman scoured the Los Angeles film schools for local talent, like director Francis Ford Coppola, writers Willard Huyck and John Milius, and producer Gary Kurtz, all of whom were eager to work for nonunion wages.) When they began to enter the industry at the turn of the decade, average weekly film attendance was approaching an all-time low (it bottomed out at 16 million tickets sold in 1971). By 1975, admissions had recovered, and box-office grosses increased by $150 million; the young auteurs of the "New Hollywood," as it was being called, were leading the major studios into the 1980s in relative prosperity. But this success depended largely on the strength of several expensively produced blockbusters that reaped windfall profits—*The Godfather* (Coppola, 1972), *Jaws* (Steven Spielberg, 1975), *Star Wars* (George Lucas, 1977), *Close Encounters of the Third Kind* (Spielberg, 1977)—the ultimate effect of whose success was to multiply the average negative cost of the films fivefold during the decade (from $2 million in 1972 to $10 million in 1980) and make directors increasingly dependent on studio financing through commercial banks for production and distribution. This "blockbuster syndrome" caused a trend toward the production of fewer and fewer films, with an attendant increase in advertising and marketing budgets to insure a film's success. Ironically, the decade in which "the film generation took over Hollywood" produced fewer feature films than any before it and witnessed the historical shift whereby the cost of promoting a film actually exceeded the cost of producing it, often by twice the negative cost.

The auteur directors of the 1970s—aka the "whiz kids," "*wunderkinder*," or "Hollywood brats"—entered the industry at a time when, as Thomas Schatz has noted, Hollywood was desperately searching for its bearings.[7] What they offered was fresh creative talent and adaptability to a system in the midst of radical change. Their youth guaranteed their ability to address the new audiovisual sensibility of an audience, like themselves, that had grown up watching television. And, as far as the studios were concerned, their relative inexperience as first- and second-time directors meant that they could be hired for much less than established talent. Because of their training, they knew more—conceptually, at least—about the production and marketing of motion pictures than the nonfilm executives, like Charles Bludhorn (CEO of Gulf & Western) and Ted Ashley (executive vice president of Kinney / National Services), who had recently taken over the industry, and this academically certified expertise was attractive to new managers who themselves were largely ignorant of the film business. Francis Ford Coppola (b. 1939) was the vanguard figure of this new breed of directors, and he quickly became the mentor of a generation of exceptional filmmakers, whose other prominent members included George Lucas (b. 1944), Steven Spielberg (b. 1947), Martin Scorsese (b. 1942), and Brian De Palma (b. 1940). At one remove from these filmmakers stand the screenwriters and sometime directors John Milius (b. 1944) and Paul Schrader (b. 1946).

Coppola received his undergraduate degree in drama at Hofstra College (now University) in 1960 and then enrolled in the graduate film program at UCLA. There, he began to work on the fringes of the industry, doing uncredited second-unit direction for Corman and eventually directing *Dementia 13* (1963), a horror film shot in Ireland in three days with the cast and crew left over from Corman's *The Young Racers* (1963). After collaborating on several major scripts both with credit (*Is Paris Burning?*, directed by René Clément, 1966; *This Property Is Condemned,* Sydney Pollack, 1966) and without credit (*Reflections in a Golden Eye,* John Huston, 1967), Coppola was able to adapt and direct *You're a Big Boy Now* (1967) from the comic novel by British writer David Benedictus. Produced by Phil Feldman for the newly merged Warner Bros.-Seven Arts for $800,000, the film was heavily influenced by the style of Richard Lester's *A Hard Day's Night* (1964) and the French New Wave, and Coppola submitted it to UCLA as his master's thesis. Although it did not break even until its sale to television, *You're a Big Boy Now* attracted enthusiastic reviews and impressed Warners sufficiently for the studio to sign Coppola to direct the $3.5 million musical, *Finian's Rainbow* (1968), starring Fred Astaire and Petula Clark. Based on a twenty-year-old Broadway hit that satirized racial prejudice in the South, the film was ultimately a box-office disappointment, but Coppola had done an extraordinary job of giving it the feel of a big-budget spectacle, and Warner Bros. blew the film up to 70mm for roadshowing.[8] Before the failure (a relative term in these times—the film eventually earned $5.5 million) of *Finian's Rainbow* became clear, the studio staked Coppola $750,000 for a small personal project, *The Rain People* (1969), based on his own script about a pregnant Long Island housewife who leaves her husband to go on an odyssey across America. Experimental in form, the film has been hailed as a feminist document before its time; many critics in 1969 applauded the distinctly individualistic nature of the film's direction, and *The Rain People* won first prize at the San Sebastian Film Festival. Meanwhile, the *Easy Rider* panic had struck at Warner Bros., and Coppola convinced its executives to bankroll his own small studio in San Francisco, American Zoetrope, to develop films for the youth market. (As Jon Lewis notes, the deal was relatively simple: Warners put up $600,000 in exchange for the right of first refusal to any and all American Zoetrope projects.)[9]

American Zoetrope (renamed Omni-Zoetrope in 1979 and Zoetrope Studios in 1980), in which Coppola was the only shareholder, was deliberately modeled on the Roger Corman unit at AIP.[10] Within a year of its opening, Coppola had produced his friend George Lucas's first feature, *THX-1138* (1971), sponsored John Milius's first script for *Apocalypse Now,* and become a guiding light of his generation. But Warner Bros. hated the rough cut of the Lucas film and demanded repayment of its investment, thus pushing Zoetrope and Coppola close to bankruptcy. To make ends meet, Coppola coscripted Fox's megahit *Patton* (1970), sharing with Edmund H. North that year's Academy Award for best original screenplay, and then he accepted a job directing a project for Paramount based on Mario Puzo's *The Godfather,* a best-selling novel about the Mafia. The studio had conceived *The Godfather* as a mainstream genre film with blockbuster potential on a modest budget ($6 million), and Coppola was chosen after at least two other directors (Constantin Costa-Gavras and Peter Yates) had turned it down. Coppola collaborated on a screenplay with Puzo, shot the film on location in New York City in the mid-winter and spring, and brought the project in for a scant $1 million over budget, an impressive achievement given the richly textured results. Assisted by a Paramount advertising blitz and saturation booking, *The Godfather* (1972) became the first great blockbuster of the 1970s, earning $85.7 million to become the fifth highest-grossing film of the decade (it was briefly the highest-grossing film in history and in 1997 still ranked among the top twenty-five).

Coppola's success in turning a studio-produced gangster film into an epic saga of a (crime) family, full of operatic intensity, won Oscars for best film, best screenplay, and best actor (for Marlon Brando's performance in the title role), as well as Academy Award nominations for direction, three supporting actor performances, sound, editing, and costumes. *The Godfather* made Coppola a power in Hollywood, and he used his leverage immediately to produce George Lucas's coming-of-age mosaic, *American Graffiti* (1973), and to direct from his own screenplay *The Conversation* (1974), a brilliant meditation on electronic surveillance and paranoia stylistically indebted to the European art film, most immediately Michelangelo Antonioni's *Blow-Up* (1966). *The Conversation* won the Palme d'or at Cannes, as well as Oscar nominations for best picture, best original screenplay, and sound, but it could not compete at home with Coppola's own remarkable sequel, *The Godfather, Part II* (1974), which won the 1974 best picture award, as well as Oscars for direction, best supporting actor (Robert De Niro), screenplay from another medium (Coppola and Puzo), art direction, and musical score. Constructed as a contrapuntal movement between the Corleone family's noble, if violent, past in the early years of the century and its presently corrupt involvement in Batista's Cuba and Las Vegas in the fifties, *The Godfather, Part II* both continues and enriches the original, portraying an America in which legitimate and illegitimate power are closely intertwined (and suggesting obliquely, in its final moments, the mob's involvement in the assassination of President John F. Kennedy). Produced for $13 million, *The Godfather, Part II* grossed less than half as much as *The Godfather,* but it brought Coppola to the pinnacle of his influence. He had directed two of the most financially successful films in industry history up to that time, both of them hailed critically as major contributions to American cinema; he personally had won five Oscars, and he had become the only director ever to be nominated for two best picture and best screenplay awards in the same year; he had founded his own studio and become a beacon to a whole generation of American filmmakers. (For the record, it should be noted that he also had written—in three weeks—the screenplay for Paramount's disastrous *The Great Gatsby,* directed by Jack Clayton, 1974). George Lucas would say of him: "Francis was the great white knight who made it,"[11] and John Milius was even more direct: "He subsidized us all. . . . If this generation is to change American cinema, he is to be given the credit, or the discredit. Whichever it may be. . . ."[12] It was in this spirit near the decade's end that Coppola produced Carroll Ballard's *The Black Stallion* (1979), coproduced Akira Kurosawa's *Kagemusha* (1980), and distributed the restored version of Abel Gance's silent epic *Napoléon* (1927; 1980) under the Zoetrope banner. He also continued to produce, distribute, and otherwise promote the work of a remarkably eclectic group of filmmakers, including Wim Wenders, Hans-Jürgen Syberberg, Jean-Luc Godard, and Paul Schrader, until the studio was sold in 1984.

Coppola's last work as an auteur in the seventies was the legendary Vietnam war film *Apocalypse Now* (1979), which he also coproduced and coscripted (with John Milius and Michael Herr). Budgeted at $12 million and shot on location in the Philippines, the production was plagued by illness, natural disasters, and other logistical problems, which nearly tripled its costs and resulted in a flawed, if brilliant film. *Apocalypse Now* is a version of Conrad's *Heart of Darkness* transposed to the hallucinated, horrific landscape of post-Tet Vietnam, whose first two-thirds may be one of

1 The last great auteur picture: Francis Coppola's *Apocalypse Now* (United Artists, 1979).

the greatest war / antiwar films ever made, but whose conclusion bogs down in a morass of pomposity and metaphysics. Widely viewed as an act of hubris or folly, the film permanently damaged Coppola's position within the industry, although it earned $40 million; it was nominated for eight Academy Awards and won two—cinematography (Vittorio Storaro) and sound (Walter Murch et al.), and it shared the Palme d'or at Cannes with Volker Schlöndorff's *The Tin Drum* (1979). Unfortunately, Coppola followed *Apocalypse Now* with *One from the Heart* (1982), an expensively stylized musical set in Las Vegas that employed an experimental production technique called "electronic cinema"—basically a method of "previsualizing" the film on video so that it could actually be edited before it was shot.[13] Although Coppola claimed that the electronic cinema method cut production time and therefore costs, *One from the Heart* cost $27 million and lost most of it at the box office, forcing Coppola to sell Zoetrope Studios in 1984 and leaving him financially at sea for the rest of the eighties.

Before that happened, however, Coppola shot two last Zoetrope films on location in Tulsa, Oklahoma, both adapted from the youth novels of S. E. Hinton. *The Outsiders* (1983) was a film of teenage rebellion thematically and formally indebted to Nicholas Ray's *Rebel Without a Cause* (1955), whereas *Rumble Fish* (1983) was a highly allusive family melodrama shot in a black-and-white style deliberately evocative of German Expressionism. *The Outsiders* was a box-office success and propelled Coppola into a deal to direct *The Cotton Club* (1984), a Robert Evans production, for release through Orion. A big-budget gangster film (projected at $20 million; final cost $47 million) set in Harlem during the 1920s and featuring the music of Duke Ellington, *The Cotton Club* received mixed reviews and broke even at the box office. Faced with enormous debt, Coppola directed two films in rapid succession for Tri-Star, *Peggy Sue Got Married* (1986), a thoughtful treatment of the previous year's *Back to the Future* formula, which became his highest-grossing film of the decade, and *Gardens of Stone* (1987), a melancholy account of a Vietnam-era military burial unit at Arlington National Cemetery, which was, almost inevitably, a commercial failure. *Tucker: The Man and His Dream* (1988), produced with George Lucas's backing for Paramount, fared little better at the box office but was perceived by critics as a major comeback for Coppola as auteur. Based on the true story of Preston Tucker, the visionary entrepreneur who successfully fought postwar Detroit to manufacture what he called "the first completely new car in 50 years" but was ruined by monopolistic collusion before he could market it, Coppola's film was crafted as both roman à clef and homage to Orson Welles's *Citizen Kane* (1941). It was followed by the successful second sequel, *The Godfather, Part III* (1990), which Coppola wrote with Puzo for Paramount to continue the Corleone saga into the present. Produced for $57 million and earning $70 million, the film received Academy Award nominations for best picture, best director, and best supporting actor. But Coppola's true redemption as far as Hollywood was concerned came with *Bram Stoker's Dracula* (1992), produced on time and just slightly under budget for $50 million for Columbia. The film grossed $200 million worldwide in its first four months. For all of its contemporary music-video patina, *Dracula* is a knowing compendium of silent film special effects; and this unlikely triumph of cinematic style over exhausted genre and jaded audience alike signaled that Coppola's auteurist impulse had not only survived the industry upheaval of the past three decades but in some sense prevailed.

George Lucas, Coppola's close friend and protégé, traveled a less bumpy road from renegade auteur to industry mainstay. Befriended as a teenager by veteran cinematographer Haskell Wexler, Lucas studied animation at the University of Southern California's cinema school, working intermittently as a cameraman for Saul Bass and editing documentaries for the U.S. Information Agency. In 1965 he produced the award-winning student film *THX-1138: 4EB/Electronic Labyrinth,* which won first prize in that year's National Student Film Festival. Lucas subsequently won a scholarship to observe production at Warner Bros., where he met Coppola, then at work on *Finian's Rainbow,* and subsequently Lucas became a production associate on *The Rain People.* Within a year, Lucas was working on the feature-length version of *THX-1138* that would become the first and only production of American Zoetrope as originally conceived. Written, directed, and edited by Lucas, with an electronic score by Lalo Schifrin, this chilling vision of an Orwellian future in which TV and drugs have subsumed the will and replaced sex, offended Warner Bros., which cut it for distribution[14] and canceled the entire Zoetrope deal with Coppola. Produced for $777,000, *THX-1138* earned only $945,000 in rentals, but it attracted considerable praise from critics and contributed to the prestige of Lucas's next project, *American Graffiti.* Produced by Gary Kurtz under the new Lucasfilm Ltd. banner for slightly more than $775,000, with Coppola as executive producer and Wexler as cinematographer, this nostalgic re-creation of early 1960s' adolescence in Modesto, California, was shot in twenty-eight days. Replete with vintage cars and period music, *American Graffiti* was hesitantly released by Universal (which spent only $500,000 for advertising, publicity, and prints, refusing to dupe them in stereo), but it struck a chord with audiences and went on to earn $55 million in domestic rentals alone. It also won nominations for five Academy Awards, including best picture, best director, and best screenplay (Lucas, Gloria Katz, and Willard Huyck), and it became the source of a sequel and a highly successful television spinoff, "Happy Days," positioning Lucas for the even more astonishing success of his next project, the now legendary *Star Wars.*

Three years in preparation, *Star Wars* was conceived by Lucas as a folkloristic "space fantasy" with the breathless pace of the cliff-hanging Saturday serials. To counterpoint the fairy-tale quality of its narrative, hyper-realistic special effects were produced at Industrial Light and Magic (ILM), a subsidiary of Lucasfilm, which was founded in Van Nuys in 1975 specifically for this purpose. Here F/X director John Dykstra, who had worked as Douglas Trumbull's assistant on *2001,* perfected a computerized motion-control system for traveling matte photography that made the process uniquely cost-effective, enabling Lucas to create hundreds of complicated stop-motion miniature sequences for *Star Wars* at a fraction of their cost in earlier films. ($2.5 million was spent on traveling matte effects for *Star Wars,* compared to $6.5 million for fewer mattes for *2001.*) With live-action sequences shot in Tunisia, Death Valley, and EMI-Elstree Studios in England, *Star Wars* also became the first film both recorded and released in six-track Dolby stereo. Still, it was produced (by Gary Kurtz) for Twentieth Century-Fox for the relatively modest sum of $11.5 million, and neither Lucasfilm nor Fox expected to earn back more than twice that in domestic rentals (because of the traditionally hard-sell market for science fiction). All were stunned when *Star Wars* made almost $3 million in its first week of limited release in May 1977. By the end of August it had grossed $100 million. The film played continuously throughout 1977, and was officially rereleased in 1978 and 1979, by the end of which it had earned $262 million in rentals worldwide. (Rereleased with updated special effects in 1997, the *Star Wars* trilogy has since grossed several hundred million dollars more.) Moreover, Lucas had designed the film with an eye toward merchandising tie-ins and retained licensing control through Lucasfilm, but he could not possibly have foreseen the bonanza of sales in toys, books, records, posters, and other *Star Wars* memorabilia that made him a multimillionaire and turned Lucasfilm Ltd. into a $30 million corporation. (Ultimately, *Star Wars*-related merchandise would gross more than the film itself—more than $1 billion retail by most accounts.) Nor could he have imagined that the film would be nominated for ten Academy Awards and receive seven (including art direction, sound, original score, film editing, costume design, and visual effects), win two Los Angeles Critics Awards, and capture three Grammies for its score.

Clearly, *Star Wars* was a cultural phenomenon as well as a movie, and its spectacular success enabled Lucas to move to Marin County in northern California and establish a state-of-the-art production facility free from the constraints of the major studios in Hollywood, fulfilling one of his personal and generational dreams. At this point, he abandoned directing to become executive producer of the blockbuster *Star Wars* sequels *The Empire Strikes Back* (Irvin Kershner, 1980) and *Return of the Jedi* (Richard Marquand, 1983), both of which were hugely profitable ($142 million and $169 million in rentals, respectively). He was also the inspiration behind Steven Spielberg's lucrative Indiana Jones trilogy—*Raiders of the Lost Ark* (1981, $116,000 million), *Indiana Jones and the Temple of Doom* (1984, $109,000), and *Indiana Jones and the Last Crusade* (1989, $116,000), as well as less successful films like Ron Howard's sword-and-sorcery epic *Willow* (1988, $28 million in rentals) and bombs like *Howard the Duck* (Willard Huyck, 1986, $10 million). ILM, which revolutionized photographic effects in the *Star Wars* era, went on to pioneer computer animation and digital effects in the late 1980s and early 1990s, becoming the major F / x studio in the industry, and Lucasfilm's Skywalker Sound division and its THX Group have had a similar impact on movie sound design and reproduction. Although he has directed only three films to date, George Lucas has exercised enormous creative and financial influence over the American film industry. He redefined the film generation's concept of authorship to become the creative CEO of the largest independent studio in the world— Lucasfilm Ltd. and its subsidiary LucasArts Entertainment Company— which have today attained the status of a multinational conglomerate.

Unlike Coppola and Lucas, Steven Spielberg did not have professional film school training, although he began making amateur films in 16mm as early as age thirteen. By the time he was in college at Cal State, Long Beach, he had produced *Amblin'* (1969), a short about hitchhikers that won festival prizes in Atlanta and Venice and led to a contract with the Universal/MCA television division. At Universal, he directed episodes of "Night Gallery," "Columbo," "Marcus Welby," and other popular shows, as well as the extraordinary TV movie *Duel* (1971), which was released theatrically in Europe. His theatrical feature debut in the United States was *The Sugarland Express* (1974), a deftly directed fugitive-couple film with Goldie Hawn that did not do well at the box office. Then came the epoch-making *Jaws* (1975), which even more than *The Godfather* and intervening blockbusters like *The Exorcist* (William Friedkin, 1973) and *The Sting* (George Roy Hill, 1973), changed the ways in which movies would be cost-projected and marketed. Adapted from Peter Benchley's best-selling novel about a great white shark that terrorizes a New England beach community at the height of the tourist season (which Universal producers Richard Zanuck and David Brown had shrewdly acquired before publication in 1973), *Jaws* was produced for only $8.5 million.

As Pye and Myles write in *The Movie Brats,* the single most salient feature of *Jaws* was "the transformation of film into event through clever manipulation of the media."[15] As sharks were raised to the level of a national fetish, "*Jaws*-consciousness" caught fire, and the film earned $100 million in domestic rentals in its initial run (and ultimately $130 million) to become the first megahit of the decade.

The success of *Jaws* permanently hooked the industry on blockbuster windfalls. (Coproducer Richard Zanuck made more money on his share of its profits in six months than his father, studio executive Darryl F. Zanuck, had made in his entire career.) Circumstances like these led to the kind of studio roulette played in Spielberg's next project, *Close Encounters of the Third Kind* (1977), in which Columbia gambled most of its working capital on a $20 million special-effects extravaganza about UFOS landing in middle America. It was rewarded with $116 million in North American rentals and eight Academy Award nominations (in the year of *Star Wars*, it took only one—for Vilmos Zsigmond's cinematography; *Jaws* had received four and won three—for sound, original score, and editing). *Close Encounters* gave Spielberg two consecutive $100 million grosses. The success went to his head; his next picture was the comedy-action film *1941* (1979), an inflated farce about war panic in post-Pearl Harbor Los Angeles, which bombed at the box office.

At this point, Spielberg began the historic collaboration with George Lucas on the Indiana Jones series that would make him the dominant commercial force in American cinema for the next twenty years and would shift audience demographics toward the younger end of the age scale. *Raiders of the Lost Ark, Indiana Jones and the Temple of Doom,* and *Indiana Jones and the Last Crusade*—for which Spielberg was director and Lucas executive producer—was a trilogy of fast-paced adventure films in the mode of the Saturday matinee serials that they both had loved as children.[16] Childhood wonder was also the theme of *ET: The Extra-Terrestrial* (1982), a project developed by Spielberg at Columbia and finally produced for Universal. This Disney-like fantasy about some suburban children who discover and befriend an alien from outer space became the highest-grossing film of all time until his own *Jurassic Park* (1993) surpassed it, earning more than $228 million in domestic rentals, and Spielberg's third Oscar nomination for best director (his second was for *Raiders*). The huge profits from *E.T.* and from *Poltergeist* (Tobe Hooper, 1982), for which he was executive producer, enabled Spielberg to form Amblin Entertainment in 1984, where he was co-executive producer for some of the eighties' most successful entertainments—*Gremlins* (Joe Dante, 1984), *Back to the Future* (Robert Zemeckis, 1985), *An American Tail* (Don Bluth, 1986; animated), and their various sequels, as well as the live-action/animated *Who Framed Roger Rabbit?* (Zemeckis, 1988), with Disney's Touchstone Pictures. Spielberg also coproduced his own work through Amblin in partnership with other studios. For example, with Warner Bros. he collaborated on *The Color Purple* (1985), an adaptation of Alice Walker's Pulitzer Prize-winning account of growing up black and female in the Depression-era South, and *Empire of the Sun* (1985), a version of J. G. Ballard's novel about life in a Japanese prison camp in China during World War II; and with Universal he produced a remake of director Victor Fleming's romantic fantasy *A Guy Named Joe* (1943) entitled *Always* (1989). These "adult" films met with varying degrees of critical and financial disappointment, inducing Spielberg to return to the world of childhood with *Hook* (1991; produced with Sony Pictures/Tri-Star), an inflated, star-studded version of *Peter Pan* that was considered a financial failure even though it earned $65 million domestically.

So gargantuan had Spielberg's reputation as a moneymaker become that only a slam-bang special effects turn like *Jurassic Park,* which became (and remained in 1997) the highest-grossing film in history, now seemed appropriate in terms of scale, although it briefly confirmed the impression that Spielberg worked best with juvenile material—at least until the release of *Schindler's List* (1993), a somber film about the Holocaust that was hailed by many as a masterpiece and won Academy Awards for best picture and best director, best screenplay, art direction, cinematography, editing, and score. The film silenced most of Spielberg's critics, who could no longer justify taking him for a lightweight. Unexpectedly for a deeply serious film shot in documentary-style black-and-white, *Schindler's List* earned about $80 million (its release *did* coincide, however, with the opening of the Holocaust Museum in Washington, D.C., and benefited from the attendant publicity), and together with the profits from *Jurassic Park* it made Spielberg one of the wealthiest individuals in Hollywood. Like Lucas at a similar point in his career, Spielberg decided to form his own studio—Dreamworks—in collaboration with multimedia barons Jeffrey Katzenberg and David Geffen in 1994. If even part of its expressed potential is realized, Dreamworks will become a major force in the image-making technologies of the twenty-first century and a pioneer in the realm of digital media.

The fourth major figure of the "film generation," Martin Scorsese (b. 1942), may have attained the goal of authorship more fully than any of his peers by consistently maintaining the quality of his art at the expense of its commercial viability. The sickly child of Sicilian immigrant parents, Scorsese grew up in New York's Little Italy and became deeply infatuated with movies. After graduating from high school, he entered a seminary with the intention of becoming a priest, but he left after a year to enroll in New York University's film department. There he earned an undergraduate degree in 1964 and an M.A. in 1966, while simultaneously making a number of award-winning student shorts. He then joined the faculty as an instructor and directed *Who's That Knocking at My Door?,* a low-budget semiautobiographical feature released in 1968. Next, he worked as an assistant director and supervising editor on *Woodstock* (Michael Wadleigh, 1970) and found work on several other documentaries the following year. In 1972 Roger Corman hired Scorsese to direct *Boxcar Bertha,* a violent Depression-era drama aimed at the audience for Corman's exploitation hit *Bloody Mama* (1970), but Scorsese's first important feature as an auteur

2 Travis Bickle (Robert De Niro) on the mean streets of New York in Martin Scorsese's *Taxi Driver* (Columbia Pictures, 1976).

was his third as a director, *Mean Streets* (1973), which became the hit of that year's New York Film Festival. Shot on location in Little Italy, this character study of a small-time hood wracked by Catholic guilt inaugurated Scorsese's relentless, moving camera style and was distinguished by the improvisatory ensemble performances of Harvey Keitel and Robert De Niro, both of whom would become the director's frequent collaborators. Despite its low budget ($500,000) and widespread critical acclaim, *Mean Streets* lost money. Scorsese then shot the 48-minute television documentary *Italianamerican* (1974) before his next feature, *Alice Doesn't Live Here Anymore* (1975), which was both more conventional and more popular in its feminist account of a young widow's struggle for self-fulfillment, earning an Oscar for its star Ellen Burstyn and inspiring the long-running television sitcom "Alice."

After *Alice Doesn't Live Here Anymore*, Scorsese had the base of critical and financial support he needed to make *Taxi Driver* (1975), his nightmare vision of urban hell in which a psychotic Vietnam veteran, brilliantly played by De Niro, becomes an avenging angel of violence and death. Written by Paul Schrader, who attended UCLA film school, with a score by Bernard Herrmann, and produced for just under $2 million by Michael and Julia Phillips (whose previous vehicle had been the upbeat caper film *The Sting*), *Taxi Driver* won several Oscar nominations and the Palme d'or at Cannes. It is a personal, obsessive work that treats some of the darkest aspects of human nature—including child prostitution, racial hatred, and sadism—with surreal intensity. The film turned a modest profit, taking in about $13 million in domestic rentals, which contributed to Scorsese's track record—even in the climate created by *Jaws*—and made him attractive to the production team of Irwin Winkler and Robert Chartoff, who were fielding a script about the romance of two young musicians during the big band era. Packaged as a musical with De Niro and Liza Minnelli as stars, this project became *New York, New York* (1977), which Scorsese made as a virtual compendium of MGM musical styles of the forties and fifties, watching hundreds of such films from his extensive video library. (Scorsese was an early and vocal supporter of video archiving as an instrument of film preservation.) A testament to Scorsese's remarkable grasp of film history, *New York, New York* became his first big-budget flop, returning less than $7 million of its $9 million investment and driving him back to the documentary with *The Last Waltz* (1978), a feature-length account of The Band's farewell concert in November 1976, shot in 35mm by seven of the generation's leading cinematographers (Michael Chapman, Laszlo Kovacs, and Vilmos Zsigmond among them). Scorsese then shot another documentary in 1978, *American Boy: A Profile of Steven Prince*, which chronicled the life of a young Jewish friend who had starred as a gun dealer in *Mean Streets* and later became a heroin addict; with *Italianamerican* it was intended as part of a six-film series on the immigrant experience (which to date has never been completed).

The documentary impulse nevertheless informed Scorsese's next feature, *Raging Bull* (1980), which many critics consider to be his finest film. Scripted by Paul Schrader and Mardik Martin, and shot by Michael Chapman in starkly rendered black-and-white, the film re-creates in pitiless fashion the fractured life of champion boxer Jake La Motta, who rose from squalor to the pinnacle of his brutal sport in the 1940s and was destroyed by his own paranoia (and a dubious morals charge) a decade later. The film won an Oscar for Robert De Niro's remarkable performance as La Motta as well as for its editing by Thelma Schoonmaker. Of all of Scorsese's psychodramas, *Raging Bull* is the most intense and unyielding; it was so graphic in its slow-motion depiction of battering in the ring that critics accused it of pandering to masochism (*Variety* called it "an exploration of Catholic sadomasochism"),[17] and, despite its quality, audiences basically stayed away.

Scorsese's first films of the eighties reflected the sense that he was no longer a player in the New Hollywood, an anxiety he reflected in several contemporary interviews with the press. (He would, for example, later note in one such interview that *New York, New York* opened in the same week as *Star Wars* and therefore was doomed to fail in the climate that made that film a megahit.)[18] *The King of Comedy* (1983) was a dark parable written by Paul Zimmerman about a fan (De Niro) who stalks and kidnaps a late-night television celebrity (Jerry Lewis) that resonates with John Hinckley's fixation on Jodie Foster (self-reflexively, via her role in *Taxi Driver*) and his attempted assassination of President Reagan on March 30, 1981. *After Hours* (1985), originally entitled "A Night in Soho," was a sort of Greenwich Village version of Joyce's *Ulysses*, complete with Catholic guilt. *King* cost $15 million and lost money. *After Hours*, shot in eight weeks for $3.5 million, turned a modest profit and won the best director's prize at Cannes. Conceived as a star vehicle for Paul Newman and Tom Cruise, *The Color of Money* (1986) seemingly brought Scorsese back into the fold; for just above $13 million he was able to give this sequel to *The Hustler* (Robert Rossen, 1961) a high-gloss, big-budget patina and direct Newman to an Academy Award-winning performance. The film returned about $25 million, and its commercial success enabled Scorsese to return to a long-cherished project, an adaptation of Nikos Kazantzakis's 1954 novel *The Last Temptation of Christ* (released in 1988), which was condemned for blasphemy by evangelical Christians for its sensuous portrayal of Jesus. However, the film won Scorsese the director's prize at the Venice Film Festival and his third Oscar nomination as best director.

In the nineties, Scorsese maintained his reputation as a serious artist who could nevertheless turn the Hollywood buck when required. After contributing "Life Stories" to the anthology film *New York Stories* (1989), which also featured segments by Coppola ("Life Without Zoe") and Woody Allen ("Oedipus Wrecks"), Scorsese directed the cheerfully violent *Goodfellas* (1990), based on the reminiscences of a New York mobster who cooperated with federal law enforcement authorities, and *Cape Fear* (1991), an apocalyptic remake of a 1962 thriller directed by J. Lee Thompson, with De Niro reprising Robert Mitchum's role as the vengeful ex-con Max Cady and Elmer Bernstein adapting Bernard Herrmann's turbulent original score. *Goodfellas* was a critical coup, winning British academy awards (for best film and best direction), the Venice Silver Lion, and another Oscar nomination for Scorsese, while *Cape Fear* became a box-office hit, earning nearly $40 million. Together, they enabled Scorsese to make *The Age of Innocence* (1993), an exquisitely precise adaptation of Edith

3 A Martin Scorsese musical? *New York, New York* (United Artists, 1977).

Wharton's Pulitzer Prize-winning novel whose $40 million budget shows at every turn. The film is virtually a paradigm of Scorsese as auteur—at once extravagant and academic, brilliant and fussy—a film that dazzles the eye but leaves the mind numb. Like his *New York, New York*, which had *Star Wars* to contend with, *The Age of Innocence* had little hope of commercial success in the year that *Jurassic Park* was breaking all box-office records. In fact, Scorsese may be the only director of "the film generation" who still passionately cares about the medium *as such*. This is as evident in his own work as from his tireless efforts on behalf of film preservation and restoration, which extend from the resurrection of 70mm epics like *Lawrence of Arabia* (David Lean, 1962) and *El Cid* (Anthony Mann, 1961) to the reissue on videocassette of classics like Michael Powell's *Black Narcissus* (1947) or Nicholas Ray's *Johnny Guitar* (1954).

Like Spielberg, Brian De Palma (b. 1940) did not attend film school, but as a physics major at Columbia University he began making 16mm shorts and eventually won a fellowship to Sarah Lawrence College sponsored by MCA. There he codirected (with Wilford Leach and Cynthia Munroe) his first feature film in 1963, released as *The Wedding Party* in 1969, notable largely as the debut feature of Robert De Niro and Jill Clayburgh. A negligible second feature, *Murder à la Mod* (1966) was barely released, but *Greetings* (1968), a satiric rendition of the Greenwich Village counterculture, was produced for $43,000 and returned more than $1 million, attracting the attention of critics and producers alike (it also won a Silver Bear jury prize at Berlin). After the experimental *Dionysus in '69* (1970), which records a performance of Euripides' *The Bacchae* from the perspective of both audience and players, De Palma produced a sequel to *Greetings* for Filmways on a budget of $120,000, the equally offbeat *Hi Mom!* (1970), which like the earlier film starred De Niro and Allen Garfield. This earned De Palma a contract with Warner Bros. to direct another counterculture comedy, *Get to Know Your Rabbit* (1970; released 1972), written as a star vehicle for Tommy Smothers. De Palma was removed from the project before it was completed, and he then worked briefly on *Fuzz* (Peter Colla, 1972). The following year he directed *Sisters* (1973) from his own script. This film about separated Siamese twins, one of whom is a psychotic murderer, marked De Palma's first attempt to model his work on Hitchcock's; it resonates with themes and plot devices from *Rear Window* (1954) and *Psycho* (1960), and it employs a lush Bernard Herrmann score. Next, De Palma shot *Phantom of the Paradise* (1974), a reworking of the Faust legend in the context of contemporary rock culture, and he moved farther into Hitchcock territory with *Obsession* (1976), a film indebted to *Vertigo* (1958) for its plot and to Herrmann's Oscar-nominated score (his last and, in his own judgment, his best)[19] for its emotional subtext.

Traces of Hitchcock could also be found in De Palma's *Carrie* (1976), which like *Obsession* was financed by a tax shelter partnership, but the film was more clearly De Palma's own in its use of split-screen montage (a staple since *Sisters*) and its cynical, gory hipness. Adapted from a novel by the then relatively unknown Stephen King, it concerns a repressed teenager with telekinetic powers (Sissy Spacek), whose dawning sexuality unleashes a bloodbath of revenge against her mother (a religious fanatic based on *Marnie*'s mother in the 1964 Hitchcock film) and her cruel high school classmates. Tapping into the nascent psycho-slasher trend, *Carrie* was a career-making hit for De Palma, earning about $15 million in domestic rentals.

After two more features, the stylish supernatural thriller *The Fury* (1978) and *Home Movies* (1979), a low-budget farce about moviemaking shot with a crew of Sarah Lawrence film students, De Palma was finally acknowledged as a mainstream auteur with *Dressed to Kill* (1980)—a film that is still probably his best. Directed from his own screenplay, the film combines Hitchcockean suspense with 1980s-style sex and gore in the story of a cross-dressing serial killer in Manhattan who murders women with a razor. *Dressed to Kill* is at once erotic, sadistic, and intensely manipulative,

4 The music-video/porn-movie sequence in Brian De Palma's controversial thriller, *Body Double* (Columbia Pictures, 1984).

as with *Carrie* raising charges of misogyny. But this time the film had the blessings of the critical establishment and a laudatory feature story in the *New York Times*. *Dressed to Kill* made nearly as much at the box office as *Carrie*. As if to demonstrate his new seriousness, De Palma wrote and directed *Blow Out* (1981), a self-consciously referential conspiracy thriller that leans heavily on allusions to Welles (*Touch of Evil*, 1958), Antonioni (*Blow-Up*, 1967), and Coppola (*The Conversation*, 1974), but never quite gels into a convincing narrative and contains, almost by way of signature, the graphically brutal murder of a prostitute. Produced for $18 million, *Blow Out* took in only $8 million and drove De Palma toward the money-making proposition of directing *Scarface* (1983) for Universal. This bloody, over-the-top remake of Howard Hawks's 1932 classic was written by Oliver Stone and substitutes Cuban refugees in contemporary Miami for Depression-era Chicago gangsters; its commercial success ($23.4 million domestic) propelled De Palma into the loathsome *Body Double* (1984), which he wrote, produced, and directed in a misguided, abortive attempt to exorcise his demons. Described by *Variety* as "a voyeur's delight and a feminist's nightmare,"[20] *Body Double* is a meretricious imitation of *Rear Window* that reeks with contempt for women and contains a murder sequence verging on sadistic pornography. Despite De Palma's protestations that the film was *about* voyeurism, most critics saw it as exploitation pure and simple, and the public simply did not care.

Since this debacle, De Palma has avoided the more lurid excesses of sadoeroticism, although he still specializes in violence of a particularly disturbing sort. His best film of more recent years is *The Untouchables* (1987), a flamboyant gangster film about the antiracketeering activities of Eliot Ness in Chicago during the thirties, with a script by David Mamet and based on the popular television series from the 1950s. This film demonstrated once again that De Palma could direct critically respectable and commercially viable entertainment, earning an Academy Award for Sean Connery as best supporting actor, as well as nominations for art direction (Patrizia Von Bradenstein et al.), costume design (Marilyn Vance-Straker), and original score (Ennio Morricone), plus $38 million in North American rentals. De Palma's career continues to be a patchwork of commercial successes (*Carlito's Way*, 1993; *Mission Impossible*, 1996) and failures (*Casualties of War*, 1989; *The Bonfire of the Vanities*, 1990; *Raising Cain*, 1992), and he continues to be controversial in his attitudes toward sex and violence, especially among feminists. De Palma has had more financial ups and downs than any member of his generation except Coppola, and his treatment of unpleasant material has often bordered on the prurient. Stylistically, he is more interesting than many of his peers because he is less predictable and consistent—he has employed some of the best cinematographers in the business (Gregory Sandor, Vilmos Zsigmond, Mario Tosi, John A. Alonzo, Stephen H. Burum), but he rarely has used the same one from film to film, and he has not built up a repertory of performers in the manner of Scorsese or Coppola. If nothing else, critics cannot accuse De Palma of following fashion, and most continue to find his work simultaneously fascinating, frustrating, and problematic.

Two other figures of the "film generation," John Milius and Paul Schrader, are known mainly as screenwriters, but they have also directed films of distinctly auteurist aspirations. Milius attended USC film school as a member of the "miracle class" that included Lucas, Randal Kleiser (*Grease*, 1978), and John Carpenter (*Halloween*, 1978).[21] He began to produce hard-boiled action scripts in the early seventies, writing or collaborating on *Dirty Harry* (Don Siegel, 1971; uncredited), *Jeremiah Johnson* (Sydney Pollack, 1972), and *Magnum Force* (Ted Post, 1972), before directing

Dillinger (1973) from his own script at AIP. Cast in the mixed genre format of *Bonnie and Clyde,* this violent gangster film focuses on the FBI manhunt for the famous bank robber who became "Public Enemy Number One" during an eighteen-month crime spree in 1933 and 1934. Milius portrays both Dillinger (Warren Oates) and his nemesis, G-man Melvin Purvis (Ben Johnson), as media-conscious public heroes obsessed with shaping their images within the popular mythology of their times. Although it was produced on a limited budget by a youth-market studio, *Dillinger* attracted a wide audience and unexpected critical acclaim, briefly putting Milius in the forefront of his generation.

For his next project, Milius had epic aspirations, although the result is more in the nature of an elegy. *The Wind and the Lion* (1975) is a quasi-historical account of the kidnapping of the wife and children of an American diplomat by a Berber chieftain in Morocco in 1904, in response to which President Theodore Roosevelt sent in the marines. With Kurosawa-like action sequences and widescreen desert cinematography modeled on *Lawrence of Arabia, The Wind and the Lion* creates a sort of Fordian myth of gunboat diplomacy entirely appropriate to its writer-director's right-wing ideology. (Contrary to his generation's liberalism, Milius is a self-proclaimed fascist, sworn to the mystical power of violence and survivalist gun culture). How fitting, then, that after producing the pretentious surfing film *Big Wednesday* (1978), Milius should make *Conan the Barbarian* (1982), coscripted with Oliver Stone, a sword-and-sorcery epic glorifying brute strength in the form of Arnold Schwarzenegger and based on the "Hyborean Age" pulp fiction cycle of Robert E. Howard. The film had box-office success ($21.7 million domestic), although it was not the hoped-for blockbuster, and it produced a sequel and numerous clones. Perhaps the ultimate statement of Milius's reactionary politics, however, was *Red Dawn* (1984), produced at the height of the Reagan era. (Two years later Reagan would act out the fantasy of *The Wind and the Lion* in 1980s' terms, sending F-111 fighters to bomb the headquarters of Libya's Mu'ammar Gadhafi in Tripoli in retaliation for a terrorist attack on American servicemen in West Berlin.) Coscripted with Kevin Reynolds, this film is a violent paranoid fantasy that envisions a militarily crippled United States invaded by the "Evil Empire" forces of the USSR and Cuba (!); its plot centers on a group of teenagers who become guerrilla fighters in the nearby mountains when their town is occupied by communist paratroopers, and it is consciously—if ironically—modeled on the World War II "partisan film" then so popular in the Warsaw Pact nations.

Afterward, Milius's films as a director plummeted ever more steeply toward the romantic fascism of *Farewell to the King* (1989), which adapts a Pierre Schoendoerffer novel as a combination of Conrad and Rambo, and the strident right-wing polemics of *Flight of the Intruder* (1991), an ill-conceived Vietnam War film about fighter pilots aboard a giant carrier during the 1972 Paris peace talks. Both of them box-office disasters, these two films seemed to confirm received industry wisdom that Milius had crossed the line from ideological artist to agenda-serving propagandist. It probably has not helped his reputation that federal prosecutors have established *Red Dawn* as a key film in the political education of convicted Oklahoma City bomber Timothy McVeigh.[22]

Paul Schrader, the product of a strict Dutch-German Calvinist upbringing, studied divinity at Calvin College seminary but became deeply enthralled by movies through a summer film course he took at Columbia University. When he graduated from Calvin in 1968, Schrader enrolled in the UCLA's graduate film program where he was a classmate of Coppola, Carroll Ballard, and B. W. Norton, Jr. (*More American Graffiti*, 1979). Schrader became film critic for the *Los Angeles Free Press,* the editor of *Cinema* magazine, and the author of the influential scholarly volume *Transcendental Style in Film: Ozu, Bresson, Dreyer* (1972). His first screenplay was *The Yakuza* (Sydney Pollack, 1975), but serious critical notice awaited his scripts for Scorsese's *Taxi Driver* and De Palma's *Obsession.* Schrader also worked uncredited on the script for Spielberg's *Close Encounters,* and he cowrote (with Mardik Martin) Scorsese's *Raging Bull;* he later wrote the screenplays for Peter Weir's *The Mosquito Coast* (1986) and Scorsese's *The Last Temptation of Christ.* As with Milius, Schrader's success as a writer during the 1970s enabled him by the end of the decade to direct films from his own scripts. *Blue Collar* (1978) and *Hardcore* (1979) both evoke the ambience of *Taxi Driver,* reflecting combined fascination and disgust with the underside of American life—in *Blue Collar,* labor union corruption in a grim factory town, in *Hardcore,* the nightworld of the $6 billion pornography industry into which the teenage daughter of a midwestern Calvinist lay minister—heroically portrayed by George C. Scott—disappears on a trip to Los Angeles. Scott's attempts in *Hardcore* to track the girl down leads him into unimagined byways of sexual prurience, debauchery, and morbidity, plunging him finally into a dark night of the soul that seems to be a paradigm for Schrader's neo-Protestant ethic. Less morally intense was *American Gigolo* (1980), a stylistically spare thriller involving kinky sex built primarily around the ambiguous persona of its star, Richard Gere. Nevertheless, the film was a hit, earning about $11 million domestic, and it spurred Schrader to direct the sensationalistic *Cat*

People (1982)—a remake of Val Lewton's classically understated horror film of 1942, directed by Jacques Tourneur, as a 1980s-style gorefest with prosthetic makeup effects by Tim Buram. Exactly why Schrader abandoned his theme of spiritual angst for the horrors of material dismemberment is unclear, but he returned again to the process of self-revelation in *Mishima: A Life in Four Chapters* (1985), for which he was awarded the best artistic contribution prize at Cannes. This film interweaves real events from the life of the famous Japanese novelist who committed seppuku in 1970 with highly stylized scenes from his fiction to create a richly textured portrait of his inner life. Schrader's later films as a director—*Light of Day* (1987), *Patty Hearst* (1988), *The Comfort of Strangers* (1990), *Light Sleeper* (1992), *Witch Hunt* (1994), and *Touch* (1997)—have been notably unsuccessful at recapturing that sense of spiritual and moral questing that characterized his early work. Although he continues to write screenplays, Schrader seems to have ended up very much like Milius—at the dead end of a cinema of ideas, where philosophy turns to rhetoric.

The idea that American directors, working within the world's most capital-intensive production context, could somehow approach the European ideal of authorship as incarnated in the French New Wave was probably doomed from the start, but it was especially intractable in the business climate of the late sixties and seventies. At that time, an industry that had been driven almost since its inception by the orderly pursuit of profit was gobbled up by new owners who saw it as a locus for high-stakes speculation and corporate tax-sheltering. They were skilled at these pursuits, and their new management style produced results—from a combined loss of $41 million in 1969–70, the eight motion picture companies earned profits of $173 million in 1972–73, thanks largely to the success of a few heavily marketed blockbusters.[23] But as this review of the careers of the most prominent "film generation" directors is meant to suggest, the extent to which film production became an investment-specific strategy during the 1970s was unprecedented, and it warped the shape of the industry for years to come, driving production and marketing costs to hitherto unimagined levels.

Starting in the late sixties, studios began recruiting untested film school-trained directors to appeal to the "youth market," which was correctly understood to be driving a national resurgence in film attendance. The institution of the MPPA Rating System in October 1968 was initially seen as a boon to experimentation by these directors, and in some ways it was. By the mid-seventies, however, ratings had worked to create a two-tiered system of production in which studios looked to make either cross-generational blockbusters like *Jaws* and *Star Wars* in the PG category (initially M, then GP, and later PG-13) or to tailor entertainment for specific market segments—children (G), adults (R), and "adults" (X). In effect, because the G-rating market share was relatively small, and the X rating was reserved mainly for pornography (or what was then perceived as such), most Hollywood films of the decade were rated either PG or R.[24] The "film generation" auteurs were neatly divided along this fault, with Lucas, Spielberg, and Milius cleaving mainly to the PG side, and Coppola, Scorsese, De Palma, and Schrader on the other. (Curiously, the two major non-"film generation" directors of the 1970s—Robert Altman and Stanley Kubrick—made a nearly equal number of films in each category.)

By the end of the decade, this linkage of auteurs with certain types of entertainment—Coppola with the *Godfather* films, Lucas with the *Star Wars* cycle, Spielberg with wondrous spectacles of all sorts—combined with the practice of saturation booking and massive national publicity to make promotion the most important aspect of exhibition and distribution.[25] Auteurism thus became a marketing tool that coincided nicely with the rise of college-level film education among the industry's most heavily courted audience segment.[26] From the cinema of rebellion represented by films like *Bonnie and Clyde, Easy Rider,* and *Medium Cool,* America's youth transferred its allegiance to the "personal" cinema of the seventies' auteurs without realizing how corporatist and impersonal it had become. And the auteurs themselves were transformed from *cinéastes* into high-rolling celebrity directors (many of them) with their own chauffeurs, Lear jets, and bodyguards. In 1968 Coppola had said, "I don't think there'll be a Hollywood as we know it when this generation of film students gets out of college,"[27] accurately forecasting the enormous impact his generation of filmmakers would have on the industry. What he could not foresee was how the change would boomerang on the new auteurs and recast their films as branded merchandise to be consumed along with T-shirts, action figures, Happy Meals, and, by the end of the decade, miniaturized and badly framed versions of the films themselves called "videos."

Notes

1 Andrew Sarris, "Notes on the Auteur Theory in 1962," *Film Culture,* no. 27 (Winter 1962–63), 00–00.

2 Andrew Sarris, *The American Cinema: Directors and Directions, 1929–1968* (New York: E. P. Dutton, 1968).

3 Anthologized in Pauline Kael, *I Lost It at the Movies* (Boston: Little, Brown, 1965), pp. 264–88.

4 Kael, "Raising Kane," *New Yorker*, February 20, 1971, pp. 43–89, and February 27, 1971, pp. 44–81; rpt. *The 'Citizen Kane' Book* (Boston: Little, Brown, 1971), pp. 1–84.

5 Quoted in Michael Pye and Lynda Myles, *The Movie Brats: How the Film Generation Took Over Hollywood* (New York: Holt, Rinehart and Winston, 1979), p. 191.

6 Ibid.

7 Schatz, *Old Hollywood / New Hollywood: Ritual, Art, and Industry* (Ann Arbor: UMI Research Press, 1983), p. 203.

8 The new aspect ratio had the unfortunate effect of cropping off the dancers' feet.

9 Jon Lewis, *Whom God Wishes to Destroy . . . Francis Ford Coppola and the New Hollywood* (Durham, N.C.: Duke University Press, 1995), p. 13.

10 George Lucas: "Francis saw Zoetrope as a sort of alternative *Easy Rider* studio where he could do the same thing: get a lot of young talent for nothing, make movies, hope that one of them would be a hit, and eventually build a studio that way." Quoted in Dale Pollack, *Skywalking: The Life and Films of George Lucas* (Los Angeles: Samuel French, 1990), p. 88.

11 Pye and Myles, *Movie Brats*, p. 86.

12 Ibid., p. 111.

13 Lewis, *Whom God Wishes to Destroy*, p. 59.

14 Ted Ashley, the executive in charge at Warners, turned the negative of *THX-1138* over to in-house editor Rudi Fehr, who cut a little over four minutes from the film. The footage was restored for re-release of the film after the success of *Star Wars* in 1977.

15 Pye and Myles, *Movie Brats*, p. 237.

16 Even so, controversy over the level of violence in *Temple of Doom* led directly to the creation of the PG-13 (Parental Guidance Suggested for Children under 13) rating code in 1984. Donald R. Mott and Cheryl McAllister Saunders, *Steven Spielberg* (Boston: Twayne, 1986), p. 87.

17 Quoted in Les Keyser, *Martin Scorsese* (Boston: Twayne, 1992), p. 112.

18 Anthony DeCurtis, "Martin Scorsese," *Rolling Stone*, 1 November 1990, p. 106.

19 Herrmann received posthumous nominations for both *Obsession* and *Taxi Driver* but lost to Jerry Goldsmith for his score for *The Omen* (Richard Donner, 1976). Herrmann's importance to the "film generation" was not only that he was a great (and somewhat neglected) composer but also that his scores were associated with the premiere auteurs of the American cinema—Hitchcock and Welles.

20 *Variety Movie Guide*, ed. Derek Elly (New York: Prentice Hall, 1992), p. 70.

21 The class of 1965–70 also included writer-directors Matthew Robbins

(*Dragonslayer*) and Willard Huyck (*French Postcards*), cinematographer Caleb Deschanel (*The Black Stallion*), editor Walter Murch (*Apocalypse Now*), composer Basil Poledouris (*Blue Lagoon, Conan the Barbarian*), writer Dan O'Bannon (*Alien*), writer-director Robert Zemeckis (*Used Cars*—later the *Back to the Future* series, *Who Framed Roger Rabbit?*, and *Forrest Gump*). Dale Pollack has described this group and its UCLA / NYU counterparts as "the cinematic equivalent of the Paris writer's groups in the 1920s" (*Skywalking*, pp. 47–48).

22 According to an article in the *New York Times,* July 5, 1995 ("The Gun Network: McVeigh's World," by John Kifner, pp. A1, A10–11), the FBI determined that McVeigh rented and watched two films obsessively during the two years of drifting that preceded the bombings—*Red Dawn* and Terry Gilliam's dystopian *Brazil* (1985).

23 Garth Jowett, *Film: The Democratic Art* (Boston: Little, Brown, 1976), p. 437.

24 The ratings were initially G (general audiences; all ages admitted), M.GP (all ages admitted, but parental guidance suggested), R (restricted; persons under seventeen admitted only if accompanied by parent / guardian), and X (no one under seventeen admitted). The creation of the X rating caused immediate concern that it would open the floodgates to a tidal wave of pornography. Some nonpornographic mainstream films—most prominently *Midnight Cowboy* (1969) and *A Clockwork Orange* (1971)—were X-rated early on for sexual content or violence. But X-rated films were soon ghettoized because local newspapers refused to carry ads for them, and the X (or "XXX"—a marketing invention of the adult film industry) became associated with pornography by default.

25 Garth Jowett and James M. Linton, *Movies as Mass Communication,* 2nd ed. (Newbury Park, Calif.: Sage, 1989), p. 59.

26 Auteurism as a marketing tool was not completely new to the industry; Alfred Hitchcock's cameos in his films from *The Lodger* (1926) through *Family Plot* (1976) were both signatures *and* marketing hooks.

27 Pye and Myles, *Movie Brats,* p. 81.

Auteurs and

the New Hollywood

Timothy Corrigan

As soon as you become that big, you get absorbed.   Francis Coppola

In the autumn of 1995 Quentin Tarantino celebrated his meteoric rise as Hollywood's latest artistic revolutionary with a guest-host appearance on "Saturday Night Live," a traditional television forum for cutting-edge personalities and entertainers. In one skit Tarantino played the host in a parody of a TV talk show called "Directors on Directing" in which he was joined by caricatures of some of Hollywood's other recent visionaries— Oliver Stone, Spike Lee, and Gus Van Sant. Despite such heady company, the discussion soon got to whether or not the directors sleep with the actresses in their films.

Alongside this somewhat lame parody of the American auteur today, the show featured a variety of other spoofs of Tarantino the auteur, including an episode of "This Is Your Life." In it, one guest recalled those critical formative days in the then 32-year-old director's life when Tarantino worked as a clerk in a video rental shop. On the one hand, the spotlighting of Tarantino here seems, if not premature, at least, for those of us who can associate this sort of personality probing of auteurs with Truffaut's famous conversations with Hitchcock, irreverently empty-headed. On the other hand, both the unabashed self-promotion of this appearance and its deliberate attempt to ironize Tarantino as auteur-celebrity tells us that if auteurism—as a description of movies being the artistic expression of a director—is still very much alive today, the artistic

5 Visionary director and pop celebrity Quentin Tarantino directs Harvey Keitel in *Reservoir Dogs* (Miramax Films, 1992).

expression of contemporary directors is fully bound up with the celebrity industry of Hollywood.

Tarantino is in many ways the quintessential 1990s' American auteur. His stunning debut, *Reservoir Dogs* (1992), was considered by reviewers (and his legions of fans) to be no less than a reworking of the language of contemporary film. *Pulp Fiction* (1994), his following feature, succeeded in rearticulating the figures of violence, communication, and the ethics of (mostly male) relationship laid out in *Reservoir Dogs.* Indeed, almost mimicking his sudden auteurist celebrity status, Tarantino's characters are relentlessly caught in the strain of identity—too self-consciously, too theatrically, always too intently driven by what one character in *Pulp Fiction* calls "a moral test of oneself." Offscreen, Tarantino plays the same dangerous game with identity politics. He works with Miramax Films to spot, promote, and distribute marginal films,[1] directs for both television (where he shot an episode of *ER,* the TV show rated number one in 1997. He also has appeared in acting roles and cameos and in the mid-nineties was the subject of three biographies. In a brief span Tarantino has become, from one point of view, a confrontational individual succeeding in Hollywood despite an uncompromising trash-art vision, and, from another, a showman quickly cashing in on an image that may be gone tomorrow.

But, lest we become too cynical too quickly, let us keep this in mind: auteurs and theories and practices of auteurism have never been a consistent or stable way of talking about movies. When auteurs and auteuristic codes for understanding film spread from France to the United States and elsewhere in the sixties and seventies, these models were hardly the pure reincarnations (as critics sometimes urged us to believe) of literary notions of the author as the sole creator of the film or of Sartrean demands for "authenticity" in personal expression.[2] Rather, from its inception, auteurism has been bound up with changes in industrial desires, technological opportunities, and marketing strategies. In the United States, for instance, the industrial utility of auteurism from the late 1960s to the early 1970s had much to do with the waning of the American studio system and the subsequent need to find new ways to mark a movie other than with a studio's signature. One might likewise recognize the crucial contribution to auteurism made by the new social formation of movie audiences in the early sixties: the massive "teenaging" of audiences in Europe and America. In the late sixties the global encounter with traditional forms of authority on campuses and on the streets opened doors in academia to nontraditional disciplines like film studies, and this in turn helped create a crucial forum and platform for a new international art cinema identified with auteurs like Ingmar Bergman, Luis Buñuel, Michelangelo Antonioni, and Jean-Luc Godard, as well as American auteurs like Arthur Penn, Peter Bogdanovich, and Robert Altman.

Since the 1970s especially, the auteurist marketing of movies whose titles often proclaim the filmmaker's name, such as Michael Cimino's *Heaven's Gate* (1981) or Oliver Stone's *Nixon* (1995), aim to guarantee a relationship between audience and movie whereby an intentional and authorial agency governs, as a kind of brand-name vision whose aesthetic meanings and values have already been determined. This industrial model of auteurism (as a trademark of sorts) reverses the studio's attempt to regulate the way movies are seen and received. But, in contemporary film culture, such a project is fraught with all sorts of contradictions and obstacles.

Auteurs: From Author to Agent

One of the chief claims and mystifications within early theories and practices of auteurism has been a tendency to valorize one or another idea of expression, mostly disconnected from its marketing and commercial implications. Despite their large differences, theories and practices of auteurism from Alexandre Astruc and Peter Wollen to Michel Foucault and Stephen Heath, from John Ford to Jean-Luc Godard, share basic assumptions about the auteur as the voice or presence that accounts for how a film is organized.[3] These different positions might, for example, locate the auteur's presence and power in style or narrative consistencies and variations (such as in Peter Wollen's account of Howard Hawks's different but structurally consistent "male universes" in *To Have and Have Not* and *Bringing Up Baby*). Even when contemporary critics make claims for the irrelevance (and sometimes the figurative "death") of the actual author / auteur (as Roland Barthes outlines in "The Death of the Author"), the figure of the auteur / author remains an important construct, a principle of textual causality like genre or narrative which asks and insists that readers and audiences see the work as whole, complete, and beyond individual differences and inconsistencies.[4]

Auteurs in this sense fill in the gaps of understanding and misunderstanding. While David Lynch can proudly muse about his films (*Blue Velvet* and *Wild at Heart*), "I don't know what a lot of things mean," audiences, faced with the quirky opacities of his films, can both bond and transcend in the name of the auteur Lynch.[5] To view a film as the product of an auteur means to read it or respond to it as an expressive organization that precedes and forecloses the historical fragmentations and subjective distortions that can take over the reception of even the most classically coded movie. The often strained attempts to make consistent or evolutionary the British and American movies of Alfred Hitchcock or the German and Hollywood films of Fritz Lang, for example, are governed by some sense of a historically transsubjective and transcendent category—the auteur— which authorizes certain readings or understandings of specific movies. In David Bordwell's analysis of auteurism as an interpretative cue,

the overt self-consciousness of the narration is often paralleled by an extratextual emphasis on the filmmaker as source. Within the art cinema's mode of production and reception, the concept of the author has a formal function it did not possess in the Hollywood studio system. Film journalism and criticism promote authors, as do film festivals, retrospectives, and academic film study. Directors' statements of intent guide comprehension of the film, while a body of work linked by an authorial signature encourages viewers to read each film as a chapter of an oeuvre. . . . More broadly, the author becomes the real-world parallel to the narrational presence who communicates (what is the filmmaker saying?) and who expresses (what is the author's personal vision?).[6]

Formalist and cognitive critiques of auteurism, such as Bordwell's, can vanquish most of the myths of expressivity in the cinema in favor of more formal and heuristic uses for the auteur. Yet, these too do not fully attend to the survival—and, in fact, increasing importance—of the auteur as a commercial strategy.

What I want to argue here is that today, even recent corrections, discussions, or deconstructions of the romantic roots of auteurism need to be taken another step toward recontextualizing them within contemporary industrial and commercial trajectories. Illustrating this need to investigate how "the author is constructed by and for commerce," John Caughie has noted that this question has mostly been overlooked since Brecht's 1931 account of *The Threepenny Opera* trial in which Brecht "brilliantly exposes the contradiction in cinema between the commercial need to maintain the ideology of the creative artist and the simultaneous need to redefine ownership in terms of capital, rather than creative investment."[7]

Certainly, such a reappraisal of auteurism—as more than just a way to account for artistic expression in a film or a category for interpretation—could and should take any national film history and industry into consideration. The international and commercial imperatives of contemporary culture have made it clear that commerce is now much more than just a contending discourse. If, in conjunction with the so-called international art cinema of the sixties and seventies, the auteur had been absorbed as a phantom presence, he or she has rematerialized in the eighties and nineties as an agent of a commercial performance of the business of being an auteur. To follow this move in contemporary culture, the practices of auteurism now must be reexamined in terms of the wider material strategies of social agency. Seen as a cultural agent, the auteur must now be described according to the conditions of a cultural and commercial intersubjectivity, a social interaction distinct from an intentional causality or textual transcendence.

Models of agency are useful here precisely because they are models of intersubjectivity that aim to undermine the metaphysics and the authority of expression and intention, the cornerstones of a stable subjectivity. They delineate a model of action in which both expression and reception are conditioned and monitored by reflective postures toward their material conditions.[8] In the cinema, auteurism as agency thus becomes a place for encountering not so much a transcending expression but the different conditions through which expressive meaning is made by an auteur and reconstructed by an audience, conditions that involve historical and cultural motivations and rationalizations. Here, even exasperating auteurs like Tarantino may strategically embrace the more promising possibilities of the auteur as a commercial presence, since the commercial status of that presence now necessarily becomes part of an agency that culturally and socially monitors identification and critical reception.

Celebrity Auteurs and the Business of Movies

The practice of the auteur as a particular brand of social agency appears most clearly and most ironically in the contemporary status of the auteur as celebrity. This idea of the auteur-star vaguely harks back to earlier versions of auteurism that ranked directors, from Orson Welles to Robert Bresson, in aesthetic and intellectual pantheons fitting the distinctions of their films. Today, however, American auteurs are, often and largely, defined by their commercial status and their ability to promote a film, sometimes regardless of its distinction.[9] In a twist on the tradition of certain movies being vehicles for certain stars, the auteur-star can potentially carry and redeem any sort of textual material, often to the extent of making us forget that material through the marvel of its agency.[10] In this sense, promotional technology and production feats become the new "camera-stylo,"[11] serving a new auteurism in which the making of a movie, like Oliver Stone's *Nixon* (1995), or its unmaking, like Stone's *JFK* (1991), foreground an agency that preempts and forecloses the value of the film in and of itself. As Godard has parodied so incisively in films like *King Lear* (1987), in today's commerce we want to know what our authors and auteurs look like or how they act; it is the text and not its author that now may be dead.

At this point, Michael Cimino and his film *Heaven's Gate* (1980) become an ideal example of this new phenomenon, a director and "his" event dramatizing the major points and counterpoints that have defined celebrity auteurism and its place in American film culture in recent years. Partly as a result of the takeover of United Artists by Transamerica Corporation, both Cimino and *Heaven's Gate* sprang from the urgency, especially in Hollywood after 1970, to make blockbuster movies. In a particularly resonant transformation of United Artists history—the studio originating with independent artists such as Charles Chaplin and D. W. Griffith and traditionally associated with independent productions like *Last Tango in Paris* (1972)—Transamerica decided that the studio should adjust to its parent company's financial logic and, in keeping with the economic potential that spurred the takeover, Transamerica urged the pursuit of block-

6a and b The name above and in the title: Oliver Stone's *Nixon*. Photos by
Sidney Baldwin (Cinergi Pictures Entertainment, 1995).

buster films and audiences. *Heaven's Gate* was to be one of the newly
reorganized studio's grandest efforts.

In this particular case, Transamerica-United Artists recruited Cimino
and in doing so anointed him as an auteur-artist: part epic genius and part
promotional commodity. Perhaps because he was, paradoxically, a relatively unknown quantity with a background, significantly, in advertising,
Cimino became the conglomerate image of the blockbuster auteur solely
on the basis of the massive *potential* of his Vietnam film, *The Deer Hunter*
(1978)—which many of the studio executives had not yet seen.

United Artists signed Cimino to the *Heaven's Gate* deal for a $7.8 million
projected cost and gave him almost unprecedented directorial freedom.

As is commonly reported, the budget ultimately exceeded $40 million, and the film became one of the most notorious financial and (undeservedly perhaps) critical disappointments in movie history. Just as the studio attempted to produce or exploit a new celebrity auteur, the film ultimately appealed to no one because of its blockbuster goal of appealing to everyone. At least as an indirect result, Kirk Kerkorian's MGM would purchase United Artists; this merger, in turn, would founder, and the studio would next end up in the hands of Giancarlo Parretti and Pathé Communications and then, following foreclosure, a French bank, Crédit Lyonnais. A telling by-product of these conglomerate quick changes would be the formation of a more specialized major studio, Orion, by a number of UA executives who moved out and reorganized in the wake of Transamerica's new agendas.

In the almost universal bashings that the film suffered, Cimino would become the scapegoat for "auteurism-out-of-control." Yet, in the end, the image of the auteur would survive the conglomerate formulas of the new Hollywood through the home video market; there, the film would be reissued in the form of a "director's cut," and it would return rental figures sufficient to make it a financial success. Both Cimino and *Heaven's Gate* would act out the often repeated contemporary formula of redemption through the extended market of home video where the making of successful movies and identifiable auteurs falls outside the promotional formulas of the studios and the critical pronouncements of the Establishment.[12]

Cimino's case and this particular episode are not, I want to argue, an eccentric or isolated event in modern American film history, but rather they form a central image and metaphor for what distinguishes film culture today. To clarify these historical and cultural contours and how they shape the contemporary auteur, I plan to use the example of Cimino to identify three related shifts in current film culture: (1) the blockbuster productions of the conglomerate studios, (2) the alternative films of Hollywood's mini-majors (like Orion), and (3) the importance of VCR technology in sustaining both kinds of moviemaking. Needless to say, other forces are at work in contemporary movie culture, but these three seem to me to shed the most light on the peculiar power and look of today's auteur.

The historical transition point for these industrial, sociological, and aesthetic changes within movie culture is the late 1960s, when the impending birth of the contemporary blockbuster would appear to encourage (almost in a contradictory fashion) the invasion of VCRs, satellite dishes, and

7 Lots of extras, lots of money: the Wild West roller rink in Michael Cimino's box-office disaster, *Heaven's Gate* (United Artists, 1981).

cable networks. Although the movie industry had been moving in this direction since 1946,[13] the economics had operated, up to this point, according to the fairly lucid logic of profitable investment: a successful movie invests efficient and controllable resources such as actors, technicians, and a director to meet the demands and desires of a hopefully predictable and relatively stable audience. During the period after 1965, however, that logic lost much of its antiquated clarity as the industry moved full force into an age of inflation and conglomeration. The most dramatic indicators of this shift are the acquisition of the major studios by multinational corporations attracted by growing movie profits. Universal was taken over by MCA (1962), Paramount by Gulf & Western (1966), United Artists by Transamerica Corporation (1967), Warner Bros. by Kinney National Services (1969), and MGM by Las Vegas businessman Kirk Kerkorian (1969).

The general consequences of these new production structures follow from their economic attractions. Responding to the inflation of costs—such as those caused by the new computer and laser technologies, better-organized union labor, and hugely inflated star salaries, a Marlon Brando or Jack Nicholson could command more than $5 million for a film in the early 1970s; Sylvester Stallone commanded $20 million per film in 1997—the corporate conglomerates became simply the next, and perhaps necessary, step in the economic expansion of the studios. Within this conglomerate structure, however, the movie industry became one of many products, albeit one with the special attraction of offering the extremely large and relatively fast turnovers that blockbuster movies like *The Godfather* (1972) or *Jaws* (1975) or *Batman* (1989) could provide. Despite failures and nods to other kinds of movies, this blockbuster vision—which during the early stages of film history was something between the exception and the accident that *The Birth of a Nation* was in 1915—would decisively guide the American movie industry through and beyond the seventies. While in the 1960s, only five American movies made profits of at least $40 million, four movies would make that much in 1979 alone (*Superman, Every Which Way But Loose, Rocky II,* and *Alien*). According to the revised logic of Hollywood economics, a movie attracts audiences through the excess of its investment in capital and technology; even a high-priced auteur, manufactured or real, might be justified and recuperated by an enormous payback. In the late 1970s the conglomerates had every reason to be encouraged. The value of George Lucas's *Star Wars,* for example, could be understood empirically. Some $27 million was invested in 1977 ($11 million of it in the shooting budget), and well over $500 million in revenues were generated by 1980—a 1,855 percent profit in three years.

As investments grew astronomically, the studios did much more than simply increase the size of the stakes that once were invested in film stars in the 1940s, 3-D technology in the fifties, or the productions of artistically acclaimed directors in the early sixties. The studios transformed the fundamental nature of the film product by forcing massive alterations in the relation of a film to an audience, since to return those massive investments meant appealing to and aiming at not just the largest possible audience (the more modest strategy of classical films or the alternative art-house audiences of early auteur film culture) but *all* audiences. No longer is investment capital directed at differentiating one audience, however dominant, from another. Instead, those investments must aim to "undifferentiate" the character and desires of different audiences, usually by emphasizing the importance of that investment as a universal value in and of itself—for example, the presence and use of computer animation or of an expensive star or director. With blockbusters, what begins as an attempt to target the teenage audience quickly becomes an attempt to absorb as many other groups as possible within that mass; it becomes the only methodology that makes sense to a conglomerate's bottom line.

Contemporary movie culture thus necessarily aligns itself with advertising, not only as a method to abate costs (by advertising other people's products on the screen: cars, soft drinks, and so on), but because blockbuster movies can themselves succeed only as an advertisement of a product that, in appealing to everyone, can never possibly satisfy an audience of different individuals.

The deleterious effects of this conglomerate and blockbuster perspective have all too often been remarked upon.[14] The most obvious result of this new corporate mind-set is the postmodern "high-concept film," a clear offshoot of the blockbuster phenomenon of the 1970s. Justin Wyatt and R. L. Rutsky argue that high-concept movies like *Top Gun* (1986) and *9½ Weeks* (1986) rely "less on a well-developed story and characters than on a big-name star, a commercial musical score, and high visual impact." High-concept movies, from *Saturday Night Fever* (1978) to *Rambo III* (1988), become advertisements that can be abstracted to easily assimilable images and regenerated through multiple viewings and the ancillary sales of various tie-ins (records, T-shirts, and other paraphernalia). "In a postmodern world," this position contends, "high concept is the 'rational' conclusion of those processes of capitalism which tend toward the abstraction and fetishization of the image, and to the increasingly rapid movements of fashion and style."[15] Not surprisingly, one of the most effective and common high-concept strategies today is the use of the name of the deserving or undeserving director / auteur as a foregrounded and often bombastic piece of advertisement. Not surprisingly, three of Hollywood's most successful auteurs in the seventies and eighties were Michael Cimino, Adrian Lyne, and Ridley Scott, all with backgrounds in the advertising business.

Contemporary film culture is not, however, a single story. Indeed, directly related to grandiose visions and projects of the conglomerate blockbuster are the more idiosyncratic, modest, and riskier movies produced by semi-independent companies or mini-majors such as Orion and Tri-Star. An immediate consequence of the corporate studio system that alienated and often cut loose former studio executives (such as Arthur Krim's departure from United Artists to form Orion), these groups represent smaller enterprises than both the conglomerates and, in many cases, the studios that existed before the takeovers. Many of these satellites are refashioned production houses that, because of more circumscribed ambitions, have been more able and willing to take risks (significantly in TriStar's case because of a financial arrangement with the HBO cable network). Others such as Robert Redford's Sundance Institute, whose successes include *The Milagro Beanfield War* (1988), Miramax Films with *sex, lies, and videotape* (1989),[16] or Avenue Pictures with *Drugstore Cowboy* (1989) have positioned themselves as alternatives to the blockbuster mentality. Offering purportedly more serious, "specialized" movies, these companies often produce films for well under the average price of a usual studio production, raise their funds by putting together packages from a variety of investors, and then make deals to have studios distribute their finished products. But just as the origin and success of these companies remains bound to the aspirations and limitations of the conglomerates, successful auteurs found here will relocate there tomorrow.

The third shift in this movie culture and one that in many ways is the linchpin between the first two is VCR technology (and other viewer-oriented, home box-office technologies such as cable and satellite dishes). These, I believe, have changed the nature and structure of the film industry as few other technologies have ever even approximated, reshuffling both classical and modernist relations between filmmaker and film audience, creating viewer-movie relations as impulsive and contingent as the contemporary production process. By January 1995, 81 percent of U.S. homes had VCRs capable of showing movies on videocassette. About two-thirds of U.S. households subscribed to basic cable TV service, while one out of three take pay cable (which includes such film channels as Cinemax, HBO, The Movie Channel, and Showtime). According to the Associated Press, there are 27,000 video rental stores in business, many of which have more than 30,000 "theatrical / entertainment" titles available. As it continues to permeate all cultures, this spectator-oriented technology becomes the bridge between the two dominant trends of corporate and semi-independent production, as well as the many films that might straddle the middle ground between blockbusters and specialty films. Both small and large movies now figure on advances and expectations from video sales and rentals; the lag time between a commercial theatrical release and its video reappearance decreases every year. The ability of a movie, even the most marginal, to find an audience or audiences— however large or small, discriminating or not—becomes a bankable likelihood. If the conglomerate blockbusters might be said to have created an empty advertisement at the center of contemporary film culture, this dispersal of movies through VCR technology has de-authorized them by opening up, according to more individualistic and private patterns, the possibilities of what is seen and how it is seen.

"Shareability" becomes a technological and economic imperative, as well as a textual, production, and financial strategy for addressing audiences outside the walls of traditional theatrical cinema and outside its traditional ways of viewing and making meaning of the movies. If auteur-ism was one of those traditional ways, it is no wonder that the most celebrated American auteur in the mid-1990s, Quentin Tarantino, re-ceived a good part of his training in a video rental shop where hierarchies are (potentially) always in question.

If in this new film/video culture, auteurs have been recuperated as promotional stars, the process is fairly indiscriminate. Placed before, after, and outside a film, and in effect usurping the work of that film and its reception, today's auteurs—from Spielberg to Tarantino—are agents who, whether they wish it or not, are always on the verge of being self-consumed by their status as stars. I am not suggesting merely some brand of egotism or self-marketing posture. Instead, institutional and commer-cial agencies now work, whatever the filmmaker's intentions, to define auteurism as publicity and advertisement or as the dispersal of the control of the auteur into the total flow of television monitors. A large part of auteurism has become, that is, both a provocative and empty display of material surface, subject to almost indiscriminate circulation, befuddling those more traditional auteurist relations based on models of writer/ text/reader. Meaghan Morris has noted (in language similar to Richard Dyer's description of on-screen stars) that today "the primary modes of film and auteur packaging are advertising, review snippeting, trailers, magazine profiles—always already in appropriation as the precondition, and not the postproduction of meaning."[17]

To respond to a movie as primarily or merely a Spielberg film, for example, is, after all, the pleasure of refusing an evaluative relation to it—a tendency that he interestingly has tried to reverse with his mimicking of a European art film, *Schindler's List* (1993). Much of that pleasure lies in being able to already know, not read, the meaning of the film as a product of the public image of its creator. An auteur film today seems to aspire more and more to a critical tautology, capable of being understood and consumed without being seen. Like an Andy Warhol movie, it can com-municate a great deal for a large number of audiences who know the maker's reputation but have never seen the films themselves.

Of the several tacks within the commerce of the auteur-star, let me isolate two broad, central, and ultimately interactive groupings: the com-mercial auteur and the auteur of commerce. The first category could include a vast range of stars who also direct (Mel Gibson, Jodie Foster, Robert Redford, Kevin Costner, and Clint Eastwood), as well as star directors like Steven Spielberg, George Lucas, Brian De Palma, and, with different agendas and at different periods in their careers, John Sayles and

Woody Allen. My argument so far would assimilate most of these names since what defines this group is a recognition, either foisted upon them or chosen by them, that the celebrity of their agency produces and promotes a text that invariably exceeds the movie itself, both before and after its release. That this category seems not to distinguish between actors and directors as controlling agents is telling since the commercial auteur is far more a product of the magnitude of the image that anchors a movie or group of movies than a text of ideas, styles, or nuances of expression.

With the commercial auteur, the Hollywood agent becomes essential to and emblematic of the mix. Movies are often created by the packaging of one or more dominant images (such as Dustin Hoffman, Robin Williams, Julia Roberts, and Steven Spielberg for *Hook*). One of the ironies here is that Michael Ovitz, the former agent of the William Morris Agency who later founded Creative Artists Agency (CAA), became one of the most powerful commercial auteurs in Hollywood because so many high-concept stars, directors, and scriptwriters were under contract to his agency. In this context, it becomes completely logical that Ovitz would later be recognized as such when he briefly signed on as president of Disney.

If auteurism can be used to promote a movie, filmmakers now run the commerce of the auteurist and autonomous self up against its textual expression in a way that shatters the coherence of both authorial expression and stardom. Motivations, desires, and historical developments—which are frequently dramatized in critical readings of films as at least semiautobiographical—now become destabilized and usually with a purpose. To what extent, for instance, has Spike Lee simply taken advantage of his image and name (commercialized and advertised by Air Jordan and beyond) to consolidate his power; to what extent has he used that image and name as a critical wedge to question his own and others' authority and power? A film like *Do the Right Thing* (1989) seems to me to place itself on the line between these two questions. Casting himself as the central character Mookie, Lee here questions, with some implicit self-criticism, the relation of fame, power, and responsibility within the commerce of a public sphere. While a more traditional auteurist position would want to describe these changes in perspective and expression according to some coherent notion of evolution, an evaluation of many contemporary filmmakers—like Lee—must admit fissures and discrepancies that consciously employ the public image of the auteur in order to then confront and fragment its expressive coherency.[18]

As a specific example of the contemporary auteur's construction and promotion of a self, I will now turn to one director at the historical heart of this dilemma and to one "semi-textual" strategy that is often taken for granted in the relation between a filmmaker, his or her films, and an audience—namely, the interview, which is one of the few extratextual spaces that can be documented where the auteur, in addressing fans and critical viewers, engages and disperses his or her own organizing agency as an auteur. Here, the standard directorial interview might be described according to the action of promotion and explanation; it is the writing and explaining of a film through the promotion of a certain intentional self, and it is frequently the commercial dramatization of self as the motivating agent of textuality. But it is this image of the auteur that the contemporary auteur necessarily troubles, confuses, or subverts through the agency of commerce.

The Economics of Self-Sacrifice: Francis Coppola as American Auteur

Certainly Francis Coppola is one of the more celebrated and bewildering examples of auteurism as it has evolved through the 1970s and 1980s. In an essay on the evolution of Coppola's career, Richard Macksey astutely made the connection between Orson Welles and the more recent child prodigy of Hollywood, Welles connected to Coppola across the heyday of auteurism as Romantic expression and independent (if not transgressive) vision. Yet, as Macksey observes, distinctions between the two filmmakers are even more compelling. On the one hand, "Welles has been a presiding model of Romantic genius, the myth of the explosive, comprehensive talent challenging corporate power and ultimately becoming the victim of its own genius." On the other hand, there is Coppola's marketing of that myth. As Macksey argues:

If he has inherited something of the Romantic artist's impatience with the system, his powers of persuasion and need to take risks have led him toward the board-room rather than the garret, back toward the old putative center of power in Hollywood (and the financial centers off-camera) rather than toward exile and "independent filmmaking." His perilous if uncanny power to enlist bankers prob-ably depends upon his temperamental inability to fold in a poker game; movie-making and risk-taking are synonymous for him.[19]

As a Romantic entrepreneur, Coppola becomes a self-exiled and stri-dently independent auteur who claims in one sentence, "I need to be a solo guy," and then for *Tucker* humbly surrenders the film to George

8 The auteur as romantic artist and industry entrepreneur: Francis Coppola on the set of *The Cotton Club*. Photo by Adger Cowans (Orion Pictures, 1984).

Lucas's "marketing sense of what people want."[20] Straddling the margins of European art cinema and the center of commercial Hollywood, he is one of the original directors of the contemporary blockbuster, *The Godfather* (1972), and the one whose experimental goals seem most threatened by the financial and commercial exigencies of his blockbuster successes. Jon Lewis explains that as "far back as 1968, four years before *The Godfather* made him the best-known director in America, Coppola predicted that his generation of film school-educated *auteurs* would someday trigger significant change in the movie business"; yet, with the needed special emphasis on finances in Coppola's version of this paradox, the

dazzling box office success of expensive *auteurist* movies like *The Godfather, Jaws,* and *Star Wars*—the very sort of movies Coppola had once believed would foster a new American *auteur* industry—led to an industrywide focus on blockbuster box office revenues. The success of *auteur* films in the 1970s did not, as Coppola had hoped it would, give *auteur* directors increased access to film financing. Instead, directors became increasingly dependent on studio financing to produce and distribute such "big" films.[21]

Indeed, this ambivalent double image as the auteur-star of gigantic productions and the auteur-creator victimized by the forces of those productions defines Coppola's central place within the commerce of auteurism, characterizing him, in Andrew Sarris's offhand portrait, as a "modern dissonant auteur."[22]

Coppola's career has followed an almost allegorical path. It begins confidently as a commercial talent with *Finian's Rainbow* (1968), transforms itself through the commerce of auteurism with *The Godfather I* and *II* and *The Conversation* (1974), suffers the contradictions of that position with *Apocalypse Now* and *One from the Heart* (1982), and more recently settles uncomfortably into the aims of the commercial auteur with *Tucker: The Man and His Dream, The Godfather, Part III,* and *Bram Stoker's Dracula.* Jeffrey Chown has described these commercial pressures and contradictions, beginning with Coppola's first appearance as the auteur-creator of *The Godfather.* "It is curious," Chown writes, "that the film that put Coppola on the celebrity map, that gave him the magic adjective 'bankable,' is also extremely problematic in terms of authorship. . . . Coppola coordinated diverse creative agents in this production, he was clearly the catalyst for the film's success, but, in a career view, his creative control and originality are far less than in other films that bear his directorial signature" (p. 59). Even Coppola's most arty and individual film, *The Conversation,* demonstrates major industrial complications within auteurism, at least as it is applied to the control of the filmmaker. For many critics, Walter Murch, who engineered the brilliant soundtrack and much of the editing of that movie, can claim the most important part in the film's creation. With *Apocalypse Now,* moreover, this most celebrated of contemporary American auteurs surrenders the choice of three different endings to a battery of advisers and miles of computer printouts that surveyed the expectations and desires of different audiences (including then-President Jimmy Carter). In the most industrial and textual sense, then, Coppola has become the willing victim of his successful name. As Chown observes of the critical slaughter that greeted *One from the Heart, The Outsiders* (1983), and *Rumble Fish* (1983): "The name Francis Ford Coppola connotes spectacles, Hollywood entertainment combined with artistic sensibility, Italian weddings, and napalm in the morning. Coppola the individual seems stifled by those expectations."[23] As with his capitulations to U.S. Army censors over *Gardens of Stone* (1987), self-destruction seems part of his "creative compromises" with the contemporary terms of auteurism.

More exactly, his commercial compromises with the agency of auteurism mean a kind of sacrificing of that self as a form of spending and expending. In 1975 Coppola summed up his perspective on auteurism:

The auteur theory is fine, but to exercise it you have to qualify, and the only way you can qualify is by having earned the right to have control, by having turned out a series of really incredibly good films. Some men have it and some men don't. I don't feel that one or two beautiful films entitle anyone to that much control. A lot of very promising directors have been destroyed by it. It's a big dilemma, of course, because, unfortunately, the authority these days is almost always shared with people who have no business being producers and studio executives. With one or two exceptions, there's no one running the studios who's qualified, either, so you have a vacuum, and the director has to fill it.[24]

Coppola's emphasis on "earned" is especially significant, since for him the expressionistic privileges of auteurism are directly related to financial actualities of investment and risk; an auteur earns his status by spending himself, and both gestures involve the aggrandizing, demeaning, and "expending" of oneself through a primary identification with the agency and exchange of money. Thus, the complement of a self that is constructed as a financial agency is degradation of that self as merely a financial product. For Coppola, "the artist's worst fear is that he'll be exposed as a sham,"[25] namely, that an audience's financial investment in his agency will be revealed as only commercial advertisement.

The image of the divided self curiously mirrors the obsessive geniuses found in Coppola's liturgical and operatic narratives. From the first two *Godfather* films through *Apocalypse Now* and *Tucker,* his visionary characters invariably pursue grandiose spectacles that reflect their desires but that either literally or metaphorically then serve to destroy them. While these spectacles frequently echo their nineteenth-century origins (lavish visuals and operatic soundtracks), the more exact terms of their agency as cinematic characters are the contemporary spectacles of industrial technology as a financial investment (for war, for corporate industry, for the business of the family). *The Conversation* is the most appropriate example. Driven by the passions of the protagonist, Harry Caul, it is a conversation through technology that leads to the absolute collapse of a sacrosanct individuality. Coppola's description of Harry, the devout and tortured Catholic, could indeed describe Coppola himself as auteur: "he's a man who has dedicated his existence to a certain kind of activity, to technology, and who in a part of his life experiences regrets and realizes that the weapon he uses for others in a certain fashion is destroying the man himself. . . . the single reason for which he is destroyed is perhaps that he has started to question all that."[26] If Hollywood's commercial industry is the financial agency that makes and unmakes Coppola the auteur, Coppola remains driven to invest and lose that self in ever-grander forms of its technological spectacle. Perhaps the most extraordinary and thus indicative example of this tendency are his technological dream projects: a giant domed theater in the Rocky Mountains, or an imagined film megalopolis where four elaborate video films would draw on Goethe's *Elective Affinities* to tell the story of Japanese-U.S. relations.[27]

Not coincidently, I think, an interview with Coppola becomes a media performance focused on the technology and the business that define and threaten him. Worried about the casual setting of his meeting with this Coppola, for instance, one interviewer found relief in Coppola's immediate identification with the technology of performance. "I needn't have worried. The minute I switched on my tape recorder, Coppola came to life. This was work. First, he corrected the position of the machine, then he fiddled with the volume and tone controls till he had them set to his satisfaction. Finally, he allowed me to question him."[28] During the interview, Coppola constructed himself as both an entrepreneur orchestrating the forces of technology and as a character lost in the improvisations of Hollywood business. He describes his expectations and frustrations about the Academy Awards, his struggles with Paramount executives to cast Marlon Brando in *The Godfather*, and he acknowledges that this most famous vehicle for his agency as an auteur had less to do with his control of the film than with submission of self and the loss of energy. "A lot of the energy that went into the film went into simply trying to convince the people who held the power to let me do the film my way."[29]

Ultimately, of course, the expenditure of energy and the loss of self are the contradictory measures of Coppola as auteur. Evaluating the ratio of his position as artist against his possible decision to actually assume the full agency of a studio (which he would do with the phoenixlike Zoetrope Studios), he casts himself according to the finances of running a large piece of technology:

if I were running a studio, it might take me 100 B.T.U.S worth of energy to bend something a quarter inch; if I stay independent and use my own resources, those 100 B.T.U.S could bend something a foot. . . . But look: The average executive of a movie studio may make $150,000 a year, and have a corresponding power, over his company. As a film artist I make much, much more than that and, consequently, have that much more power over my company. . . . Perhaps the wisest thing to do is to use all my energies to make a film that grosses some stupendous amount, then go out and buy a major company and change it from the top.[30]

Appropriately, for this elaborate characterization of himself at the turning point in his commercial career, Coppola begins this 1975 interview by claiming, "this is my last interview."[31] Within this glossy, high-tech conversation with an interviewer for the glossy high-tech magazine *Playboy*, he must naturally be given up from the start.[32]

Attempting to synthesize his relation to his movies in the manner of the Big Picture, Coppola becomes the presiding genius of the film of himself. However, this genius is represented not in expression or productive control but in expenditure and loss: loss of control, loss of money, loss of vision, and loss of self. His renowned posture as a risk-taker thus becomes a bombastic effacement of any distinguishing differences between his intentions and the films. In 1982 Jonathan Cott asked Coppola about the publicity gained from adopting this posture of loss during the production of *One from the Heart*:

The real answer, from my point of view, is that I just say what the facts are; in this case, that I'm working on a film, I'm told that the money's gone, and that if I want to go ahead, I'll have to risk something of my own. And by that point, I'm so far into it that I say okay. . . . And then that tends to be the story that the people who write about me want to go for. If they ask me, "If the picture's a flop, will the company go out of business?" and if I say yes, it's because that seems to be the case. But it's not the idea that I want to push out into the public. In fact, I regret that I'm treated more as a charlatan or a con man than as a professional person, and to be honest, my feelings are hurt. I feel that I'm not reckless or crazy. It's just that I'm primarily interested in making films more than amassing money, which is just a tool. If someone suddenly gave me a billion dollars, for instance, I'd only invest it in my work. I will say yes to anything that seems reasonable to me, and sometimes I get in a little deep because I want to participate so much.[33]

A key moment in this interview, as with other Coppola interviews, is the appeal for sympathy, not the distance of authority. It elicits a kind of social and psychological identification between Coppola and an imagined interlocutor, like that between a spectator and an actor-victim in an epic movie. For Coppola, the auteur communicates from one heart to another. For him, specifically, his is a self-portrait of the auteur as persecuted and dismissed by the operatic conglomerates who have made him a powerless vehicle in their success. "I've done so much for them, and yet they resent even putting me in a position where I don't have to go to one of them with my hat in my hand and have them tell me what movies I can or cannot make."[34] At other times, with astonishing dexterity, he rhetorically moves between an image of himself as the powerful agent of a financial and technological machine and an image of the completely insignificant individuality that inhabits that agency:

You know what I think? I think people are afraid of me, basically. They're afraid if I ever got like too much power, I'd change their lives, and they're right! . . . I'm only a minor representative of the times. I may be a schmuck, but you can be sure that some other people somewhere are going to start doing the same kind of stuff, and the world is going to change. . . . As for myself, I'm not worried. What the hell! If I don't do it with this film, I'll go and invent some little gadget that will make billions![35]

The sympathetic enlistment of an audience has become the path for locating multiple subjectivities ("other people somewhere") within an agency that disallows the authority or stability of any single organizing perspective. In this action, Coppola puts into play the central problem of contemporary auteurism as an interpretive category; while it remains a more powerful figure and agency than ever, it is invariably forced to disperse its authority in terms of its commercial agency.

In the end, Coppola within the commerce of auteurism remains a most utopian figure, for whom the spectacle of self-destruction becomes a way back to self-expression. For him, the destruction of the authority of the auteur can mean the resurrection of a world of private auteurs, an intimate yet expansive network of electronic communication. Speculating on the future of new technologies that regenerate themselves through money made and money spent on them, he proclaims a home-video exchange that somehow retains the aura of auteurist agency, the expressive "I" becoming a third-person plural: "Everybody will use it, everybody will make films, everybody will make dreams. That's what I think is gonna happen. You'll ship 'em over to your friend, and he'll ship one back. . . . I think that, very shortly, there's going to be a whole new approach to things, and the designers and the architects and philosophers and artists are going to be the ones to help lead the society."[36]

The Age and Aging of Auteurs

There are many kinds of auteurs in contemporary film culture. And there are many strategies through which a moviemaker can employ the agency of auteurism and by which audiences can use it as a way of understanding films. Both European and Asian filmmakers, for instance, have complicated this category for many years, and one only has to look to German or British filmmakers such as Rainer Werner Fassbinder or Peter Greenaway to get a sense of the range of such revisions of auteurist practices. In the United States the "displaced Hollywood auteur," more so than ever before, raises questions and issues that need to be addressed. From Wim Wenders to Peter Weir, this group of global filmmakers situated in Hollywood suggests contradictions and alternative descriptions to the history and coherence of auteurism. Even in my emphasis on the massive role that television has had in redefining auteurist practice today, I feel that I have only sketched the first stages of its impact.

In "The Unauthorized Auteur Today," Dudley Andrew has added another dimension to auteurism that my model does not—or I think cannot—take into account. Following Gilles Deleuze's suggestions about the relation between an auteur's signature and a temporal "duration," Andrew puts forward the idea that the largely spatial description of the commerce of auteurism laid out here—one which plays across public and private space—forgets or is asked to forget about a crucial dimension of temporality as a figure of the auteur. "The signature," Andrew posits, "embeds within it—as a hypertext—a genuine fourth dimension, the temporal process that brought the text into being in the first place. . . . The auteur marks the presence of temporality and creativity in the text, including the creativity of emergent thought contributed by the spectator."[37] One common characteristic of new American auteurs like Tarantino or David Lynch is the instantaneity of their careers. The brief duration of auteur celebrity these days parallels the rapid turnover of postmodern consumerism. Without being nostalgic, I would at least like to follow Andrew and urge this notion of duration as the missing element in the account of the modern American auteur.

Duration is a distinguishing characteristic of another sort of auteur today—the auteur who creates a figure of time as enduring, evolving within the commerce of its expression. Coppola ultimately might belong and be reclaimed here, as certainly do Martin Scorsese and Robert Altman.[38] These temporal auteurs demand that we encounter them across the historical vicissitudes of a commercial agency in which they have sometimes triumphed and sometimes faltered, temporal vicissitudes that, in today's climate, take on a variety of shapes and figures. Contemporary auteurs appear "immediately"; they "historically remake" themselves; there are "futuristic" auteurs and "nostalgic" auteurs; some auteurs become identified with the histories of different nationalities and some with

the temporalities of different genders. In the final analysis, the achievement of auteurs today will reflect far more than their individual films and
be found in a figure of temporality that their work dramatizes.

In whatever shape and in whatever agency, then, auteurs are far from
dead. In fact, they may be more alive than at any other point in film
history. This particular interpretive category has of course never addressed audiences in simple or singular ways. Yet, within the commerce of
contemporary culture, auteurism has become, as both a production and
interpretive position, something quite different from what it may have
been in the 1950s or 1960s. Since the early 1970s, the commercial condition-
ing of this figure has successfully evacuated it of most of its expressive
power and textual coherency; simultaneously, this commercial condition-
ing has called renewed attention to the layered pressures of auteurism as
an agency that establishes different modes of identification with its audiences. However vast some of their differences as filmmakers may be,
they each, it seems to me, willingly or not, have had to give up their
authority as authors and begin to communicate as figures within the
commerce of that image. For viewers, this should mean the pleasure of
engaging and adopting one more image of, in, and around a movie without, perhaps, the pretenses of its traditional authorities and mystifications.

Notes

1 Tarantino's best-known effort as an executive producer is probably *Killing
Zoe* (1994).
2 See Jim Hillier's introduction to *Cahiers du Cinema: The 1950s*, ed. Jim Hillier
(Cambridge, Mass.: Harvard University Press, 1985), pp. 1–17.
3 A collection of the major debates and documents about auteurism can be
found in *Theories of Authorship*, ed. John Caughie (London: Routledge, 1981).
4 See Roland Barthes, *Image–Music–Text* (New York: Hill and Wang, 1977),
pp. 142–48.
5 This point suggests, I believe, one way to discuss the way that cult follow-
ings of certain directors and films has become increasingly popular. In some
cases a kind of identification and appropriation of the image of the film or
filmmaker can replace the need for interpretive meaning.
6 David Bordwell, *Narration in the Fiction Film* (Madison: University of Wisconsin Press, 1985), p. 211.
7 *Theories of Authorship*, ed. Caughie, p. 2. See also Ben Brewster's "Brecht
and the Film Industry," *Screen* 16 (Winter 1976–77): 16–33.
8 Charles Taylor, for instance, has argued a model of human agency that

foregrounds "second-order desires" where the "reflective self-evaluation" of "the self-interpreting subject" has as its object "the having of certain first-order desires." Similarly, Anthony Giddens suggests a materialist model of expression as self-reflexive action: the motivation of expressive action, the rationalization of that action, and the reflective monitoring of action concomitantly interact to map the structure of expression as a reflective social discourse that necessarily calls attention to the material terms of its communication. In both cases, agency becomes a mode of expression (or more exactly enunciation) that describes an active and monitored engagement with its own conditions as the subjective expresses itself through the socially symbolic. See Charles Taylor, *Human Agency and Language: Philosophical Papers* (Cambridge: Cambridge University Press, 1985), pp. 43, 28, 15; and Anthony Giddens, *Central Problems in Social Theory: Action, Structure, and Contrast in Social Analysis* (Berkeley: University of California Press, 1983).

9 A direct and oversimplified example of this position, which responds to the contemporary status of the auteur but fails to reflect on its larger cultural and critical implications, is Joseph Gelmis's *The Film Director as Superstar* (Garden City, N.Y.: Doubleday, 1970). "Over half the movie tickets sold today," he notes, "are bought by moviegoers between the ages of sixteen and twenty-five. They know what a director is, what he does, and what he's done" (p. xvii).

10 Steven Bach's *Final Cut: Dreams and Disasters* (New York: William Morrow, 1985) is a particularly graphic and detailed account of the contemporary auteur as a self-promoting celebrity. A sophisticated analysis of this tendency around the globe (specifically in Germany) is Sheila Johnston's "A Star is Born: Fassbinder and the New German Cinema," *New German Critique* 24–25 (Fall / Winter 1981–82): 57–72.

11 The phrase "camera stylo" is from Alexandre Astruc's "The Birth of a New Avant-Garde: *le camera-stylo*," *Ecran Français* 144 (1948).

12 According to Robert Sklar, the uncut version of the movie was "rescued by home video from the limbo of lost films." "Homevideo," *Cineaste* 14, no. 4 (1987): 29.

13 In "The New Hollywood," Thomas Schatz lays out the different stages that describe Hollywood's evolution toward blockbuster production. See *Film Theory Goes to the Movies*, ed. Jim Collins, Hilary Radner, and Ava Collins (New York: Routledge, 1993), pp. 8–36.

14 Mark Crispin Miller states the negative view most strongly in his "Hollywood: The Ad," *Atlantic*, April 1990, pp. 41–54.

15 Justin Wyatt and R. L. Rutsky, "High Concept: Abstracting the Postmodern," *Wide Angle* 10, no. 4 (1988): 42.

16 Disney's 1993 purchase of Miramax is one of several indicators that this industrial field remains very much in flux, showing that independent voices can become quickly assimilated while new ones appear on the horizon.

17 Meaghan Morris, "Tooth and Claw: Tales of Survival and *Crocodile Dundee*," in *Universal Abandon?: The Politics of Postmodernism*, ed. Andrew Ross (Minneapolis: University of Minnesota Press, 1988), pp. 122–23. See also Richard Dyer, *Stars* (London: British Film Institute, 1979).

18 David Bordwell recognizes this fragmentation of the auteur but sees it more as simply a variation on the traditional auteur-narrator: "The popularity of R. W. Fassbinder . . . may owe something to his ability to change narrational personae from film to film so that there is a 'realist' Fassbinder, a 'literary' Fassbinder, a 'pastiche' Fassbinder, and so on" (*Narration in the Fiction Film*, p. 210). I believe that mobilizing these different agencies within an auteurist category has important implications.

19 Richard Macksey, " 'The Glitter of the Infernal Stream': The Splendors and Miseries of Francis Coppola," *Bennington Review* 15 (1983): 2, 3.

20 Robert Lindsay, "Francis Ford Coppola: Promises to Keep," *New York Times Magazine*, July 24, 1988, pp. 23–27.

21 Jon Lewis, *Whom God Wishes to Destroy* (Durham, N.C.: Duke University Press, 1993), pp. 21, 22.

22 Andrew Sarris, "O Hollywood!, Oh Mores!" *Village Voice*, 5 March 1985, p. 5.

23 Jeffrey Chown, *Hollywood Auteur: Francis Coppola* (New York: UMI Research Press, 1981), p. 175.

24 William Murray, "*Playboy* Interview: Francis Ford Coppola," *Playboy*, July 1975, p. 68.

25 Ibid., p. 65.

26 Gabria Belloni and Lorenzo Codelli, "Conversation avec Francis Ford Coppola," *Positif* 161 (1974): 51.

27 Not surprisingly, the ambivalent identification of the artistic self within the commerce of auteurism and its promise of the great spectacle of self becomes fraught with all the religious guilt of sin and the weight of self-sacrifice: "I am more interested in technology than I am in content. This, in some circles, is the same as admitting one is a child molester and likes it. The truth is that I am interested in a content I can't get at. I yearn to be able to move into a world where story and content is available to me; where my ideas connect into a pattern that could be identified as a story. But I truly cannot get there." Francis Coppola, "The Director on Content," *Washington Post*, August 29, 1982, p. 3D.

28 Murray, "*Playboy* Interview: Francis Ford Coppola," p. 54.

29 Ibid., p. 59.

30 Ibid., p. 68.

31 Ibid., p. 54.

32 Jon Lewis's comparative discussion of Coppola's Zoetrope studios and George Lucas's Lucasfilm studios suggests how one's failure and the other's success reveals two distinct and different philosophies of the New Hollywood.

Whom God Wishes to Destroy, pp. 38–40. I would add that in many other ways Lucas seems much more directed toward what I call commercial auteurism than toward the strains of working within the commerce of auteurism.

33 Jonathan Cott, "Francis Coppola," *Rolling Stone,* March 18, 1982, p. 24.

34 Ibid., p. 76.

35 Ibid.

36 Ibid.

37 Dudley Andrew, "The Unauthorized Auteur Today," in *Film Theory Goes to the Movies,* ed. Collins, Collins, and Radner, p. 83.

38 Altman's *The Player* (1992) is one of the most incisively bitter and lucid depictions of today's production and auteurist practice as I have described it.

From Roadshowing to Saturation

Release: Majors, Independents, and Marketing/

Distribution Innovations

Justin Wyatt

While a considerable literature addresses the aesthetics of the last golden age of Hollywood—1968–1975—accounts of changes in the industrial parameters (production, distribution, and exhibition) of those years are rare. Yet a strong relationship exists between the period's aesthetic and industrial changes. Aesthetic "products" are presented to suit the conditions of the overall market, and as the marketplace shifts across time, so does the product.

During the 1970s, film distribution practices for both the majors and minors in Hollywood altered substantially, revealing a steady shift toward the mass saturation release pattern still in place in 1997.[1] Through a review of these distribution strategies, I will illustrate the interdependence of aesthetic and industrial changes in the New Hollywood. Such an analysis also uncovers the relationship between the major studios and the independents. The innovations in film distribution can be traced directly to the margins of Hollywood, to those companies operating outside the studio system.

Searching for an Audience: Distribution, Production, and the Hollywood Recession

After the success of an increasing number of big-budget films released during the 1950s and early 1960s (such as *The Ten Commandments*, 1956; *Around the World in 80 Days*, 1956; and *Ben-Hur*, 1959), the studios began to focus on the large-scale, roadshow picture as the anchor of their distribution schedules. In many respects the roadshow picture was just another attempt to present film as different from television—or, to be more precise, as bigger, grander, and more spectacular than television. Roadshowing, with reserved seating, limited showings (usually ten or twelve per week), and lavish film subjects, transformed the act of moviegoing into a special occasion—an event. Of the components of the roadshow, a name cast and high production values (often linked to a large budget) were commonly accepted as the most significant for a successful engagement.[2] Studios furthered the practice through developing full-scale sales and publicity departments specializing in roadshow presentations.

Since these films could take months to open nationwide, roadshowing allowed studios to refine sales and advertising approaches. This method of distribution met several challenges, so the added time came in handy. Most obviously, studios incurred the greater expense of a hard-ticket engagement, which also necessitated limited (usually one matinee and one evening) shows per day. Roadshow engagements required additional box-office personnel, ushers, and support staff, creating larger "house nuts" (the cost of running the theater, which is routinely subtracted from the box-office gross before a percentage of the box-office revenues are returned to the studios). The increase in house nuts decreased studio profits on these films. Such a shortfall was exacerbated by the limited number of screenings.

Despite seemingly successful roadshow experiences with *The Bridge on the River Kwai* (1957) and *Lawrence of Arabia* (1962), Columbia Pictures executives, for example, felt so constrained financially by the traditional roadshow schedule of ten shows per week that, with the release of *A Man for All Seasons* in 1965, the studio increased the roadshow schedule to sixteen performances each week. Attendance at *A Man for All Seasons* also was bolstered by group sales that were attracted to the "event-quality" of the roadshow. Fund-raising groups were able to buy blocks of tickets at a 10–20 percent discount;[3] groups then would add an amount to raise funds for their cause or activity.

While these fund-raising activities helped roadshows to recoup a greater amount in a shorter period, the increased number of roadshowed films created a more competitive market for these "events." As Twentieth Century-Fox sales vice president Abe Dickstein commented, "If there's one thing that will kill the roadshow business, it's presenting films on a reserved-seat basis that just don't warrant this special attention."[4] To remedy this problem, studios resorted to elaborate advertising and publicity opportunities to establish a particular film in the marketplace. Columbia, for instance, was concerned about the lag between the completion of principal photography on the roadshow musical *Funny Girl* and its 1968 release date; the lengthy gap of almost a year between production and release could diminish press interest. Columbia and producer Ray Stark provided an array of impressive publicity materials during the interim: a five-minute collage of stills set to Barbra Streisand singing "I'm the Greatest Star," a fashion short made for women's clubs and merchandisers, a ten-minute documentary for theatrical release, a CBS-TV show titled "Barbra in Movieland," and a lengthier documentary, *This Is Streisand*, for the film's European release.[5]

Roadshows eventually began to target specific audiences.[6] United Artists helped to pioneer the "ethnic roadshow" through *Exodus* (1960) and *Cast a Giant Shadow* (1966), both films concerning Israel's battle for independence. Targeting the substantial Jewish population in New York City, UA set *Cast a Giant Shadow* as a roadshow attraction only in theaters there. The attempt to connect with a Jewish audience was furthered by scheduling extra holiday matinees during Passover.[7] Paramount, on the other hand, was credited with developing the youth roadshow with Franco Zeffirelli's *Romeo and Juliet* (1968); focusing on the film's young stars, Paramount scheduled a series of youth screenings and tie-ins with publications such as *Seventeen*.[8]

By 1968 the roadshow had become significant to every studio's yearly distribution slate. As *Variety* analyst Lee Beaupre reported, "Roadshow is the name of the game now being played by the majors. At current count 12 pix are expected to go out on hardticket in the last four months of this year, as compared to only 10 reserved-seaters during the previous 20 months."[9] While the emphasis on spectacle produced some large-scale hits, the bottleneck of expensive roadshows by the end of the decade proved problematic for the studios. As Darryl F. Zanuck, then head of Twentieth Century-Fox, commented in 1968: "We've got $50 million tied up in these three musicals, *Doctor Doolittle*, *Star!*, *Hello, Dolly!*, and quite frankly if we hadn't had such an enormous success with *The Sound of Music*, I'd be petrified."[10] Fox did, in fact, realize a substantial loss on these films, posting overall losses of $36.8 million in 1969 and $77.4 million in 1970. Other studios—MGM, Warner Bros., United Artists, Paramount—also suffered significant losses around the start of the 1970s. Conditions became so dire that from 1968 through 1971, total industry losses exceeded $500 million.[11]

This crisis indicated the poor fit between the youth audience, an increasingly important part of the filmgoing public, and Hollywood's focus on the large-scale roadshow film. This situation proved to be beneficial to independent companies, as some of the most financially successful youth films were produced outside the majors. Avco Embassy released *The Graduate* (the top-grossing film of 1968), and a small production company, BBS, produced *Easy Rider* for Columbia. Independent companies thrived by exploiting market segments ignored by the majors. These smaller studios were able to prosper in this environment through two methods: (1) by working with subject matter untouched by the majors and (2) by operating outside the traditional realm of the majors in terms of distribution. The independents were able to forge a market identity through these methods, but in the long run such innovations—both aesthetic and institutional—were subsumed by the major studios.

Independent Opportunities:
Markets and Marketing Beyond the Majors

The margins of the major studios extended considerably in the era following the Paramount decision, a period during which the studios essentially became distributors.[12] With projects being set up on a case-by-case basis, and the old studio system unraveled, the majors began to lease their lots and cut back considerably. Distinctions between the "Big 5" (Loews/ MGM, Fox, Warners, Paramount, RKO) and the "Little 3" (Columbia, United Artists, Universal) became blurred with these industrial changes. Smaller companies (for example, American International, Roger Corman's New World) maintained a presence through offering a clearly differentiated product such as the exploitation or teen film. Those companies presenting "typical" Hollywood fare, such as National General, Allied Artists, and Associated Film Distribution (AFD), primarily marketed their projects around stars and found the competition too fierce for long-term survival.

Art-house films thrived in urban centers, signaling a renewed interest in foreign nations and an apparent audience desire to view film as akin, under certain circumstances, to the high arts of literature, music, and drama. The art film benefited from the decreased production among the majors; as the industry trade paper *Variety* commented in 1953, "There is a feeling among the film importers that, with a general product shortage in the offing, imports from abroad—in both subtitled and dubbed versions— are heading into a somewhat brighter future."[13] While these art films were primarily foreign, a small number of independent American films also served this market throughout the period. By the early 1960s the art cinema had matured through the efforts of companies such as Cinema V, run by Don Rugoff. Cinema V, often cited as a model for such later 1990s' independents as Miramax and New Line, operated in part by establishing an identity based on its otherness from the major studios. This otherness—through products and business practices—greatly aided the continued presence of the art cinema.

Cinema V grew from Rugoff and Becker theaters, a chain started in 1921 by Don Rugoff's father, Edward, and Herman Becker. When the elder Rugoff died in 1952, the chain had just opened its first theater on the Upper East Side of New York, the Beekman.[14] After Becker's demise in 1957, Rugoff gained sole control of the company, and he began a quick expansion in the burgeoning world of art-house exhibition, building one of the first twin theaters, Cinemas I and II, in 1962, and controlling such prestigious theaters as the Paris, Baronet, Plaza, and Sutton, all on the up-scale Upper East Side. Added to other Cinema V houses in New York—the Murray Hill, Gramercy, Art, and Paramount theaters—Rugoff established himself as the premier exhibitor of art-house product. At the time, Rugoff claimed that the expansion responded to a greater visibility and awareness of art-house product in the United States: "Now I meet people who don't know I'm in the business and they start talking about movies as an art form. Much more so than plays or books or anything else. I think the movies have created more excitement in recent years than any other form of art. . . . I don't know if there's going to be enough product to go around. But I think that new theaters are going to stimulate the growth of moviemaking."[15]

Simultaneously, with the backing of a consortium headed by composer Richard Rodgers, Cinema V branched into distribution, starting with the British comedy *Heavens Above!* in 1963.[16] Sticking with a slate of art films, Cinema V was responsible for distributing such influential independent projects as Shirley Clarke's *The Cool World* (1963), Michael Roemer's *Nothing But a Man* (1964), Larry Peerce's *One Potato, Two Potato* (1964), and François Truffaut's *The Soft Skin* (1964). Structurally, the ties between distribution and exhibition were crucial to the success and longevity of Cinema V. With control of the art houses in New York divided between Rugoff, Walter Reade, Jr., and Dan Talbot (New Yorker Films), Rugoff was assured a market for his Cinema V product.

Despite several successes during the decade, though, the market for art film was product-driven. Given the discerning filmgoers at the core of the art-house audience, attendance for each film was key to the market. As a result, Rugoff eventually encountered some downtime in distribution, which by the early 1960s was remedied by a shift in demographic focus to the youth audience. Rugoff claimed by the mid-1960s that the teen and college set was "the future for Cinema V."[17] This strategy proved successful through two 1966 films: Karel Reisz's antiestablishment comedy *Morgan!* and Bruce Brown's surfing documentary *The Endless Summer.* With the surfing film, Rugoff scheduled an intermission even though the entire picture ran only ninety-one minutes; the entrepreneur believed that audience members would want the break to discuss the action and surfing styles depicted up to that point. This unusual practice illustrates the care with which Rugoff approached each project—seeking to maximize the uniqueness and marketability of each film.

In this manner Rugoff and Cinema V responded to the difference of each film from mainstream fare and then sought to exploit this difference as much as possible. This approach responded directly to the perceived lack of promotion given to imported and independent films during the period. As *Variety* described this problem, while imported films compare favorably in storytelling, human interest, and artistic merit, "they simply don't promote. . . . Their product is unsold in the land of sell. They remain hopelessly inferior competitors in the department of make-known and make-attend."[18] *Variety*'s call for "special treatment" and "slow-build releases" was met by Rugoff, who clearly appreciated niche marketing.

Print advertising for the films became the primary focus for this treatment. Following a graphically simple, visually distinctive approach to visual marketing often associated with Saul Bass's film credits and advertising designs, Rugoff believed in simple, recognizable logos to distinguish each project. Even for those films playing at Cinema V theaters from other distributors, Rugoff reserved the right to reject the ad campaigns, and he reportedly discarded 90 percent of them in favor of his own designs.[19] Rugoff worked closely with the Diener, Hauser, Greenthal advertising agency in designing the Cinema V logos. Among the most memorable were a trio of silhouetted surfers on the beach for *The Endless Summer* (1964), an embracing couple for *Elvira Madigan* (1967), and, most provocatively, a hand with a scantily clad girl replacing the raised middle finger for Robert Downey's satire of the advertising industry, *Putney Swope* (1969).[20] Once an image was shaped for the campaign, Rugoff "wild-posted" the ad across major cities playing the film, so that a visual reminder of the picture would appear in both ordinary and extraordinary settings.

Whenever possible, Rugoff was aware of the publicity benefits connected to these images. For the Dušan Makavejev film, *WR: Mysteries of the Organism* (1971), Rugoff fashioned an ad featuring a photograph of a young woman eating a banana. With the New York play date in December 1971, the photo and critical quotes were subordinated to a larger story attending the refusal of three Boston papers to advertise the film.[21] One paper, the *Record-American,* offered to run the ad if the word "organism" was omitted (because it seemed too close to "orgasm"). The *Globe* flatly refused to advertise the title at all, as did the *Herald Traveler,* which had a policy not to advertise any X-rated films. After Rugoff used the Boston ban to advertise the picture, a series of pieces appeared in the New York papers on the principles of cinema advertising, all of which mentioned the film. That Rugoff used the controversy to promote the film was hardly a secret or a surprise to his fellow exhibitors. As one observed at the time: "Latest ad ploy [for *WR*] comes after a number of varied campaigns, none of which seemed to have captured the essence of [the film's] sexual-political satire." The attempt to create media controversy, a "media moment," around a current release was replicated many times by Cinema V.

The successful exploitation of the youth audience, albeit still within the context of the art cinema, proved instrumental to another lucrative market, one more amorphous and marginal—the adult film. Despite the reluctance of the independents, the MPAA Rating System instituted in 1968 acted as an economic benefit, effectively segmenting the marketplace for film. The "X" rating became synonymous with stronger adult (later pornographic) content. The consequences of the X rating for the majors were dire. Approximately 50 percent of theaters across the country refused to play X films, and as many as thirty large city newspapers, along with many television and radio stations, refused to advertise them.[22]

The MPAA Rating System, the growth of the art and underground cinema, "X" as a marketing tool, and the confusions around censorship, all served to nurture the adult film as a marketplace phenomenon. Growing from the tradition of art films that seemed to stretch the boundaries of free expression (*La Dolce Vita,* 1961; *And God Created Woman,* 1956; *Room at the Top,* 1958), the first adult features were supplied by European producers. Films from Sweden and Denmark were screened with subtitles in art houses, thus giving them an aura of class and sophistication. The economic power of the adult film became stronger toward the close of the 1960s, ranging from the foreign imports, such as *I Am Curious (Yellow),* breaking capacity records in New York in 1969 and placing number one on *Variety*'s Top Grossing Films Chart (November 26, 1969), to the sexploitation films, such as *The Stewardesses, I, A Woman,* and *Therese and Isabelle.* (The possibility of any soft-core independent film reaching the top of the *Variety* chart currently is ludicrous, with the contemporary market for this product primarily dominated by videocassettes and cable.)

The increased visibility of the adult feature toward the end of the decade accompanied greater freedom in content. In 1968, for example, a case before the Maryland state censorship board demonstrated that exploitation features had progressed from upper torso nudity to full frontal nudity in features such as *Walls of Flesh* and *Savage Blonde.* As a trade paper reported on this transformation, "Time was when a nudie made exclusively for the exploitation house had the male retain some dress, usually a pair of shorts, and the female her panties."[23] The issue was extended even further with the U.S. distribution of the Swedish film *I Am Curious (Yellow).* Featuring simulated sexual intercourse in medium and long shot, the film offered a mix of political, sexual, and social satire with a story centering on the "radical" life-style choices of Lena, a young Swede unbound by sexual and social conventions. Banned outright in Norway, Vilgot Sjoman's film was censored heavily in Britain, France, and Germany.[24] Grove Press picked up the American distribution rights, only to have the film impounded by the U.S. Customs Service. Until the U.S. Circuit Court of Appeals, Eastern District, ruled that the film could be shown uncut, Grove Press continued to maximize publicity around the movie through releasing a paperback copy of the script, with over 250 stills for those who preferred looking at the pictures. When the film finally was cleared for release, Grove was able to extract strong terms from exhibitors as a result of the media controversy; the distributor asked for $50,000 in advance and a 90/10 split favoring the distributor after recoupment of the advance.[25] The New York opening in March 1969 was phenomenally successful, with an opening week of $91,785 at two small theaters, the Cinema 57 Rendezvous and the Evergreen.[26] Within six months of release, the film grossed $4 million in fewer than twenty-five theaters. *I Am Curious (Yellow)*'s performance illustrates the power of publicity and the ability of one film to exploit the national dialogue over sexual freedom and expression.

Given that *I Am Curious (Yellow)* is a dry, didactic film, distinguished only by its sexually frank material, its box-office performance was extraordinary. As the reviewer for *Time* wrote: "If it were not for the sex scenes, *Yellow* would probably never have been imported. It is simply too interminably boring, too determinedly insular and, like the sex scenes themselves, finally and fatally passionless."[27]

By the end of the 1960s the market for adult film had branched into three distinct areas: adult dramas that incorporated increasingly explicit sex scenes and subject matter; softcore pornography often using X as a ratings attraction; and hardcore pornography that was limited to large cities and linked to hardcore bookstores and strip clubs. Softcore was distinguished from hardcore by the degree of "realism": soft involved simulated sex, while hard included insertion and orgasm shots. Reflecting the increased competition from mainstream cinema, the Presidential Commission on Obscenity and Pornography estimated that adult film receipts dropped 10 percent to 20 percent from 1969 to 1970 because of "increased competition from sexually oriented motion pictures playing outside the exploitation market."[28]

The majors responded to the more liberal climate by distributing a larger number of explicit films often made "palatable" by their association with adjoining art forms. Adaptations of *Goodbye, Columbus* (1969); *Portnoy's Complaint* (1972), and *Rabbit, Run* (1970)—all based on "serious" works of fiction—arrived on screen with the trades cynically describing their appearance as an attempt by the major studios "to broaden sexploitation."[29]

Cinema V and Continental continued to buttress their release schedules with adult-oriented dramas: *WR: Mysteries of the Organism, Ulysses* (1967), *Putney Swope,* and *Trash* (1970). These films all contained explicit material, yet were perceived as part of an "art cinema." The combined distribution and exhibition arms of both Cinema V and Continental aided these films that were alternately too strong, too esoteric, and too "foreign" for the majors, and too arty and too serious for the porn market.

The art and adult markets demonstrate the viability of independent cinema during the turbulent era in which the major studios were experiencing a considerable recession. These markets were able to thrive by stressing their difference from mainstream filmmaking and by targeting youth as a market segment. In addition, both markets prospered by focusing on advertising and separate exhibition opportunities rather than competing with the majors for circuit theaters. These lessons would be underlined even further with the burgeoning family market that was being served by independent distributors. Through acknowledging the innovations of these markets outside of their primary focus, the major studios eventually were able to regain their prominence. This comeback illustrates how innovations in the independent film world eventually filter through to mainstream studio moviemaking.

Distribution Shifts:
Toward Four-Walling and Saturation Releases

While the markets for art and adult cinema were crucial for the independent companies of the period, perhaps the most significant innovation in distribution during the era can be traced to another segment marginalized by Hollywood, the family film. The practice of four-walling, in which the distributor rents the theater outright, was fostered greatly in the mid-1960s by American National Enterprises (ANE).[30] Located in Utah, ANE concentrated on family wilderness adventures starting with *Alaskan Safari* (1965). While four-walling had been done on occasion as far back as the silent era, traditionally the distributor and exhibitor split box-office receipts according to a predetermined formula (often a sliding percentage of the gross, favoring the distributor). The four-wall arrangement placed a greater onus on the distributor, since the exhibitor received the theater rental up front. The upside of the deal for the distributor was that, for a high-grossing film, it retained the vast majority of the box-office revenue.[31]

The form of marketing attached to four-wall projects, as practiced by ANE and similar companies such as Pacific International Enterprises and Sunn Classics, was fundamentally different from films distributed more traditionally. As the four-wall engagements were limited to one or two weeks, the films needed immediate audience awareness and interest.[32] The four-wallers relied heavily on saturation television ad campaigns targeted precisely at specific audience segments through market research conducted beforehand. Distributors of four-wallers needed to gauge their audience specifically. As Frederick Wasser recounts, "Surveys taught the producers that their prime customers were lower-income families with earnings in the ten to twelve thousand dollar range, two or three children, and limited schooling."[33] These companies fostered a market identity around the concept of family, attempting to establish an image in which the company name would become synonymous with family entertainment. The Walt Disney Company was often cited as a role model for this goal.[34]

While the studios might have downplayed the success of these companies as the consequence of aggressive marketing and their ability to serve a starved but small family audience, the case of Tom Laughlin's film *Billy Jack* (1971) could not be so easily dismissed. Tapping into concerns voiced by the young, *Billy Jack*'s eponymous hero (played by Laughlin)

9 Tom Laughlin assumes control in *Billy Jack* (Warner Bros., 1971).

was a half-breed and a self-appointed guardian of Indian rights. Most of the film's action revolved around the hero's attempts to help a school on the reservation against the violent and racist residents of a neighboring town. *Variety* assessed the film's box-office potential on its release in 1971: "Film is strictly for selected situations where its chances are uncertain, but reception may benefit by feature also reflecting some of the troubles of present-day youth."[35]

As released initially by Warner Bros., *Billy Jack* garnered little interest. Laughlin, who believed that Warners did not adequately support the film in the market, filed suit, charging that Warner Bros. "didn't give the picture the proper 'push.'"[36] Settling out of court, Warners agreed to a reissue and a renegotiated distribution deal with Laughlin. The revised deal stipulated an even split of both profits and costs for the reissue.[37]

While the new arrangement covered Warners in terms of advertising expenses, Laughlin was able to realize a much greater share of the revenue than under his previous deal. His approach with the reissue was to four-wall the film. Armed with much data about the specifics of each engagement, including demographics of the region, educational characteristics, population density, and ethnic composition, the goal was to saturate the airways with commercials aimed at different demographic groups.[38] Separate ads foregrounded romance, countercultural / anti-Vietnam War aspects, action, and martial arts—a varied campaign designed to reach a broad spectrum of moviegoers. A large number of theaters were rented within the region of the television signal, maximizing the convenience of attending the film. As Laughlin commented, "A filmgoer should be able, after being dunned relentlessly by the campaign, to fall out of bed and find a theater where *Billy Jack* is playing."[39] Rereleased on May 9, 1973, with a generous ad expenditure of $250,000 in Southern California under the four-wall approach to mostly second- and third-run neighborhood theaters, *Billy Jack* grossed $1.02 million in the first week in sixty-two theaters from Santa Barbara to Bakersfield.[40] The gross represented a record box-office return for the region. Weekly figures were similarly impressive in such major markets as New York ($1.45 million), Philadelphia ($710,000), and Chicago ($600,000). Eventually the film reached more than $30 million in domestic film rentals.

Throughout the country the film was sold on a week-to-week four-wall basis. Exhibitors were reimbursed for their theater rental before the engagements started, and a uniform admission price ($2.50 for adults, $1 for children) was enforced at all engagements. While previously there had been a large number of four-wall family films, Laughlin's movie marked the first time that a PG film had been four-walled. The success of *Billy Jack* indicated that the four-wall method of distribution could be used successfully beyond the family market.[41] In addition, *Billy Jack* demonstrated that a failure in initial release could be reversed by the four-wall approach and, more specifically, by saturation television advertising targeting specific demographics. The extensive use of market research to gauge potential audience interest and demand enabled Laughlin and marketeer Max Youngstein to aggressively pursue many audience segments. The approach was built like a "military campaign," aiming for an audience from several independent segments rather than narrowly exploiting a single segment.[42]

Following Laughlin's lead, studios began to experiment with four-walling as a distribution method. At first, their attempts were limited to films that had proven financially disappointing in initial release: in 1974 Warners rereleased the Robert Redford film *Jeremiah Johnson,* Universal chose the May / December romance *Breezy* starring William Holden, and Avco Embassy attempted to resurrect *The Day of the Dolphin.* On the *Dolphin* rerelease, *Variety* described the effort: "Armed with newfound marketing ploys such as saturation video campaigns, four-walling, and tested and retested advertising campaigns, distributors [are] try[ing] anything short of shanghai-ing people off the street to improve grosses."[43]

Distributors eventually broadened the range of films released under four-wall situations. For instance, six months after its initial release, Warner Bros. four-walled its blockbuster *The Exorcist* as a means to rejuvenate the film late in its release.[44] Within the independent film world, distributors also began to four-wall films outside the family genre: arthouse films (the 1953 Italian opera film *Giuseppe Verdi*, rereleased in 1974), religious / inspirational films (*The Hiding Place*), and even gay porn (Wakefield Poole's *Boys*).[45]

While four-walling grew as a market practice, a backlash developed within several industry sectors. Distributors began to complain that exhibitors' terms were inflated far beyond the house nut covering the theater's operating expenses. Laughlin stated that the theaters added "so much air" to the rental price that four-walling would start to become a losing proposition.[46] Exhibitors, along with the National Organization of Theater Owners (NATO), feared that four-walling would become the primary arrangement between exhibitors and distributors. The concern was that exhibitors would be cut off from profits usually gained from hits (where the percentage split between exhibitor and distributor would benefit both if the film grosses well). As Paul Roth, NATO president, described four-walling, "That's not a marketing technique . . . it's an illegal sales technique, that would put the larger exhibitors who could afford to charge less rent, at an unfair advantage."[47]

Union projectionists demanded higher salaries for four-wall films, arguing that they should receive their slice of the four-wall windfall.[48] In the New York City case of *The Exorcist*'s four-wall engagements, the projectionists were able to secure a wage differential for the run, leading to the prediction that in the future four-walling would routinely lead to such "hidden" costs and thus no longer be as attractive to distributors.

Richard Moses, senior vice president for MGM marketing, argues that unrealistic theater rentals were not the only reason for the eventual decline of four-walling. He puts the blame on the distributors, which selected the wrong pictures for the venture.[49] Just as roadshows relied on event status, four-walled films needed to be *perceived* as special events. Admittedly, the hard-sell television approach may have created an illusion of difference that the films themselves did not warrant. Films with an exploitable premise, especially those with sensationalistic subject matter (witness the 1974 four-wall Sunn Family lineup: *Chariots of the Gods, Myste-*

rious Monsters, and *The Outer Space Connection*) that could be conveyed quickly and efficiently through the visuals in a television commercial, yielded most readily to the four-wall selling method.

Four-walling's overexposure can be gauged by the release pattern for Laughlin's next effort, *The Trial of Billy Jack* (1974). Surprising the industry greatly, Laughlin chose not to exclusively four-wall the picture, but instead relied on saturation bookings for the sequel. Given the extraordinary success of the original, Laughlin was able to extract strict terms from exhibitors for the sequel: a 90 / 10 split toward Laughlin and a minimum cash guarantee adjusted for a low house nut for each theater.[50] In cases for which Laughlin could not set deals on these terms, he chose to four-wall the film in a limited number of cities, as he did in New York, Philadelphia, and Phoenix.[51] Laughlin maintained that he did not want to four-wall the film in every location since exhibitors had been substantially padding their rental terms given the tremendous success of many four-wall films. For the vast majority of engagements, Laughlin negotiated standard, albeit stringent, contracts. *The Trial of Billy Jack* opened very wide, at 1,100 theaters, plus 180 four-wall engagements, on November 13, 1974. Box-office after the first week was $10.5 million; amazingly, given the terms of his deal, Laughlin was able to recoup the total negative and advertising costs within seven days.[52]

The major studios retained the most significant marketing components of the four-wall approach in their move to saturation releases: the heavy reliance on television advertising supported by market research and the pattern of wide distribution linked to the advertising. The shift to audiovisual marketing was crucial. Its significance can be noted by changes in the motion picture advertising budgets. From 1972 through 1974 the proportion of the average film's ad budget paid to newspapers dropped from 58 percent to 44 percent, while the amount paid to television in the same period jumped dramatically from 15 percent to 42 percent.[53] Addressing a convention of newspaper advertising executives in 1975, Jonas Rosenfield, Fox's advertising, publicity, and promotion vice president, blamed the poor reproduction of ads, discriminatory rate practices, and lengthy deadlines for the shift. Rosenfield used the example of the Fox action film *Dirty Mary and Crazy Larry* (1974) as a paradigm for effective film marketing. Produced at a negative cost of under $1 million, Fox allocated $2 million for national television advertising. The film grossed $32.5 million. Rosenfield commented that, if the marketing had focused on print instead of television, $6 million would have been the expected gross.[54]

This alteration in marketing was fostered by advertising agencies, particularly Grey Advertising, which had expertise in targeting specific demographic groups through network and local television ad buys. Columbia worked extensively with Grey Advertising in developing the campaign for the 1975 wide release of *Breakout,* a Charles Bronson action thriller about a split-second rescue from a high-security jail. Aiming to make every potential customer (particularly males 18 to 49) aware of the film prior to release, Columbia ran ads to hit 92 percent of the television households in the country and 84 percent of the entire 18–49 male target group.[55] To achieve this result, Grey used a technique known as "roadblocking" in which a *Breakout* ad ran the same hour on a designated night on every television station in the key New York market.[56] The goal was to exhaust the box office within two weeks rather than allow the film to play out across several months.

The Marketing Legacy of *Jaws*

While saturation release and television-oriented marketing were designed to diminish the effect of a critical response to a film (a hit-and-run sales tactic), the stakes were raised considerably that same year with Steven Spielberg's *Jaws.* The difference between *Breakout* and *Jaws* can be seen if we examine the marketing assets of the two films. While *Breakout* offered its star (Charles Bronson) as insurance of success, *Jaws* sported a presold property (Peter Benchley's best-selling novel released in paperback less than six months before the film's release). The marketing campaign for *Jaws* created a strong visual identity for the film: the image of the naked swimmer tiny in the ocean against the huge shark, jaws open and ready to attack its prey. Producers Richard Zanuck and David Brown, realizing the benefits of the book tie-in, actively promoted the print version of *Jaws,* even though they had no direct financial interest in those sales. Zanuck and Brown wanted to clearly establish the paperback cover image in the marketplace by the time of the film's release months later. As Zanuck commented, "We adapted the artwork from the book to the artwork in the film promotion. By the time we sneaked the film in Dallas, we didn't even need to name it in the ad. We put in the logo of the shark's teeth and the swimming girl and 3,000 came out in a hailstorm."[57]

Following from the earlier saturation and four-wall experiments, *Jaws* relied on television advertising to support opening wide (at 409 theaters) on June 19, 1975. Universal boasted "the biggest national TV spot campaign in industry history," substituting national for local television buys. Given the wide release, Universal realized cost advantages by buying national prime-time spots rather than buying a similar schedule on a market-by-market basis. Since the national television spot replaced the usual local television commercials, Universal demanded that exhibitors contribute to the national advertising based on potential local earnings; exhibitors therefore were charged between $175 and $400 depending on the number of theaters in a given market, a one-time charge as part of their deal with Universal.[58] The match of television advertising, presold property, and strong playability (*Time* aptly referred to the film as "an efficient entertainment machine") created the record opening-week gross of $14.31 million. By September 5, *Jaws* became the largest-grossing film in motion picture history up to that time, surpassing *The Godfather.* Not willing to toy with success, the foreign release of the film replicated the domestic pattern with heavy television advertising and wide simultaneous release. For example, *Jaws* was booked into more than a hundred theaters in Britain alone for the United Kingdom opening on December 26.[59]

The commercial performance of *Jaws* demonstrated many lessons to the film industry, not only reinforcing the power of large-scale national television advertising and wide release, but also illustrating the potential of a presold property and an early summer release date (it opened in June and played through Labor Day). By setting a new box-office record in such a short span of time, *Jaws* also redefined the concept of a blockbuster or breakthrough hit. In later years $100 million became a benchmark for the blockbuster. Spielberg's film also heralded the era of the high-concept film, that is, cinematic products with a narrative that could be boiled down to a single sentence and then marketed visually through print advertising and especially commercials.[60] These projects were oriented around marketing assets: stars, genre, an exploitable premise, all targeted toward specific demographics with the potential to cross over into other audience segments. The commercial equation was refined through several other films in the second half of the 1970s. These projects effectively extended the marketing and distribution opportunities of these high-concept projects.

A year after *Jaws,* Twentieth Century-Fox copied the strategy for its promotion of the horror film, *The Omen.* Launching the picture with a tie-in novelization and two weekends of nationwide sneak previews, the distributor also sank the equivalent of the film's negative cost ($2.8 million) into advertising. While the film was not based on a best-selling book, the producers generated their own $1.50 Signet paperback novelization by

Omen screenwriter David Seltzer. Based on the sneak previews and extremely positive word-of-mouth, the paperback sold out in three hours in New York and Los Angeles.[61] The film's distinctive logo—three 6's inside the O of the title—was emblazoned across the paperback and all print advertising to establish the movie's identity. While its R rating, almost by definition, limited grosses to a modest degree, *The Omen* opened at 526 theaters and garnered rentals of $28 million, making it the third-highest grosser for the year.

Producer Dino de Laurentiis, obviously admiring the box office of *Jaws*, offered another man-against-beast tale, a remake of *King Kong*, for a Christmas 1976 release. Depending on the success of the original film and the forty-three years between the two versions, the remake of the original 1933 classic was a presold property, bringing an immediate point of identification for potential audience members. De Laurentiis also heavily merchandised the film, choosing not to focus exclusively on the children's market. The promotional campaign for the film included licenses with obvious tie-ins (GAF Viewmaster pictures and 7-Eleven Slurpee cups) as well as more imaginative and esoteric ones (Jim Beam "King Kong" cocktails from a "commemorative King Kong" bottle). With a massive television advertising campaign running from December 12 until Christmas Day on national television (the film opened on December 17), de Laurentiis released the film to 961 theaters—more than twice the size of the *Jaws* opening. Given the similar themes, marketing angles, and distribution approach, the industry anticipated similar grosses between the Spielberg film and *King Kong*. As a *Variety* columnist asked after *Kong*'s opening weekend, "Can *Kong* best the Great White Shark of *Jaws* and become the biggest and fastest grossing film in the history of the biz?"[62] Although *King Kong* failed to reach that goal, the opening was robust (a $6.97 million weekend).[63]

The case of George Lucas's *Star Wars*, released initially on May 25, 1977, shifted the equation of the breakthrough hit further through an emphasis on merchandising and a box-office gross that eclipsed even *Jaws*. Lacking the presold properties of *King Kong*, *Star Wars* had to combat several marketing problems to establish a market identity. First, market research tests showed that the title alone connoted a science fiction film about a conflict in outer space, a concept that appealed to a limited audience (males under twenty-five years of age).[64] To counter that notion, Twentieth Century-Fox through its advertising positioned the film as space fantasy, with a strong element of human relationships and adventure; the studio's campaign centered on human characters in the foreground and space hardware in the background. The ad line ("A long time ago in a galaxy far, far away") was designed to evoke "a future world with the romanticism of fairy tales, myths, and real heroes."[65] Initiating a trend that became common practice during the next decade, Fox opened the film on May 25, before the usual onslaught of June openings. In certain engagements 70mm Dolby prints were used, and the initial release was limited to only forty-two theaters in thirty cities. The distributor hoped to establish positive word-of-mouth before widening the film's market in late June.

Breaking house records in many engagements, *Star Wars* proved to be even sturdier at the box office than its predecessor. By November 1977, six months after its release, the Lucas film replaced *Jaws* as the highest-grossing film ever. While its maintenance in the market obviously resulted from extremely enthusiastic word-of-mouth, the film also benefited from Fox's merchandising strategy. Fox executive Mark Pepvers commented, "George Lucas created *Star Wars* with the toy byproducts in mind. He was making much more than a movie."[66] (In fact, Lucas sought to control all merchandising rights for the film, but the final contracts specified an equal revenue split between Fox and Lucas after Fox's administrative costs were covered.)[67]

Part of the film's phenomenal success as a licensing property has been its diverse set of characters and its creative hardware. The film's completely novel environment and characters were so striking that Kenner Toys was able to go beyond the figures in the film and add new characters to their *Star Wars* line in keeping with the film's mythological world.

Superman, released for Christmas 1978, similarly exploited its merchandising potential. Producer Ilya Salkind was able to maximize the film's licensed products in large part because of the conglomerate structure of Warner Communications Inc. (WCI). Ten years earlier, the conglomerate purchased DC Comics, the publisher of the Superman adventures since 1939. This deal facilitated the licensed product developed in conjunction with the film released by Warner Bros. Another WCI subsidiary, the Licensing Corporation of America, was assigned the task of allocating rights to different product companies; major companies involved with the promotion included Bristol Meyers, General Foods, Pepsico, Lever Bros., and Gillette. Warner Books issued eight separate *Superman*-related titles, and Warner Records released not only a soundtrack album but two singles.[68] Happy about the possibilities for cross-promotion of the film, Salkind enthused, "They're all part of the same conglomerate. That was a big advantage of going with them—besides their being a very good company.

We had all the areas of merchandising and books and records and everything else with the same conglomerate. If we had dealt with another major we would have had a very difficult situation. . . ."[69]

By the end of the decade a number of other high-concept films were released—*The Deep* (1977), *Close Encounters of the Third Kind, Grease* (1978), *Jaws 2* (1978), and *Star Trek: The Motion Picture* (1979)—all of them foregrounding marketing assets, wide distribution, and aggressive marketing. The shift toward these marketing-oriented projects has led industry analyst Lee Beaupre to designate the 1970s as "Hollywood's Marketing Decade."[70] The effects of these marketing changes were far-reaching. By 1976, data illustrated that box office was being generated from a wider base of theaters, reflecting the move toward saturation releases; from 1969 through 1975 the number of theaters accounting for the *Variety* key city box-office chart grew by 54 percent.[71] With increased television advertising, costs for marketing the average film increased dramatically in the second half of the 1970s; by 1980, print and advertising expenses averaged about $6 million per film (at a time when the average negative cost—all production expenses before prints, advertising, etc.—was more than $10 million). With advertising costs escalating even faster than increases in production costs, the break-even level for films during this period became inflated. According to producer Salkind, a gross of less than $80 million for *Superman* would have been a "flat out disaster," while estimates before the release of *Close Encounters* placed the break-even point at a gross of close to $70 million.[72] With the higher box-office potential of breakthrough hits, a single film could make a studio the market share leader for the year. *Jaws* led Universal to a 25 percent market share in 1975 (the next highest studio had only 14 percent), *Star Wars* gave Fox a 20 percent market share in 1977, and the combination of *Saturday Night Fever* and *Grease* helped Paramount dominate with a 24 percent share in 1978.[73]

The pattern of wide releases, television advertising, and spiraling marketing costs continued into the 1980s, and, indeed, it still dictates the distribution plans of the major studios. These innovations can be traced back to the independent companies (including Tom Laughlin working as equal partners with Warner Bros.) and their aggressive advertising and seeking of new distribution arrangements from the mid-1960s through the mid-1970s. Through these independents working in art, adult, and family cinema, the studios were able to gauge the significance of new marketing techniques and novel forms of distribution. In turn, by adopting these innovations for their own product, the majors rebounded in the mid-1970s, experiencing greater box office for individual films than ever before.

The era of the high-concept film also eventually limited the diversity of American filmmaking both within the studios and within the ranks of the independents. Focusing on marketing-oriented projects and searching for larger and larger hits, the studios looked to "tent-pole" films, a single one of which on its own could support a studio's yearly distribution schedule. Such efforts substantially limited the production of films that could not be described in a single sentence and that did not contain inherent marketing hooks. With their greater precision in targeting audience members through market research, studios were also able to exploit those markets, including the art cinema, that traditionally had been the domain of the independents. As with all aspects of the independent-mainstream relationship, this situation does not remain static. Indeed, by the end of the 1970s two of the most influential independents—Miramax and New Line Cinema—were in their infancy and would soon be ready to rejuvenate the world of commercial independent filmmaking.[74]

Notes

1 A saturation release pattern involves the widespread distribution of a film across North America at a single point in time rather than a tiered approach favoring a slow rollout concentrated at first in large cities.

2 "Click Roadshows," *Variety*, August 21, 1968, p. 17.

3 Ron Wise, "Fundraisers Miss Modest Film Ducats, An Easier Sell Compared to Legit," *Variety*, January 17, 1973, p. 24.

4 Lee Beaupre, "Dickstein's 'Star' Strategy," *Variety*, September 4, 1968, p. 2.

5 Lee Beaupre, "Third Roadshow Study: 'Funny Girl' Getting a Specialized Sell to Party Agents," *Variety*, August 23, 1968, p. 25.

6 As Beaupre describes it: "With the hindsight exception of *2001*, most roadshows cater to family audiences or, less frequently, 'serious' adults." *Variety*, August 14, 1968, p. 5.

7 "'Ethnic Roadshow' Policy Gives Marcus Epic Special N.Y. Dates," *Variety*, May 23, 1966, p. 21.

8 Lee Beaupre, "Clutch of Roadshows," *Variety*, August 14, 1968, p. 5.

9 Ibid.

10 John Gregory Dunne, *The Studio* (New York: Limelight Editions, 1968), p. 242.

11 Teresa Grimes, "BBS: Auspicious Beginnings, Open Endings," *Movie* 31-32 (1986): 57.

12 For a review of this transitional period, consult Jim Hillier, "Forty Years of Change," *The New Hollywood* (New York: Continuum, 1994), pp. 6–17.

13 "Foreign Pix Hope for More U.S. Dates with Curtailment of 'B' Productions," *Variety*, September 30, 1953, p. 3.

14 Todd McCarthy, "Exhib/Distrib Donald Rugoff Dies at 62," *Variety*, May 3, 1989, p. 7.

15 "Art-House Boom," *Newsweek*, May 28, 1962, p. 102.

16 McCarthy, "Exhib/Distrib Donald Rugoff," p. 7.

17 "Rugoff Cinema V Shakes Moribund Tone," *Variety*, October 5, 1966, p. 5.

18 Robert J. Landry, "Unsold in the Land of Sell," *Variety*, April 24, 1957, p. 5.

19 McCarthy, "Exhib/Distrib Donald Rugoff," p. 7.

20 In a move that reverses the usual protocol, Cinema V placed an industry ad thanking their agency (Diener, Hauser, Greenthal Co.) for their images; the ad was titled, "They make images to sell dreams." *Variety*, September 10, 1969, p. 47.

21 "Rugoff's N.Y. Times Ad for 'Organism' Chides Hub's Mixed-Reason," *Variety*, December 29, 1971, p. 1.

22 Stephen Farber and Estelle Changas, "Putting the Hex on 'R' and 'X,'" *New York Times*, April 9, 1972.

23 "From Nearly to Total Nude Pix," *Variety*, October 2, 1968, p. 1.

24 "'Curious': Sexy-Dull Shocker," *Variety*, March 19, 1969, p. 7.

25 Kent E. Carroll, "Some Fear to Be 'Curious,'" *Variety*, April 9, 1969, p. 5.

26 "Sex Dominates B'way First-Runs," *Variety*, March 19, 1969, p. 9.

27 "Dubious Yellow," *Time*, March 14, 1969, p. 98.

28 "Porno Study: 600 U.S. Film Sites Comprise Playoff For Sexploitation," *Variety*, October 7, 1970, p. 7.

29 Aubrey Tarbox, "High-Brow Novels to Screen," *Variety*, March 12, 1969, p. 8.

30 For a history of the distribution innovations introduced by American National Enterprises, see Frederick Wasser, "Four Walling Exhibition: Regional Resistance to the Hollywood Film Industry," *Cinema Journal* 34 (Winter 1995): 51–65.

31 Lee Beaupre, "Utah Gospel of Four-Wall Sell," *Variety*, November 7, 1973, p. 5.

32 For an industrial history of Sunn Classics, see Gary Edgerton, "Charles E. Sellier, Jr. and Sunn Classic Pictures: Success as a Commercial Independent in the 1970s," *Journal of Popular Film and Television*, October 1982, pp. 106–18.

33 Wasser, "Four Walling Exhibition," p. 55.

34 "'Four Wall' Sun International Reversing Distribution Field," *Independent Film Journal*, March 18, 1974, p. 7.

35 "'Billy Jack' as Labor of Love," *Variety*, November 10, 1971, p. 4.

36 "'Billy Jack': Happy Denouement of a Warners Courtroom Drama," *Variety*, November 7, 1973, p. 5.

37 Ibid.

38 Richard Moses, "The Rise, Fall and Second Coming of Four-Walling," *Variety*, January 8, 1975, p. 22.

39 Richard Albarino, "Billy Jack Hits Reissue Jackpot," *Variety*, November 7, 1973, p. 1.

40 Richard Kahn, "The Day Film Marketing Came of Age," *Variety*, October 30, 1979, p. 38.

41 Ibid.

42 Albarino, " 'Billy Jack' Hits Reissue Jackpot," p. 1.

43 "Motivational Research in Promotion," *Variety*, June 26, 1974, p. 7.

44 " 'Exorcist' Four-Walling Causes Trade Disruption," *Independent Film Journal*, March 4, 1974, p. 6.

45 See, for example, "Outsider Hits," *Variety*, November 6, 1974; "Hudson's Four Wall Deal," *Variety*, November 5, 1975, p. 3; Addison Verrill, "Amateurs Bring in Bonanza," *Variety*, February 2, 1972, p. 7.

46 Richard Albarino, " 'Billy' Sequel's Grand $11-Mil Preem," *Variety*, November 20, 1974, p. 61.

47 "Four-Walling, 'A Burgeoning Menace,' Stirs Exhib Wrath," *Independent Film Journal*, August 7, 1974, p. 12.

48 Frank Segers, "Pix Unions Scaling Up Four Walls," *Variety*, June 12, 1974, p. 1.

49 Kahn, "The Day Film Marketing Came of Age," p. 38.

50 " 'Billy Jack' Sequel Four-Walls New York, Philly, Phoenix Areas," *Variety*, October 23, 1974, p. 76.

51 Ibid., p. 4.

52 Albarino, " 'Billy' Sequel's Grand $11-Mil Preem," p. 61.

53 Larry Primak, "Jonas at the Pic Distributors Wall," *Variety*, February 5, 1975, p. 5.

54 Ibid.

55 "Col's Break Out Campaign to Launch Its Own Breakout," *Variety*, May 14, 1975, p. 6.

56 Ibid.

57 Jim Harwood, "Anticipated Success Mutes Squawks on Cost, Rental Terms," *Variety*, June 4, 1975, p. 7.

58 Harlan Jacobson, "U Introing Nat'l Coop Ads on 'Jaws,' " *Variety*, April 16, 1975, p. 3.

59 "CIC Shifts from Usual Foreign Marketing Pattern in 'Jaws' Bow," *Variety*, December 3, 1975, p. 3.

60 For an analysis of the economics and aesthetics of the high-concept film, see Justin Wyatt, *High Concept: Movies and Marketing in Hollywood* (Austin: University of Texas Press, 1994).

61 "Big Ballyhoo Gamble on 'Omen'; Fox Hopes Antichrist Is Saviour," *Variety*, June 30, 1976, p. 4.

62 Addison Verrill, " 'Kong' Wants 'Jaws' Boxoffice Crown," *Variety*, December 22, 1976, p. 1.

63 Stuart Byron considers several reasons for *King Kong*'s ultimate commercial 'disappointment' (although he remarks that the film still earned a $40 million profit) in "Industry," *Film Comment*, March-April 1978, pp. 72–73.

64 Olen J. Earnest, "*Star Wars*: A Case Study of Motion Picture Marketing," *Current Research in Film: Audiences, Economics and Law*, ed. Bruce A. Austin (Norwood, N.J.: Ablex, 1983), p. 9.

65 Ibid., p. 11.

66 "E.T. and Friends Are Flying High," *Business Week*, January 10, 1983, p. 77.

67 Dale Pollock, *Skywalking: The Life and Films of George Lucas* (New York: Harmony Books, 1983), p. 194.

68 "Merchandising New Abracadabra of Cinematic Showmanship," *Variety*, August 23, 1978, p. 6.

69 "Ilya Salkind Defines Disaster," *Variety*, August 30, 1978, p. 40.

70 Lee Beaupre, "Grosses Gloss: 'Breaking Away' at the Boxoffice," *Film Comment*, March-April 1980, p. 69.

71 "New Era of Custom-Fit Pix Releases," *Variety*, March 17, 1976, p. 48.

72 "How Close the Encounter to a Profit," *Variety*, November 9, 1977, p. 5.

73 A. D. Murphy, "North American Theatrical Film Rental Market Shares: 1970–1989," *Variety*, January 17, 1990, p. 15.

74 For an analysis of Miramax's and New Line's significance to the world of independent filmmaking of the last two decades, consult the section "The Outsider: The Era of the 'Major' Independent," in Justin Wyatt, "Economic Constraints / Economic Opportunities: Robert Altman as Auteur," *Velvet Light Trap* 38 (Fall 1996): 59–65.

Money Matters:

Hollywood in the Corporate Era

Jon Lewis

The system stinks. It's fed by greed and ego. . . . [Hollywood has] been changing and always in the same direction, which is more about money and much less about what movies are. I hate it, I hate it. But you can't ignore it. As much as you keep reminding yourself with the mantra, "It's about the movies; it's about the movies," it's about the money. Joe Roth, *Chairman, Disney Film Division*[1]

Moviemaking is a business. The long list of credits at the end of a Holly-wood film remind us of that. At stake in every major film release is less the merit of the picture as a work of art than its measurable value as a product. The amount of money invested in each production and the staggering revenues (earned first at theaters and then in video, on cable and network television, and finally in the various merchandising, ancillary and foreign distribution, and exhibition deals) that attend the release of every significant Hollywood movie has so upped the stakes that most of the major players these days know a lot more about money than they do about making movies. Since the early 1980s—a period routinely referred to in the trades as "the corporate era" in Hollywood—increasing deregulation and a dramatic reinterpretation of antitrust guidelines, the introduction of junk-bond financing and its use in leveraged mergers and acquisitions, and the growing consolidation of assets and power by large corporations within the deeply incestuous and collusive industry subculture have dramatically altered the way business is conducted in Hollywood.

The roots of this newest of new Hollywoods can be found in two roughly coincident events: the 1948 Supreme Court decision in *U.S.* v.

Paramount Pictures, Inc. (the so-called Paramount decision) and the Hollywood blacklist (set in motion by the Committee on Un-American Activities of the House of Representatives in the fall of 1947). The Paramount decision culminated a ten-year court battle between studio Hollywood and the United States Justice Department. In 1938 the Justice Department filed a suit contending that the studios' control over the development, production, post-production, distribution, and exhibition of motion pictures violated federal antitrust laws and amounted to an unreasonable and unlawful restraint of trade. In 1948 the Supreme Court decided in the government's favor and the studios were forced to sell off their interests in the theatrical exhibition of motion pictures, effectively (albeit temporarily) eliminating the distribution/exhibition guarantees that had supported the studio system for decades.

Industry complicity with the House Committee on Un-American Activities came about just as the old studio system was unraveling. By the fall of 1947, the studios had come to expect—and had begun to prepare for—a decision in the government's favor in the Paramount case. Industry guilds (unions) had organized the industry workforce and had waged a series of successful job actions and strikes. Studio executives were particularly anxious and the House Committee exploited the situation.

When the House committee (which included committee chair J. Parnell Thomas of New Jersey, John McDowell of Pennsylvania, Richard Vail of Illinois, and Richard Nixon of California) moved to indict Alvah Bessie, Lester Cole, John Howard Lawson, Albert Maltz, Edward Dmytryk, Samuel Ornitz, Herbert Biberman, Dalton Trumbo, Ring Lardner Jr., and Adrian Scott (the so-called Hollywood Ten) for contempt of Congress, it had significant support in the House but hardly a mandate from the press or from the people (so far as polls revealed at the time). The Hollywood Ten, on the other hand, had the support (or so they were led to believe) of the various industry guilds and much of the mainstream press. The MPAA (the Motion Picture Association of America, the mainstream industry's self-governing apparatus) maintained a fairly low profile during the proceedings, although it seemed safe to assume at the time that the organization would oppose government regulation. But just five days before the House was scheduled to process the contempt citations, MPAA president Eric Johnston issued a stunning public statement condemning the Hollywood Ten, thus paving the way for the indictments, incarcerations, and blacklists to follow.

At the time, Johnston's statement to the press and the industry's subsequent capitulation to federal government pressure to purge subversion within its ranks seemed illogical and unnecessary. After all, the House committee seemed interested in regulating not only Hollywood employment but film content as well, a situation that seemed to undermine studio autonomy. But the blacklist—whose full force came into effect at the very moment the Paramount decision seemed to spell the end of the old studio system—proved useful to industry management. By monitoring those who worked and those who did not, the studios ruthlessly policed the post-Paramount decision workforce and as a result effectively stripped the various industry guilds and unions of their bargaining power.[2] As a result, the unions fell victim to the patriotic fervor of the Red Scare, so much so that the Ronald Reagan-led Screen Actors Guild joined management in implementing the industry-wide ban against some of its members.

Looking back, one is drawn to the tragic story of who got work and who didn't, but those stories just scratch at the surface of the net industrial effect of the blacklist. So far as the studios were concerned, the lesson learned during those years—the lesson learned in the very act of complying with government interference—had little, finally, to do with patriotism or ideology or free speech. Instead, in the very process of exploiting the power and threat of an industry-wide blacklist, the studios not only succeeded in the short term, rigorously controlling the pool of talent dependent on their largesse for work, but in doing so they developed a long-term strategy that persists (in a far more complex form) today. The so-called free market engendered by the Paramount decision got a whole lot less free during the blacklist era because the studios discovered that they could circumvent the spirit of the antitrust decree and control the industry guilds if they just learned to work better together.[3]

In November 1968 the MPAA instituted a radically new motion picture ratings system (G, M, R, and X—later refined to G, PG, R, and X; G, PG, PG-13, R, and X; and finally G, PG, PG-13, R, and NC-17). The new system was adopted by the studio membership at a crucial moment in Hollywood history. Box office revenues were stuck in a decade-long decline. Studio executives who had learned to work together to establish a new relationship with exhibitors after the Paramount decision—executives who colluded in order to reestablish control over the Hollywood workforce during the blacklist era—broke ranks and began to explore short-term solutions to their box office problems. Some of these short-term strategies worked; certain films were released and screened because the studios were willing to ignore long-standing codes of industry conduct. The abandonment of the gentleman's rules that had prevailed in Hollywood for over half a century threatened the long-term stability of the industry.

The new code became *necessary* because the studios had given up adhering to the old one. *The Moon Is Blue* (1953), *Baby Doll* (1956), *Room at the Top* (1958), *Never on Sunday* (1960), *Lolita* (1962), *Kiss Me Stupid* (1964), *The Pawnbroker* (1964), and *Blow-Up* (1967), all mature-content films that did well in an otherwise dead box office, reached America's screens only after their studio distributors circumvented the letter and spirit of the old code.

MPAA President Jack Valenti, a former advertising executive and LBJ administration insider, sold the new code to the public with continued reference to the changing times. But first and foremost the new ratings system was a business proposition. The studios needed to update their product lines and the new code was a means toward that end. The code not only supported a product overhaul—American films after the fall of 1968 look and sound different from those produced before then—it also promised to better insulate the studios against local efforts to interfere with the production, distribution, and exhibition of their product. Moreover, the new ratings system was a significant and collective move toward establishing a stake in a series of new markets formerly dominated by foreign and American independent films and helped the studios to better differentiate their product from the far more strictly regulated shows on television.

Local film censorship boards—in Dallas, Chicago, and Memphis, for example—had long been a problem for the studios. The Catholic Legion of Decency (and then NCOMP, the National Catholic Office of Motion Pictures) and other grassroots organizations sported their own ratings systems and reached and influenced their own large and loyal constituencies. From the silent era to the late 1960s, the various local and grassroots/national censorship codes complicated the production process at the studios. To release a picture nationwide, the studios were compelled to submit scripts and preview films to a number of independently operated censorship boards and then forced to negotiate and/or capitulate to gain what amounted to a series of seals of approval. The development, production, and post-production costs prompted by local board input were significant and especially painful in times of box office decline.

A series of successful Supreme Court challenges to censorship prosecutions in the 1950s and 1960s—*Roth* v. *United States* (1957), *A Book Named John Cleland's "Memoirs of a Woman of Pleasure (Fanny Hill)"* v. *Attorney General of*

Massachusetts (1966), and *Redrup* v. *New York* (1967)—encouraged the studios to develop a new national production standard that they eventually finessed into an industry-run system of self-regulation. By the time the ratings system was adopted by the MPAA, studio executives were confident that their better financed, better produced, more up-to-date studio pictures would easily carry the day.

But the new production code *did not* solve the studios' problems at the box office, at least at first. And worse, the very Supreme Court decisions that significantly diminished the authority of local censorship boards eliminated one big problem for the studios by creating another. Between 1968 and 1973, the first spate of G, M, R, and X studio pictures performed poorly at the box office. Over that same period, features produced and distributed by the burgeoning hard-core porn movie industry recorded astonishing box office revenues.

On February 25, 1970, less than two years after the MPAA adopted its new ratings code, a front-page article in *Variety* highlighted the mainstream industry's ongoing frustration at the box office: "The depressed state of first run business among some of the New York cinemas has been given a conversational going over within the trade itself. And there is slightly more inclination to the interpretation that the Code Ratings are having a bearing on those b.o. ups and downs . . . [though] it cannot be said with authority that an X can make a picture and a G break it, it is known that some non-major distributors plant an X on their pictures on their own and view this as a move toward profitability." Between 1968 and 1970, six hundred theaters in New York City alone stopped showing studio fare in favor of, as *Variety* so glibly described them, "budgetary lowbrow exploitation specials [skin flicks]."[4] Three early seventies' skin flicks, *Deep Throat* (1972), *Behind the Green Door* (1972), and *The Devil in Miss Jones* (1973), outearned big-budget studio films not only on a screen-by-screen basis but in total box-office revenue nationwide; all three would have been top twenty films for their year of release had *Variety* not limited its year-end lists to "legit" pictures.

In 1971, the year before the release of *The Godfather* and the beginning of the vaunted Hollywood auteur (and box-office) renaissance, Gulf & Western / Paramount contemplated shutting down its film unit and selling off its motion picture lot for real estate development. The success of hard-core—the apparent failure of the collusive strategy to circumvent federal censorship of their product—served as a reminder of just how desperately the studios *needed* a new Hollywood and how, at least in this instance, they needed further federal regulation to get there.

10 Robert Altman's auteur classic, *Nashville* (Paramount Pictures, 1975).

Finally, in 1973 the studios caught their break, and they have not turned to look back since. On June 27 five pivotal U.S. Supreme Court decisions (*Miller* v. *California, Paris Adult Theater* v. *Slaton, U.S.* v. *Orito, U.S.* v. *12 2,000' Reels,* and *Kaplan* v. *California*) established "community standards" as the measure of what could and could not be shown at "legit" theaters nationwide. The June 1973 decisions effectively criminalized hard-core and forced the vast majority of the theaters that specialized in screening X, XX, and XXX movies to reconsider the industry product or else risk prosecution by ambitious, politically conservative district attorneys.

The artistic freedom fostered by the new ratings code and the deservedly celebrated industrial swing to more ambitious auteur pictures from 1972 through 1979 owe far more to the slow resolution of the hard-core controversy in the courts than to some loftier notion regarding young and smart studio executives supporting young and smart university-educated auteurs because they made better movies. Hollywood was "a better place" in the early seventies in large part because of the competition posed by the hard-core industry, which in the space of five short years taught the majors a lesson, not so much (or at least not only) about the vagaries of unlimited, unregulated artistic freedom, but instead (or also) about how to market a product and how to use artistic freedom as a means toward better identifying that product in advance of release. Once the studios figured out this new marketplace—once production of more explicit but decidedly not hard-core films was systematically and collectively instituted—the studios retrenched and essentially abandoned the auteur theory in favor of greater management control over the development, production, and release of motion pictures in the United States.

In 1978 the federal government commissioned a task force to study the new Hollywood. The commission's findings (published in 1979) confirmed what most players in the business already understood all too well. "The industry is clearly oligopolistic," the report concluded, "major studios appear to be controlling the market to restrict competition and lessen output so as to maintain tight control over employees and an exceedingly low buyer profit."[5] Unsurprising as these findings were, the commission report proved prescient; within a year, the Paramount decision and antitrust regulation, in the film business at least, were rendered moot.

The industry story that set the stage for this new, deregulated Hollywood broke on April 25, 1979, when the industry trade journal *Variety* published a piece on a series of fairly confusing stock moves by MGM CEO Kirk Kerkorian. In accordance with Securities and Exchange Commission rules, Kerkorian filed notice of his intention to sell 297,000 shares of his MGM stock. With the proceeds from the sale, Kerkorian secured a $38 million loan, which he then used to finance the purchase of a block of stock (amounting to 24 percent of the publicly trading shares) in a rival studio, Columbia Pictures Industries (CPI). By the end of business on April 25, 1979, Kerkorian had become the largest shareholder at two of the six major studios.[6]

Exactly what Kerkorian wanted with CPI is not completely clear. He already owned one studio and seemed to have no idea how to run it. Kerkorian purchased MGM from the Bronfman family (which controls Seagram) and Time, Inc. in 1969 in an acrimonious leveraged buyout. For several months MGM management opposed Kerkorian's bid, a strategy that proved a nuisance for Kerkorian (who had to purchase 40 percent of the company's stock at a premium before closing the deal) and the beginning of the end for MGM. From the moment he took over the studio, the Las Vegas real estate tycoon was in financial trouble. To meet debt obligations stemming from the buyout, Kerkorian's first moves as MGM CEO were to downsize the studio operation and to sell off assets. Within the first few years of owning the studio, Kerkorian cut the MGM work force from 6,200 to 1,200, hired the notorious, cost-conscious TV executive James Aubrey to run the studio, and alienated everyone who had a deal pending with the company. By the time he purchased the block of stock in CPI, MGM could hardly be considered a *major* studio any longer.

When news of the stock purchase hit the trades, Kerkorian came forward to insist that he had no interest in running Columbia Pictures; rather, he claimed, he was making an investment in a company that, like another company he owned and controlled, just happened to make movies. Predictably, management at CPI and attorneys in the antitrust division of the U.S. Justice Department saw the stock purchase in a far different light. To force Kerkorian to divest interest in one of the two companies, the Justice Department filed suit in federal court and set in motion a confrontation between federal regulatory agencies and Hollywood ownership, the outcome of which would change the movie business in dramatic and so far it seems irreversible ways.[7]

The Justice Department suit called for Kerkorian to divest interest in one of the two film companies. But as early as August 7, 1979, the opening day of the trial, things began to go wrong with the government's case. Citing Kerkorian's standstill agreement with CPI—his pledge not to buy any additional stock for three years—Judge Andrew Hauk challenged the Justice Department to prove "actual Kerkorian intent to meddle in Columbia's affairs,"[8] something that the government attorneys were unprepared and unable to do.

The key to the Justice Department's argument was that Kerkorian's purchase of such a large block of CPI stock amounted to a hostile move on the company and that his potential dual ownership (of both MGM and CPI) created a diminution of competition, one which promised significant disadvantages to exhibitors. But when pressed by Judge Hauk, none of the exhibitors brought in by the Justice Department to testify could cite a single example in the four months after Kerkorian's stock move to support such a contention. In fact, the exhibitors testified that they had little choice with regard to product, no matter who owned or controlled the major studios.

Columbia CEO Herbert Allen proved of little use to the government case either. While other Columbia executives like Fay Vincent (who would eventually leave Columbia to become the commissioner of Major League Baseball) and Frank Price were wont to vent their frustration with Kerkorian in the trades, Allen tended to be more philosophical, appreciating the ways in which public relations (and maintaining one's cool) are important in any battle for control of a company. When called to testify, Allen downplayed the significance of Kerkorian's stock purchase, glibly remarking that while he would have preferred a standstill agreement that held for ten years as opposed to just three, he felt confident that Kerkorian had no "current anti-competitive scheme." Allen then pointed out that the primary problem was not with Kerkorian's dual ownership, but rather with how that dual ownership was perceived by talent, by independent producers, by others in less stable and powerful positions within the industry. "In the motion picture business," Allen quipped, "perception is often more important than reality."

The most telling blow to the government's case came when Judge Hauk called on two expert witnesses, University of California professors of economics Robert Clower and Fred Weston. Both academics challenged the government's argument regarding the antitrust implications of Kerkorian's *stock purchase*. Clower's testimony proved particularly damaging. He argued that even an outright merger of MGM and CPI "would not significantly lessen competition in any line of commerce," and he added that even if two of the more successful studios were to *merge*, "you still should have five or six major distributors, thus a reasonably competitive environment."[9]

Judge Hauk concurred and on August 22 decided in Kerkorian's favor. Hauk admonished the government attorneys for pursuing the case in the first place: "How on earth the government can arrive at the thought that there will be *a diminution of nonexistent competition* is beyond me."[10]

Whether or not Kerkorian appreciated the historical significance of the decision is anybody's guess; what he did understand was that the victory put him one step closer to taking over Columbia. In September 1980 Kerkorian made a tender offer to purchase an additional 1 million shares of CPI stock (via call options to be exercised once the standstill agreement expired). The proposed purchase would raise his total stake in CPI to 35 percent. The CPI board labeled the move "an outrageous assault" and, though they should have known better, "a blatant violation of anti-trust laws."[11]

To block Kerkorian's apparent takeover attempt, Columbia issued more stock through a convertible debenture.[12] The stock move promised to put the shares in the friendly hands of in-house studio producer Ray Stark (who, Kerkorian alleged, was running the studio anyway),[13] thereby diminishing Kerkorian's stake in the company. Kerkorian responded by declaring the standstill agreement null and void, and then, to cull shareholder support, he turned to the press and accused CPI management of all sorts of corporate misadventures, including the stock deal with Stark and a $1 million underwriting fee paid by CPI to its CEO Herbert Allen's brokerage firm.

At the time, it was fair to assume that Kerkorian would eventually mount a proxy fight for control of the company, a fight he seemed poised to win. But a fire at the MGM Grand Hotel in Las Vegas proved a significant sidetrack, and eventually CPI succeeded in buying him off. In lieu of yet another standstill agreement, Kerkorian agreed to finally sell his Columbia stock, netting a cool $137 million.

Four months later, Kerkorian made an offer to buy Chris Craft Industries' 22 percent stake in Twentieth Century-Fox. Ultimately outbid by Marvin Davis (who management preferred at first but lived to despise as he sold off pieces of the studio to service his debt), Kerkorian emerged a few years later as a principal in Saul Steinberg's leveraged move on the Disney Company (which led to yet another huge greenmail payoff).[14]

Kerkorian spent the better part of the 1980s trying to unload MGM (and the eventually annexed United Artists), first to Ted Turner, then to producers Peter Guber and Jon Peters, then to an Australian multinational (Quintex), and finally (and successfully) in a highly questionable deal with Giancarlo Parretti, who, amid accusations of fraud and mismanagement (leveled against both Parretti and Kerkorian), was, in less than a year's time, forced to relinquish control of the company to the French bank Crédit Lyonnais, which maintained control of the studio until July 1996[15] when Frank Mancuso engineered a management buyout financed by— who else—Kirk Kerkorian.[16] Though he generally goes unmentioned in contemporary film history, Kerkorian was the 1980s' most powerful and at times most perplexing player. Through the course of the decade, he held significant stock positions at MGM, Columbia, and United Artists; negotiated distribution deals and thus integrated MGM Filmco with United Artists, Paramount, and Universal (with whom he released MGM features through the jointly held Cinema International); and made tender offers for Columbia, Fox, Disney, and United Artists.

As a consequence of the 1979 antitrust decision in the Kerkorian-CPI case, a flurry of corporate restructurings and takeovers occurred throughout the eighties and the first half of the nineties. These featured an elite set of multinational corporate players—for example, Rupert Murdoch, Seagram's Edgar Bronfman, Jr., Sony's Norio Ohga—each with enough money and ambition to exploit the increasingly deregulated entertainment industry. The majors now exploit synergies in a wide variety of ancillary markets: cable and network television, videocassettes and laser discs, toy manufacturing, book and magazine publishing. Most of the studios now own their own theaters, and though they do not use them to show their own product exclusively, they could do so anytime they want to or need to. In seeming preparation for the next new Hollywood—one in which the film market may well be even more dependent on "home box office" and more integrated with the vaunted information-entertainment superhighway—the studios have begun to establish strategic relationships with computer software providers like Microsoft and cable and telephone hardware providers like TCI (which owns almost one-fourth of the voting stock in Turner Broadcasting), Viacom (which owns Paramount), Bell Atlantic (which nearly merged with TCI in 1994), Nynex (a partner in a number of Viacom/Paramount ventures), and US West (a partner in Time Warner Entertainment).

The ability to coordinate and exploit different media outlets have enabled the big studios to better insulate themselves against potential box-office disappointments. Skyrocketing production costs (especially above-the-line salaries—in the seven- and eight-figure range for star performers and directors), the increasing market reliance on bigger and better special effects, and increases in short-term interest rates have virtually shut down the B-movie industry.[17] But the same conditions pose little threat to these new media conglomerates. So much money can be made in such a variety of markets, all under the purview of the parent company, that it is hard to imagine any film actually losing money these days. Despite creative bookkeeping that often defies rationality, even would-be blockbusters like *Waterworld* (1995) or *Last Action Hero* (1993)—well-publicized flops at the domestic box office—earn out their investment because the studios have access to first-money revenues in so many ancillary markets. (Studio distributors are routinely first in line to receive theatrical and home box office revenues.)

The astonishing profitability of the "film business"—or, more accurately, the diversified entertainment industries of which motion pictures are a part—continues to support increased conglomeratization and monopolization. To understand how the companies themselves view this new Hollywood, consider the following three paragraphs excerpted from the 1989 annual report published by the then newly formed Time Warner. Note the references to "profound political and economic changes," ostensibly referring to Reaganomics and deregulation (which followed the Kerkorian decision) and the conspicuous omission—the conspicuous irrelevance—of such outmoded "old Hollywood" notions as divestiture and free trade:

In the Eighties we witnessed the most profound political and economic changes since the end of the Second World War. As these changes unfolded, Time, Inc., and Warner Communications, Inc., came independently to the same fundamental conclusion: globalization was rapidly evolving from a prophecy to a fact of life. No serious competitor could hope for any long term success unless, building on a secure home base, it achieved a major presence in all of the world's important markets.

With this goal in mind, Time and Warner began discussions on joint ventures. The more we talked—the more we learned about each other—the more obvious it became that the most significant and exciting possibility was a synthesis that would lift us to a position neither could achieve alone.

In a season of history when technology has combined with political and social change to open vast new markets, we are a company equipped to reap the greatest benefits.[18]

The merger of Time and Warner took shape at a moment of record prosperity in Hollywood; the entertainment business had become increasingly well-integrated, federal regulation (FCC, FTC, and Justice Department interference) had been all but eliminated, and everyone who could afford to make a product was making money. It is important, then, to look at the Time-Warner merger not as an isolated event, nor as a landmark deal later aped by other companies. Instead, the merger should be viewed in light of the larger picture of this newest of new Hollywoods, one in which Time Warner was a logical consequence.

In April 1989, while Time and Warner Communications Inc. (WCI) were negotiating the details of their planned merger, the Gulf & Western Corporation, the parent company of Paramount Pictures, announced its decision to sell off its financial unit in order to consolidate its interests in the entertainment business. The announcement was newsworthy but hardly surprising. In 1980 Paramount's film and television divisions accounted for only 11 percent of Gulf & Western's annual revenues. But by 1989 the entertainment division accounted for more than 50 percent. Throughout the 1980s Paramount boasted the best average market share, 16.2 percent, two of the best franchise properties, the *Indiana Jones* and *Star Trek* films, and a lucrative television production unit.

Less than two months after its consolidation as an entertainment conglomerate, Gulf & Western CEO Martin Davis took out a full-page advertisement in *Variety* to announce the company's new name: Paramount Communications Inc. (PCI).[19] The new studio/entertainment conglomerate was well set-up to do business in the 1990s; its holdings crossed genres and industries, it had the ability to reproduce a single product in various forms and formats, and it had the distribution and exhibition network to exploit profits at every stage, in every market.

A cursory look at PCI's principal holdings in the entertainment business in 1989 reveals just how diversified, just how powerful the company had become: Paramount Pictures, Paramount Television, Paramount Home Video, Famous Music, Madison Square Garden (which included the New York Knicks and New York Rangers, a cable television station, an attractive piece of New York City real estate, and the Miss Universe Pageant), Simon and Schuster, Prentice Hall, Pocket Books, and majority interests in the TVX broadcasting group (including WTFX-TV in Philadelphia, WDCA-TV in Washington, D.C., and KTXA-TV in Dallas), two major theater chains (United Cinemas International and, in partnership with Warner Communications, Cineamerica Theaters, with its subsidiaries Mann and Trans-Lux), and the USA cable television network.[20] But despite the seeming wisdom in selling off its financial unit and concentrating on its most lucrative markets, the industry trades at the time seemed interested only in what Davis planned to do with all the cash netted from the sale and who might make a move on PCI if he failed to find a place for it soon.

The amount of cash in question was substantial by any standards: $3.5 billion. Industry experts speculated that PCI would go after MCA (which owned Universal). Subsequent rumors had PCI interested in acquiring the Chicago-based Tribune Company, ABC, NBC, CBS, and least likely (but most interestingly) Time Inc., or merging with Viacom, a company that owned a cable delivery system and a number of pay television stations that could provide PCI with access to and a profit stake in yet another media market.[21]

On June 6, 1989, Davis put an end to all the speculation and announced his decision to mount a hostile takeover of Time. From that date until September 20, 1989, the battle between WCI and PCI dominated the trades.[22] That either company might be allowed (by the FCC or FTC) to merge with Time prompted *Variety*'s Richard Gold to glibly conclude that in the future "all of show business [will be] controlled by two or three conglomerates."[23]

Paramount's bid to acquire Time—a hostile but astonishingly high $10.7 billion, $175/share offer[24]—put the planned merger between Time and WCI on ice. Speculation in the trade press identified a number of possible scenarios, the most likely of which was that Time, in order to quash the PCI offer, would tap into its $5 billion line of credit and simply buy WCI outright. In doing so, the merger with WCI would ostensibly take a somewhat altered shape, and Time would become far too debt-laden for PCI to buy. Moreover, the combined companies—Time Warner—would be in a strong position to take over PCI, itself vulnerable because of its enormous cash reserves and minimal debt load. Others guessed that Paramount would succeed in its hostile takeover of Time, leaving WCI an attractive takeover target.

Complicating matters were some of the secondary players—major stockholders not participating in the various merger negotiations (and not liking it). Principal among this group was Herbert Siegel, CEO at Chris Craft Industries, who owned a 17 percent share of WCI. Industry analysts speculated that Siegel, who despised WCI chairman Steve Ross, might sell his stock to PCI just to spite his longtime enemy. Another possibility—one even more complex and interesting—was that Siegel might sell his stock back to Ross and use the money to make a move on Paramount.

From the start, the proposed merger with WCI met with little enthusiasm in the financial press because it did not seem to offer Time shareholders much of anything, at least in the short run. The poor public relations—Time's inability to control the spin in the financial press evaluating the merger—encouraged the widely held assumption that Time's shareholders might force the company to accept the Paramount offer.

Rumors then began to circulate that WCI chairman Ross had entered into negotiations with the French media giant, Hachette. A deal with Hachette made sense, and it caught Time's attention. If Time accepted the Paramount offer, Ross needed to prevent a hostile move on WCI, perhaps by PCI-Time. But it is certainly possible that Ross met with the executive team at Hachette in order to force Time's hand; given the astonishing stock bonus tied to the proposed Time-WCI merger, Ross had plenty of motivation to coerce Time into dealing with him.

Paramount's challenge to the Time-Warner merger betrayed the pervasive collusive spirit that had prevailed in Hollywood since the blacklist. Gold, again writing in *Variety*, expressed concern about how "Paramount's new acquisitive ferocity" might "escalate the traditional level of competition between the Hollywood majors and alter the gentlemen's rules by which the majors play the game." Hollywood insiders, none of whom would speak on the record to Gold, spoke anxiously about the evolving, less friendly, new Hollywood. "This is a major rift in what has become an extremely incestuous industry," one executive opined, "with corporate consolidation coming you have to wonder about how studio politics are going to settle." A second executive added: "Hollywood is a small town. Everyone is friendly . . . there hasn't been a war like this since the days of the moguls."[25]

Those industry players involved in development and production—writers, directors, producers, actors—watched the PCI / WCI battle with trepidation. As industry analysts pointed out in the trades, consolidation on the scale of either a Time-Warner or Time-Paramount merger promised to dramatically increase studio debt loads. The first wave of conglomeratization in the early 1980s had accompanied and had supported a marked decline in production—an industry-wide strategy to make fewer, bigger movies. In 1989, with a second wave clearly under way, it seemed highly possible that companies involved in these megamergers and acquisitions might take on so much debt that they would stop producing new movies altogether.

The studios already had a taste of what it might be like to stay in the entertainment business and not make movies (for a while at least). In 1980, when the Screen Actors Guild (SAG) went out on strike, the studios refused to capitulate. Universal went so far as to invoke the force majeure clause in its contract with talent, effectively locking out not only the actors but other industry guild members as well. As the strike dragged on, and new film and television production remained shut down, the studios nevertheless continued to record profits; they had become so well diversified that film—new films, that is—had become only a very small part of the overall operation. In light of the new studios' continued profitability without them, SAG eventually capitulated, and since then the guilds have been mostly ineffective. Given the variety of markets—the variety of forms and formats—into which a single product (produced as much as several years earlier) could be recycled, the studios came to understand that even extended walkouts holding up production for months, perhaps even years, would have little effect on the balance sheet so long as there were ancillary markets into which they might rerelease products chosen from their vast film and television libraries.

On June 16, 1989, Time announced a significant restructuring of its proposed merger with WCI; to counter PCI's hostile bid, Time proposed to buy WCI outright for $14 billion. The revised plan did not require shareholder approval, which was a good thing, because after the new deal was announced in the press, Time's stock entered free-fall, dropping nine-and-a-half points on June 16 and another six points the following day. Michael Price of Heine Securities, a mutual fund company that owned more than 1 million shares of Time stock, spoke for all of Wall Street when he predicted: "if Time buys Warner at this price, the stock falls off a cliff."[26]

One June 26, Paramount sweetened its offer to $12.2 billion. The offer was summarily rejected. When news of the bid and its rejection was made public, Paramount's stock went up and Time's went down. For those who play the market, the message was clear. Investors were betting that Time would succeed in its efforts to buy WCI, and as a result Paramount would become the target of a takeover.

When the second Paramount offer was rejected by the board, three of Time's biggest shareholders—Robert Bass, Jerry Perenchio, and Cablevision Inc.—filed suit in chancery court in Delaware. By early July, Paramount had filed a suit of its own, but the court rejected any and all attempts to block Time's acquisition of WCI. In a 79-page decision issued on July 14, 1989, Chancellor William Allen wrote: "There [is] no persuasive evidence that the board of Time has a corrupt or venal motivation in electing to continue with its long term plan even in the face of the cost that course will no doubt entail for the company's shareholders in the short run. In doing so, it is exercising perfectly conventional powers to cause the corporation to buy assets [in this case a major studio] for use in its business." The decision not only struck a blow to shareholders' rights (to influence corporate decisions, especially decisions of the magnitude of Time's acquisition of WCI), it regarded competition and monopoly control in the film business in much the same terms as the Kerkorian decision had ten years earlier.[27]

When the chancery court decision was announced, Time's stock price again fell dramatically by another $12 per share. But while stockholders failed to reap the benefits of the merger, the officers of the combined companies benefited from the moment the court decided in their favor. For example, when the Time-Warner deal was finally signed, Steve Ross received a $200 million stock bonus.[28]

The new Time Warner formed in 1989 sported assets amounting to nearly $25 billion and annual revenues estimated at $7.6 billion.[29] Its holdings spanned a number of related entertainment industries, including film and television studios (involved in development, production, and distribution), movie theaters, magazine and book publishing, cable television delivery systems and pay television stations, recording industry operations, and theme parks. At the time of the merger Time Warner's publishing division included book publishers Little, Brown and Co., Warner Books, and Time Life Books; comic book manufacturer DC Comics (which produces, distributes and licenses Superman, Batman, and Wonder Woman; and the mass market magazines *Time, Fortune, Sports Illustrated, Sunset,* and *Parenting.* Its Publishing Services Division distributed books and magazines for other publishers and owned the Book-of-the-Month Club. The company's music division was the world's most lucrative and extensive and included Warner Brothers Records (and its subsidiaries Reprise, Sire, and Paisley Park), the Atlantic Recording Group (and its subsidiaries, Interscope and Rhino), Elektra Entertainment (and its subsidiaries, Asylum and Nonesuch), and Warner Music International. Its wholly owned subsidiaries ATC and Warner Cable Communications Inc. made Time Warner the second-largest cable provider in the nation (behind TCI), and its cable software division sported some of the better premium channels: HBO, Cinemax, and the Comedy Channel. First-run films were screened at Cineamerica Theaters (co-owned with Paramount). Finally, Time Warner owned the Licensing Corporation of America (LCA), which managed and protected the copyright on all Warner Bros. characters (Batman, Bugs Bunny, etc.). Products bearing these trademarks are available for purchase all over the world at Warner Bros. chain of retail stores.

In 1989, just before the merger was finalized, the Warner Bros. film division released *Batman.* The film grossed in excess of $250 million domestically. But for Time Warner, the domestic gross was just a very small part of the film's overall worth to the company. Batman is a DC Comics character, licensed by LCA. The merchandising subsidiary has taken its cut from the profits of every T-shirt, cup, book, or action figure sold (at a Warner Bros. store and elsewhere). The film has appeared on HBO and Cinemax and has been delivered into homes across the country by means of cable systems owned by Time Warner. When the film was released on video and laser disc, it bore the Warner Home Video label, and the popular soundtrack CD came out in two versions, both from companies owned by Time Warner. Coverage—constant reminders about the film (as an event, as a franchise)—appeared in Time Warner magazines like *Time, Life,* and (at least in time for the first sequel) *Entertainment Weekly.*

But despite its size and the range of its influence and control, since 1990 Time Warner has performed disappointingly on Wall Street. Saddled with massive debt (which insulates it against possible takeover but restricts its ability to exploit its assets), Time Warner stock has failed to keep up with the market, and rumors have circulated, especially since Steve Ross's death and Gerald Levin's ascension to the CEO post in 1993, that a management shake-up, from the top down, might be the only antidote to the company's poor performance.[30] Facing pressure from major shareholders

11 The quintessential new Hollywood product: Time-Warner's *Batman* (Warner Bros., 1989).

to downsize and constant criticism in the financial press to improve the stock's performance and scale down the conglomerate's debt load, in the summer of 1995 Levin defied the shareholders and experts and instead began negotiating a deal that promised to make Time Warner a whole lot bigger and its debt load a whole lot less manageable.

While Levin was fashioning a plan to retain power, the big news in Hollywood was Disney's $19 billion acquisition of Capital Cities / ABC, at the time the second-largest takeover in U.S. business history. The acquisition followed on the heels of another big deal, Edgard Bronfman, Jr.'s purchase of MCA from Matsushita. Both deals reveal the studios' growing reliance on market synergies and multinational capital and the government's unwillingness or inability to enforce antitrust regulations in Hollywood.

The Capital Cities / ABC acquisition gave Disney, in then-corporate president Michael Ovitz's words, "amazing vertical clout." Disney is now a company that not only manufactures a product (that is routinely reproduced in a variety of forms and formats) but one that also owns a network of venues into which to distribute that product: for example, ABC and the Disney Channel, the theme parks and retail stores.

In many ways the acquisition was merely a sign of the times, a move that Disney CEO Michael Eisner had to make in order to maintain the conglomerate's strong position in the evolving international marketplace. "In communication," argues Alan Schwartz, the investment banking chief at Bear-Stearns who advised Eisner on the deal, "if your company does not grow and consolidate with others, then you run the risk of lacking the outlets to justify spending for a quality product. By consolidating, you can sell your product to your network, to your cable channel, as reruns, as merchandise, as overseas product. There are now so many more ways to get paid. . . ."[31]

For Disney, the deal not only gave the company a television network (like Fox), but a second key cable television channel, ESPN. The acquisition of ESPN was strategic—perhaps even the key to the entire deal—because it enabled Disney to package the sports network with its own Disney Channel, which should prove very attractive to fledgling cable providers abroad, especially in Asia where relatively strict programming guidelines (regarding sexual and political content) have made it impossible for Time Warner (with its flagship cable station HBO) to participate.

The synergy between ESPN and "team Disney" extends beyond Eisner's apparent goal of internationalizing the pay TV division. In a 1995 *New York Times* interview, Eisner revealed plans to open a chain of ESPN sport bars (with ESPN's on-line service available at every table), ESPN interactive games, and an ESPN theme park at Disney's new boardwalk complex near EPCOT in Orlando. Additionally, Eisner planned to exploit the popular ESPN logo by imprinting it on various product lines: "When we lay Mickey Mouse or Goofy on top of products, we get pretty creative stuff. ESPN has the potential to be that kind of brand."[32] Of course, ESPN merchandise will be made available to the public at Disney's chain of retail stores, the gift shops at its theme parks, and through its mail-order catalog.

The effect of the Disney acquisition of Capital Cities / ABC on product supply—on the amount of capital that will be made available to the several Disney film units (Touchstone, Buena Vista, Hollywood Films, Miramax)—is not yet clear. By all accounts, Eisner had little trouble raising the money to make the deal; indeed, he put it together very much out of the spotlight on the phone in a single afternoon. Thus, it is likely that the increased debt load will affect film production minimally. But it is fair to suspect that films with sports themes (that might have useful tie-ins to ESPN) or adaptations of new and old ABC-TV properties will stand a better chance of getting made than might have been the case before the acquisition.

12 The diversified new Disney: a. The once and future lion king is born (Buena Vista, 1994). b. Producer Tim Burton and the puppet stars of *The Nightmare Before Christmas.* Photo by Elizabeth Annas (Touchstone Pictures, 1993).

The Disney-Capital Cities / ABC deal, much like Time's purchase of Warner Communications in 1989 and the Viacom acquisition of Paramount in 1993, was less a watershed moment than a symptom, a logical consequence of the industry's ongoing adjustment to, and revision of, the Paramount decision. Case in point: within weeks of Eisner's dramatic move—which briefly made Disney the largest media conglomerate—Gerald Levin announced the purchase of 82 percent of the outstanding shares in Turner Broadcasting System[33]—a stock deal worth roughly $7.5 billion. With the purchase, Time Warner projected annual revenues for the combined companies in excess of $19.8 billion—surpassing Disney's $16.4 billion.[34]

Because it posed antitrust problems—even a superficial survey of the acquisition reveals that Time Warner Turner is both a vertical and horizontal monopoly; the company has a stake in virtually every aspect of the production process (from development to exhibition) and controls much of the hardware and software providing customers with access to that product—and, more importantly, because it seemed to violate federal guidelines, the deal was held up by the FTC. Federal regulations prohibit one company from owning cable systems that reach more than 30 percent of the nation's households.[35] Given the combination of Time Warner's and TCI's cable systems (owned by John Malone, a major Turner shareholder), Time Warner Turner will reach more than 50 percent.

But, apparently, federal regulations can be flexible in such cases. Both Time Warner as a corporation and Malone as an individual contribute strategically to political campaigns, and for their time and money they expect something in return. To insure government approval, just as news of the deal was leaked to the press, Levin dispatched two teams of Time Warner attorneys to Washington, D.C.—one to call in favors in Congress and another to negotiate a compromise with the FTC.

Though the acquisition was meant to consolidate Levin's power, the purchase of Turner Broadcasting met with unanimous disapproval from the financial press. Wall Street journalists acknowledged that the Turner acquisition increased Time Warner's holdings, but few of them could get past the fact that in the process Levin had boosted the conglomerate's already mammoth debt load to $17 billion.

Joining the financial press in its criticism of the deal, a variety of Turner Broadcasting's major shareholders, including the cable companies Comcast and Continental Cablevision as well as Time Warner partner US West, publicly expressed their disapproval.[36] So far as these unhappy shareholders were concerned, the deal (at least as it was laid out when Levin first went public about the acquisition) unfairly favored TCI chairman John Malone.

To fully understand shareholder opposition to Time Warner's acquisition of Turner Broadcasting, one must first consider Malone's stake in the deal and the growing anxiety within the industry over his role in the management of the combined companies. And to fully appreciate why there was so much concern about Malone, it is instructive to examine his role in the 1993 bidding war to acquire Paramount between Viacom and QVC. Back in 1993 it was convenient for journalists to concentrate on Barry Diller, the former Paramount studio chief (fired by Paramount CEO Martin Davis) and chairman of QVC, a home shopping network,[37] and Viacom's Sumner Redstone, and the stratospheric cash offers they made for the studio. Malone, who bankrolled Diller and eventually made a coercive move on the studio himself, stayed out of the media spotlight. But other players in the industry and later the financial and mainstream press came to understand if not admire the way he worked the deal.

Though Malone was not all that well known outside the cable industry in 1993, by 1996 he had become something of a corporate celebrity, due in large part to Vice President Al Gore.[38] In two different speeches the vice president called Malone, alternatively, "the Darth Vader of the cable industry," and "[the man who runs] the cable Cosa Nostra."[39] Malone is, as Gore suggests, widely perceived as a dark, reclusive, ruthless force by other players in the cable industry. Because of that reputation, a lot of media executives prefer to cooperate with him rather than compete.

In 1993 Malone had little difficulty financing Diller's bid not only with his own money (through TCI and another of his holdings, Liberty Media) but with funds culled from a group of major media companies anxious to play along: Cox Enterprises, Comcast, Condé Nast and Bell South. As Malone accumulated financing (and partners), Redstone was forced to follow suit. He recruited help from Blockbuster Video's Wayne Huizenga and Bell South's competitor, Nynex. Before Diller, Malone, and Redstone began bidding on PCI, Davis had already tried to negotiate deals with Sony (which owns Columbia Pictures) and General Electric (which owns NBC). The list of interested parties says less about the value of Paramount in 1993 than it does about the sorts of strategic alliances—the constellations of monopoly capital—brought together in the very process of mounting acquisitions of this magnitude.

In 1993 Malone was QVC's largest shareholder and ultimately Diller's steadiest financial backer. But once it became clear to Malone that Davis would never sell the studio to Diller, Malone went behind Diller's back with an offer of his own, a secret bid that would have given TCI (independent of QVC) a 17 percent stake in Paramount. Malone's offer at first intrigued Davis because—unlike Diller or Redstone—Malone promised to leave Paramount management for the most part intact. But there was one significant condition. In deference to Diller, who stood to lose the bidding war and in the process a business partner as well, Davis had to agree not to go public with regard to the Malone bid.

Davis now claims that Malone's secret counteroffer troubled him from the start—that he told Malone that he could not consider an offer so transparently in conflict of interest with the QVC bid. According to Davis, in order to diminish such a conflict, Malone offered to recuse himself from QVC's decision-making process. Davis then reminded Malone that, as the major stockholder in QVC, there was no way he could do so without publicly severing ties with Diller. Malone decided against such a public betrayal, withdrew his offer, and contented himself with the fact that he stood to win whether Davis went with QVC or a severely overextended and debt-ridden Viacom.

Redstone ultimately acquired Paramount and added the studio to his media empire. But, thanks to Malone, he did so at a significantly higher price than he could afford. A few months after the purchase, Viacom was forced to sell some of its assets to meet its debt obligations. To the surprise of no one in the industry, these assets were purchased by Malone.

Once the deal closed, Davis told the press that he was glad Viacom got the studio. Moreover, Davis speculated that Malone had supported the QVC offer, even though—or, more accurately, because—he knew Davis would never accept a bid from Diller. So far as Davis is concerned, it was Malone's plan from the start to step in with his own bid and to present it as a viable alternative to the professional embarrassment of selling out a former subordinate.[40]

In what may well serve as a warning to Turner and Levin in the not so distant future, Davis offered the following glib assessment of the TCI chairman: "Let me tell you about Malone. The guy is a genius—but then so was Al Capone. He's the smartest guy you'll ever meet. He's brilliant. But he stands for nothing, and has contempt for anyone that walks and talks. I'll tell you something else . . . . He has nothing good to say about anyone, whether it's Barry Diller or Ted Turner. He told me I couldn't trust one of his own lawyers for Christ's sake."[41]

In 1995 Malone owned a 21 percent stake in Turner Broadcasting and had veto power over all significant deals. To get him to cooperate, the deal first structured by Levin and Turner offered Malone's TCI a long-term preferential arrangement. In exchange for his agreement not to block the acquisition, Malone was contracted to receive Turner Broadcasting channels CNN, TBS, TNT, and the Cartoon Network for twenty years at a 15 percent discount to air on all TCI systems and options to buy Turner's stake in two sports networks (then in development). Levin additionally agreed to purchase, at some future date, Malone's satellite operation (ostensibly in exchange for even more stock in the combined companies). Malone's

Liberty Media was contracted to receive 57 million shares of Time Warner stock, amounting to approximately 9 percent of the combined companies, worth at the time of the announced merger $2.3 billion. And finally, Malone refused to sign a standstill agreement; in fact, he told the press on several occasions that he had not ruled out the possibility of buying more shares and of taking a majority position in the company.

Comcast and Continental Cablevision—two other cable television outfits controlling a smaller block of Turner stock—were effectively frozen out of the merger negotiations. They received no such sweetheart deal, and though they stood to benefit from the acquisition because the value of their Turner stock would increase, Levin's long-term deal with Malone promised to make it even more difficult for them to compete with TCI in the future.

Another apparent winner in the deal was Ted Turner, who gained control over almost a third of the combined companies' voting stock. "Imagine courting and marrying the prettiest girl in town," financial columnist Christopher Byron felicitously wrote assessing the dynamics of Levin and Turner's relationship after the merger, "and bringing her home only to discover that she's some hairy dude named Ted who thinks he's the guy and you're the girl."[42] The deal with Levin allowed Turner to maintain control over Turner Broadcasting operations and added to his responsibilities the management of Time Warner's Home Box Office (HBO). Turner's new title—vice chairman of Time Warner—gives him two seats on the Time Warner board and an additional 64 million shares of Time Warner stock, worth $2.6 billion at the time of the announced deal. Though he has no stock position in the combined companies whatsoever, the acquisition also benefited Michael Milken, the erstwhile felon[43] and megadealmaker, who received $50 million for brokering the deal, topping the old record of $40 million paid to Michael Ovitz for setting up the MCA / Matsushita stock purchase.

Though they are in the minority, some analysts believe that Levin has finally made the right move and that Time Warner will benefit as well. The purchase promises to add $147 million to the company's cash flow and should increase its annual revenues and market share. Turner's film and cartoon libraries greatly enhance Warner Bros.' already considerable archive, and the combined companies can lay claim once again to the title of biggest and richest media conglomerate.

Within days of the announced acquisition, US West filed suit in federal court to block the deal, citing a condition of its partnership agreement with Time Warner signed less than four years earlier. In 1992, to raise $3.5

billion in cash to serve the corporation's mounting debt, Levin sold shares in Time Warner Entertainment (TWE: the Warner Bros. film studio and Time Warner's cable television channels) to US West, Itochu, and Toshiba. At the time, the arrangement, especially with US West, seemed smart. Time Warner needed a partner in the telephone business in order to expand its access to the burgeoning home box office market. (The key to the industry's future may well lie underground. Access to the cables owned and operated by local telephone and cable television companies are essential—hence, Malone's attempt to merge with Bell Atlantic and Redstone's alliance with Nynex.) Ostensibly because of its strategic importance to Time Warner's future, US West, unlike the two Japanese partners, secured veto power to protect itself against any future arrangement that might compete with (and diminish the profitability of) the TWE partnership, which, at least according to US West's attorneys, is exactly what the Turner acquisition promised to do.

When US West first stepped forward to block the Turner acquisition, many in the business press believed that it would prevail in court, or, more likely, that Levin would have to pony up, perhaps even cede control over Time Warner's cable systems in order to get US West to drop the suit. In 1996 Time Warner cable television, a division of TWE, served 11.7 million homes and accounted for almost 50 percent of TWE's total assets. Had Time Warner lost control of its cable holdings, it is hard to imagine what Levin could have told shareholders and the board to defend the Turner deal.

US West eventually lost its case in court, and Levin never had to make the deal. But Time Warner's victory before a federal judge may only serve to make US West even harder to deal with. "The victory will likely bring more headaches," Christopher Byron predicted in *Esquire* shortly after the US West suit failed, "as Time Warner's prize assets (its entertainment division) remain under shared control of a now publicly defeated, resentful and distrustful partner that will be scrutinizing—and perhaps looking to sabotage—any new deal the Turner merger may lead to."[44]

On July 17, 1996, the FTC conditionally approved Time Warner's purchase of Turner Broadcasting. After lengthy negotiations between Time Warner attorney Robert Joffe and William Baer, the director of the Bureau of Competition at the FTC, Time Warner Turner agreed to comply with three fairly substantial changes in the deal as it was first set up by Levin and Turner: (1) in order to prevent Time Warner Turner's CNN (and CNN's sister stations CNN Headline News, CNN International, and CNNFN) from completely controlling cable television news (and public opinion), the cable systems owned and controlled by the combined companies have to carry an additional all-news station produced and distributed by someone other than Time Warner Turner (marking the first time the government has mandated what sort of programming a cable system must carry);[45] (2) because it so clearly disadvantaged competing cable systems and potentially destabilized pricing in the cable market, the sweetheart deal discounting Turner programming on John Malone's TCI systems was nixed; and (3) Time Warner Turner will not be able to "bundle" or package its product; HBO, Cinemax, CNN, CNNFN, CNN Headline News, TNT, TBS, etc., must be made available to all systems for a fair price whether or not the respective systems want all or just one of the products.

Though the rhetoric cited in the trades and the financial press suggests that the FTC was primarily interested in maintaining freer access to the airwaves and freer trade in the cable industry, the regulatory agency's primary target throughout the negotiations was Malone. But while the FTC-Time Warner compromise gets rid of the twenty-year programming discount and further limits Malone's power within the combined companies,[46] Malone still has plenty of room to maneuver. By special arrangement, his windfall in Time Warner stock will be exempt from any IRS / capital gains tax liability. The compromise agreement also contains a provision that will enable Time Warner to renegotiate with TCI six months after final approval from the FTC. Though both parties will be restricted to a deal that holds for five (instead of twenty) years, nothing in place limits the future discount arrangement. Moreover, should Levin be ousted, nothing in the FTC compromise prevents Malone from taking the top spot at the conglomerate.

Federal Trade Commission approval of Time Warner Turner was the biggest of three big media industry stories, all of which broke within a few days of each other in mid-July 1996. The Tribune Company, which controls television stations and newspapers nationwide, announced a strategic alliance with the computer information service company, America Online. The deal with America Online gives the Tribune Company a strategic ally should (as many predict) television cable systems be tied to home computer on-line services in the future.

The other story—actually two related stories—far more closely paralleled Time Warner's blockbuster deal. On the very day that the FTC-Time Warner accord was announced, Rupert Murdoch's News Corporation made two significant moves: (1) to expand its holdings in TV—especially in syndicated programming—Murdoch purchased New World Communications (and its extensive television library) from Ronald Perelman for $2.48 billion, and (2) Murdoch made official his intention to produce a 24-hour news channel to compete with CNN. New World Communications gave Murdoch more product to sell into cable and gave him the most extensive television library in the industry. The cable news channel gave Murdoch an important media synergy in news reporting; moreover, it gave him greater access to and potentially greater influence over public opinion.

At the time, Murdoch's all-news channel seemed certain to benefit from the FTC compromise compelling Time Warner Turner to carry an alternative news station on its cable systems. Murdoch's only competition for the slot came from MSNBC, an all-news station launched earlier in 1996 by new partners NBC and Microsoft. In July, when the FTC agreement was first announced, the industry line was that, despite Turner's problems with Murdoch (and with Murdoch's all-news channel director Roger Ailes), Time Warner Turner would opt to carry the Fox News Channel. MSNBC, after all, had lobbied actively against the Turner acquisition, and the widely held assumption was that Levin was anxious to send MSNBC a message. But in September, Turner prevailed and the combined Time Warner Turner cable systems signed with MSNBC. As of September 1996 Murdoch was pretty much locked out of the New York market, at least the significant part of the market controlled by Time Warner Turner.

In the fall of 1996, industry journalists speculated (with I think laudable if not reasonable optimism) that the fallout from the purchase of Turner Broadcasting might lead to the breakup of Time Warner and that such a breakup might finally stem the tide of corporate gigantism in the entertainment industry. It seemed at the time certain that Levin would be forced to liquidate assets in order to service the debt that financed the acquisition of Turner Broadcasting. Rumored to be up for sale were Turner subsidiaries New Line and Castle Rock, film production companies, as well as Turner's professional sports teams: the Atlanta Braves and the Atlanta Hawks.[47] At the 1996 Time Warner shareholders meeting, several of those present boldly took the microphone and asked Levin if he planned to resign. In addition to widespread disappointment with management, shareholders offered support to a plan to break up the company. Had such a fire-sale of assets gained board approval, former Time Inc. shareholders would finally have seen the money they were promised when Ross and Levin negotiated the merger of Time and WCI back in 1989.

Pre-acquisition speculation posted the breakup value of the company at over twenty dollars per share higher than Time Warner's trading price.

According to *Forbes* senior editor Matthew Schifrin, the Turner acquisition diluted the break-up value of Time Warner from $63 to $57 per share, still a 39 percent premium over the undervalued $41 per share price in October 1996.

But even a dramatic re-structuring and a large-scale sale of assets did not guarantee Levin's future or Time Warner's eventual profitability. Financial analysts remained concerned about US West's ability to hold up developments at TWE, and cautioned that even if Levin offered to buy out US West, such a change in ownership at TWE would likely be considered a *sale* by the IRS. Should Levin reconsolidate control over TWE by repurchasing US West's shares, Wall Street experts cautioned, Time Warner Turner could owe the IRS more than $1 billion in capital gains taxes.

Warner Music, long a strong performer, was, at the time of the Turner acquisition, in a rare downturn; first quarter 1996 cash flow fell 16 percent. Projections had movie profits falling 33 percent industrywise with Warner studio profits also down one-third. Many in the industry blamed the downturn at Time Warner on Levin, who in the mid-nineties forced the exit of two key and popular executives: Mo Ostin, who ran the record company, and HBO visionary Michael Fuchs.

As of September 1996, Murdoch's News Corporation and Disney were in the process of *expanding* their holdings in the entertainment market. The very decline in theatrical revenues that so discouraged Time Warner investors in 1996 fueled corporate expansion at these better-managed and less debt-ridden companies. The future may indeed be one in which the entertainment industry will be controlled by two or three conglomerates; in the fall of 1996 it was uncertain whether or not Time Warner would be one of them.

Though there is a seductive, narrative quality to these financial wheelings and dealings, we must not miss the political ramifications of corporate maneuvers of this magnitude. Rumor has it that in 1995 Turner—who was sorely in need of cash to keep Turner Broadcasting going—decided to make a deal with Levin because he feared a hostile takeover from Rupert Murdoch.[48] Turner, who is left-of-center politically, feared Murdoch's conservative political agenda; moreover, he feared what adding CNN to Murdoch's tabloid newspaper empire might mean in terms of the shaping of American public opinion.

But trouble, again in the person of John Malone, looms at the periphery. Like Murdoch, Malone is a right-wing ideologue. Should Malone maneuver his way to the top spot at Time Warner Turner, the consequences of the monopoly proposed by the alliance of these two media giants would then not only concern the domination of an economic market, troubling in and of itself, but it would portend an ideological domination of America's airwaves and print media as well.

So far, Malone has made decisions that seem to suggest that what he really wants is more money, not more political power. For example, when Roger Ailes, a former Republican Party operative who worked for Nixon and Reagan, took control of CNBC and proposed a 24-hour news channel to compete with Turner Broadcasting's CNN, Malone refused to carry the station on TCI systems. Although he shares more in common politically with Ailes than with Turner, Malone seems to have thought first about his role and responsibility as a major shareholder in Turner Broadcasting and thus snubbed and bullied Ailes, who quickly and quietly backed off— testimony to Malone's clout. But what Malone would do in control of Time Warner Turner is, I think, cause for concern.

"Time Warner would have a powerful combination of businesses if someone could get them to work together," argues Michael Wolf, a partner at the investment firm Booz, Allen and Hamilton, which specializes in media industries. "They need stronger management and the right incentives."[49] Malone may well be the man for the job, the man who finally might enable the entertainment conglomerate to realize its full potential, or at least to keep pace with the market. There is no doubt that Malone is an outstanding manager; even those who have dealt with him—and if they have, they tend not to like him much—concede that he is a brilliant and ruthless businessman, which is to say that he may well be the only man currently in line to take over Time Warner who could handle the job. But as one executive (who wisely opted for anonymity) noted, only half in jest, "Malone wants to control the world."[50] When we follow the money, in Hollywood and in big business in general these days, such megalomania—and the potential to realize megalomaniacal goals—has become disconcertingly commonplace.

September 1997

When I finished the first draft of this essay in the fall of 1996, the financial press was rife with speculation that Gerald Levin would be forced to resign as CEO at Time Warner. At the time, journalists suggested three potential successors: Turner, former Time Warner second-in-command Nick Nicholas (who was forced out by Levin in a boardroom coup shortly after Steve Ross's death), and Malone. It is now a year later and Levin is still in control. Despite all the talk (and all his stock), Turner is clearly not running things. Nicholas is no longer mentioned as even a dark horse candidate for Levin's job, and Malone is talking about getting out of the communication business altogether.

After ten months of bitter fighting—in boardrooms, on the airwaves, in print—in July 1997 Levin and Murdoch reached a truce; Time Warner's New York City cable system now provides space for the Fox News Channel. Although he has never commented on the deal (for the record, at least), it is fair to assume that Turner is unhappy. In a court deposition leaked to the press in October 1996, Turner called Murdoch "a joke," "a scumbag," "a pretty slimy character," and "a disgrace to journalism." In the *New York Post*—a tabloid owned by the News Corporation—Murdoch printed a cartoon depicting Turner in a straitjacket above a caption that read: "Is Ted Turner veering dangerously towards insanity—or has he come off the medication he takes to fight his manic depression?" When Turner got a look at the cartoon he challenged Murdoch to a public boxing match.

The Levin-Murdoch alliance caught the industry by surprise, but it made good business sense. When Time Warner opted to exclude the Fox News Channel from its New York City cable service, Murdoch threatened to take his headquarters (and some nine-hundred jobs) across the river to New Jersey. New York City Mayor Rudolph Giuliani interceded and publicly sided with Murdoch, offering a city-run channel to Fox. Although the courts rejected Giuliani's plan, Levin had reason to believe that unless he made peace with Murdoch, Time Warner might lose its lucrative New York City cable franchise.

When Levin announced the deal to the Time Warner board, Turner (who no doubt had been briefed if not consulted) remained uncharacteristically, but understandably silent. While the deal with Murdoch was no doubt a source of some embarrassment to Turner, it met with unanimous approval on Wall Street. Given how many shares of Time Warner stock Turner owns, it is (at this writing at least) to his advantage to keep quiet about Murdoch and let Levin run the company.

It is important to emphasize here by way of conclusion that Levin's present security as CEO at Time Warner came not at the expense of a competitor but rather as a consequence of a competitor's good fortune. While the New York City cable deal expanded the subscriber base for Murdoch's Fox News Channel from twenty-two million to almost forty million, it supported at the same time a significant surge in the value of

Time Warner stock. As former New York City Deputy Mayor (and Levin confidant) Richard Powers so aptly pointed out in the press, when the conglomerates get along, "It's win, win, win for everybody."[51]

Notes

1 Neal Gabler, "Brother, Can You Spare a Million?" *Esquire*, September 1995, p. 140.

2 An example: Dave Hilberman, a longtime Disney animator who led the successful 1941 strike against the company, was ratted out by Walt Disney to HUAC in 1947. As a result, Hilberman, who was never a member of the Communist Party, was blacklisted and put under surveillance (and regularly hounded) by the FBI for fifteen years. Also, at least one employer that Hilberman later worked for—a Disney competitor—was added to the *Red Channels* list of undesirable and un-American companies and went out of business. After labeling Hilberman a Communist infiltrator and agitator, Disney made the following closing statement to HUAC: "I know I have been handicapped out there in fighting [the Communists], because they have been hiding behind this labor set-up, so that if you try to get rid of them they make a labor case out of it. We must keep the American labor unions clean." For more on Disney, the unions, and HUAC, see Mark Eliot, *Disney: Hollywood's Dark Prince* (New York: Birch Lane, 1993), pp. 178–97, and Holly Allen and Michael Denning, "The Cartoonists' Front," *South Atlantic Quarterly* 92, no. 1 (1993): 89–117. A final note: in 1993 Hilberman finally got his revenge by participating in Eliot's scathing unauthorized biography of "Uncle Walt," mentioned above, a book that has done much to revise the historical record on Disney.

3 A useful analogy can be found in contemporary professional baseball. When a collective bargaining agreement and various court decisions established free agency and salary arbitration for players, industrywide collusion—eventually proved in court by the players' union—temporarily enabled team owners to ignore these changes and punish the players who tried to use the new rules to their advantage.

4 "Trade Ponders: X the Key to B.O.?," *Variety*, February 25, 1970, p. 1.

5 Chris Hugo, "The Economic Background: Part II," *Movie* 31-32 (1986): 84.

6 "Needed Cash for Bank Balances, Kerkorian Sold 2% of His MGM," *Variety*, April 25, 1979, p. 4.

7 The CPI antitrust suit was not the first time the U.S. Justice Department had gone after Kerkorian; federal attorneys had interrogated him years earlier about his relationship with reputed gangster Meyer Lansky, who, they alleged, helped finance Kerkorian early on in his career. Kerkorian refused to testify against Lansky. In 1979 rumor had it that the government's decision to challenge the CPI deal had less to do with antitrust than retaliation.

8 James Harwood, "Trial Begins on Col. Stock Buy," *Variety*, August 22, 1979, p. 5.

9 "Economic Professors Unalarmed if Distribs. Shrink or Combine," *Variety*, August 15, 1979, p. 7.

10 James Harwood, "Dept. of Justice Draws a Defeat," *Variety*, August 22, 1979, p. 53. The italics are mine.

11 "Vincent Memo Defends Price as Pivotal Studio Power: Hits KK's Claims of Stark Control," *Variety*, October 8, 1980, p. 36.

12 A convertible debenture is an unsecured corporate bond that can be converted into common shares at the option of its owner. In this case the convertible debenture allowed Allen to issue additional shares in order to put them in the *friendly* hands of in-house producer Ray Stark.

13 David McClintick's book on the David Begelman scandal, *Indecent Exposure* (New York: Dell, 1983), confirms Kerkorian's allegations. According to McClintick, when Begelman, then Columbia Pictures president, was caught forging checks and embezzling money, Stark was powerful enough to get him (briefly) reinstated despite opposition from then CPI president (and Allen and Company employee) Alan Hirschfield.

14 Companies are said to pay *greenmail* to corporate raiders when they buy back stock at a premium in order to prevent an unwanted takeover. *Forbes* estimated Kerkorian's personal wealth at $4.36 billion, much of that earned in greenmail payoffs. See "The Superrich," ed. Graham Burton, *Forbes*, July 15, 1996, p. 188.

15 While MGM/UA was owned by Crédit Lyonnais, it was managed/advised by the powerful Hollywood talent agency Creative Artists Agency (CAA).

16 Several companies bid on MGM, including Polygram, Morgan Creek, and the News Corporation, but Kerkorian's $1.3 billion—the exact amount Giancarlo Parretti paid him in 1990—carried the day. As of late 1997, MGM executives were said to be happy; one of them, quoted in *Entertainment Weekly*, succinctly explained why: "The bottom line is I get to keep my job." Beyond the absurdity of Kerkorian buying back—or financing the buyback—of a studio he tried for almost a decade to unload, it is fair to wonder what Crédit Lyonnais executives were thinking. In 1989, when it was forced to foreclose on Parretti, Crédit Lyonnais filed suit against Kerkorian for misrepresenting the value of the property and concealing certain facts about his deal with Parretti. Apparently the French bank has learned a little since taking over the studio; in Hollywood it pays to have a short memory. See Casey Davidson, "Deals," *Entertainment Weekly*, July 26, 1996, p. 11.

17 Between 1978 and 1988 ten independent film companies—American International, Filmways, Weintraub, Cannon, De Laurentiis, New World, Lorimar, Vista, New Century, and the Atlantic Releasing Corporation—went out of business. With the possible exception of the exploitation pictures made by direct-to-video production companies there are no B-movies anymore.

18 Janet Wasko, *Hollywood in the Information Age* (Austin: University of Texas Press, 1995), pp. 48–49.

19 *Variety*, June 7–13, 1989, p. 15.

20 "Paramount Communications Inc. Is Our New Name" stated an advertisement in *Variety*, June 7, 1989, p. 15.

21 "Pending Sale of Associates First Indicates G&W May Be Considering Viacom Merger," *Variety*, May 10–18, 1989, p. 3. The Time Warner merger and Paramount's attempt to block it is discussed at greater length in Jon Lewis, "Trust and Anti-trust in the New New Hollywood," *Michigan Quarterly Review* 35, no. 1 (1996): 97–102.

22 Richard Gold, "Size Is the Ultimate Prize as Showbiz / Media Corps Fight for Supremacy," *Variety*, June 14–20, 1989, pp. 1, 6; "Will Par-Time-WCI War Victimize Creatives?" *Variety*, June 21–27, 1989, pp. 1, 4; "Intense Propaganda Fight Mars Par vs. WCI War," *Variety*, June 26–July 4, 1989, pp. 1, 5; "Par's Block Looks Like a Bust as Court Backs Time Director's Stand," *Variety*, July 19–26, 1989, pp. 1, 6; Richard Gold and Paul Harris, "Time Marches On, Grabs Warner, Outpaces Par," *Variety*, July 26–August 1, 1989, pp. 1, 6; "Time Inc. Buyout Attempt Puts Dent in Paramount Communications Qtr.," *Variety*, September 20–26, 1989, p. 9.

23 Gold, "Size Is the Ultimate Prize," p. 1.

24 Bids are considered "hostile" when they go against the expressed wishes of management. Such bids, though, often are very good for individual shareholders, each of whom is offered a premium, for his or her stock.

25 Gold, "Size Is the Ultimate Prize," pp. 1, 6.

26 Gold, "Will Par-Time-WCI War Victimize Creatives?" p. 4.

27 In *Forbes*, Matthew Schifrin projected that the decision to turn down the PCI offer had cost Time shareholders "60% of what they could otherwise have realized." See "The Mess at Time Warner," *Forbes*, May 20, 1996, p. 170.

28 Gabler, "Brother, Can You Spare a Million?" p. 38.

29 See Wasko, *Hollywood in the Information Age*, pp. 48–52. Wasko's sources are, primarily, Time Warner's annual reports.

30 Schifrin, "The Mess at Time Warner," p. 170.

31 Ken Auletta, "Awesome," *New Yorker*, August 14, 1995, p. 31.

32 Bill Carter and Richard Sandomir, "The Trophy in Eisner's Big Deal," *New York Times*, August 6, 1995, sec. 3, pp. 1, 11.

33 Time Warner already owned the remaining 18 percent.

34 Mark Lander, "Time Warner and Turner Seal Merger," *New York Times*, September 23, 1995, pp. 1, 18.

35 These regulations may be altered or waived altogether if pending media industry legislation, endorsed unsurprisingly by the Republican Right, is passed. Given the political influence of Rupert Murdoch and John Malone, both of whom contribute strategically to conservative Republican campaigns, it is fair to predict even further deregulation.

36 Lander, "Time Warner and Turner Seal Merger," p. 1.

37 After Gulf & Western CEO Charles Bluhdorn died in 1983, Martin Davis took over the company. Davis was much more fiscally conservative than his predecessor, and in one of his first moves he promoted marketing expert Frank Mancuso to the top spot at Paramount. In doing so, he not only slighted Diller, who had been the studio chairman under Bluhdorn and was widely acknowledged as a creative executive rather than a careful one, but reneged on an agreement that, over time, had promised Diller more autonomy and power. Diller then filed suit for breach of contract. The dispute was settled out of court, and Diller quickly landed on his feet—in the top spot at Fox—but he never forgave Davis for treating him so shabbily. Soon after Diller left, his second in command, Michael Eisner, resigned to take the CEO position at Disney. Eisner's exit led to a dramatic display of distrust for Davis's leadership as Jeffrey Katzenberg, Bill Mechanic, Helene Hahn, Richard Frank, and Bob Jacquemin—top executives in the creative and legal departments—left Paramount to join Eisner at Disney. The mass exodus made Davis look bad; and the subsequent Disney turnaround—which has been executed by the former Paramount brain trust—only made things worse. By the time that QVC made its bid for Paramount in 1993, both Diller and Davis had major scores to settle with each other.

38 See Gary Samuels, "Gore-ing Malone's Ox," *Forbes*, September 9, 1996, pp. 52–56.

39 Bryan Burrough, "The Siege of Paramount," *Vanity Fair*, February 1994, p. 131.

40 The chancery court decision on Time Warner made clear that companies no longer had to take short-term profit into account when making decisions of this magnitude.

41 Burrough, "The Siege of Paramount," p. 70.

42 Christopher Byron, "Time of Troubles," *Esquire*, August 1986, p. 51.

43 In March 1989 Michael Milken was indicted on ninety-eight felony counts resulting from a government sting operation engineered by then Assistant U.S. Attorney Rudolph Giuliani as part of his ongoing investigation into insider trading in the securities industry. Using testimony from two other (arguably bigger) felons, David Levine and Ivan Boesky, Giuliani eventually forced Milken to plea-bargain (down from a potential 520 years in jail and a fine of $1.8 billion). Milken, astonishingly, given his personal wealth and team of attorneys, did time—he was sentenced to ten years but got out considerably quicker—and he is barred from working in the securities business for life. Milken came to prominence in 1983 when he helped Roy Disney Jr. finance a move on his father's old company (mostly through the proposed sale of junk bonds). Upon his release, Milken soon resumed his role as the broker of choice in film mergers and acquisitions. The only difference from his former role is that these days his fee has gone up.

44 Byron, "Time of Troubles," p. 50.

45 Previously, the only programming mandate enforced by the FCC was that all cable systems must provide, as part of their service, network and local independent stations available over the airwaves.

46 Pending an IRS ruling, TCI's investment in Time Warner Turner will be limited to 9.2 percent. All these shares will be spun off into a new company owned and controlled by shareholders in TCI's subsidiary, Liberty Media. See Bryan Gruley and Eben Shapiro, "Time, FTC Staff Agree on Turner Deal," *Wall Street Journal,* July 18, 1996, p. A3.

47 Ibid., p. A14.

48 Connie Brunk, "Jerry's Deal," *New Yorker,* February 19, 1996, p. 59.

49 Schifrin, "The Mess at Time Warner," p. 174.

50 Connie Brunk, "Jerry's Deal," p. 61.

51 The primary source for this last section is John Cassidy, "Brotherhood of Man Dept.: Where Was Ted Turner When Levin and Murdoch Made Up?" *New Yorker,* July 21, 1997, p. 23.

Cinema and Culture

A Rose Is a Rose?

Real Women and a Lost War

Tania Modleski

If there ever was a purely masculine genre, it is surely the war film.[1] That women in the genre represent a threat to the male warrior is revealed in a timeworn convention: a soldier who displays a photograph of his girl-friend, wife, or family is doomed to die by the end of the film. The convention is so well-known that it is parodied in *Hot Shots* (1991), a spoof of the very popular film *Top Gun* (1986). Playing off the nicknaming of fighter pilots, one of the early scenes in *Hot Shots* shows the hero's sidekick slamming a locker door on which is taped a family photo. He then intro-duces himself to the hero as "Dead Meat."

Feminist critics of the war film, most notably Susan Jeffords, have con-vincingly argued that the genre is not only *for* men but plays a crucial role in the masculinizing process so necessary to the creation of warriors. Through spectacle (bombs bursting in air) and sound (usually heavy rock), pro-war fantasies like *Top Gun* mobilize the kind of aggression essential to the functioning of men as killing machines. So, too, often enough, do antiwar films. Indeed, it is frequently noted that films like *Platoon* (1986) not only do not effectively protest war but actually partici-pate in and extend it, to the point where the spectator him- or herself becomes the target of the warrior-filmmaker's assault. As Gilbert Adair, in a thoughtful critique of *Platoon,* puts it: "It is surely time that film-makers learned that the meticulously detailed aping of an atrocity *is* an atrocity; that the hyper-realistic depiction of an obscenity cannot avoid being con-taminated with that obscenity; and that the unmediated representation of violence constitutes in itself an act of violence against the spectator."[2]

Moreover, just as it is the goal of war to crush opposing viewpoints and violently secure the opposing side's assent to the conquerer's truth, films like *Platoon* "bully" us into "craven submission," as Adair puts it, by pointing "an accusatory finger" and asking, " 'How do you know what it was like unless you've been there?' " (p. 169).³ Such questions are often literally addressed to women in the Vietnam films: "Who the hell are you to judge him?" the uncle, a vet (Bruce Willis), rebukes his niece (Emily Lloyd) in Norman Jewison's *In Country* (1989) when she expresses dismay at the racist remarks she finds in the diary that her father kept before being killed in the war. Thus, since "being there" has so far been out of the question for women (who are prohibited from combat), their authority on any issue related to war is discredited from the outset, and insofar as they may be inclined to question or oppose war (except in and on the terms granted them by men), they find themselves consigned to the ranks of the always-already defeated.⁴

Given the extent to which war has been an exclusive masculine preserve, it is not surprising to find the critics themselves desiring generic purity, expressing discomfort when bits of "feminine discourse"—for example, melodrama and the love story—are used for non-ironic purposes, thereby "contaminating" the war film. One of the earliest films about the Vietnam War, *Coming Home* (1978), which David James condemns because it "rewrite[s] the invasion of Vietnam as erotic melodrama," is a case in point.⁵ While commentators like John Hellman have found much to praise in other films' rewriting of the invasion in terms of *male* genres— the western (*The Deerhunter*, 1978) and the detective genre (*Apocalypse Now*, 1979)—*Coming Home*'s incorporation of female genres provokes derision.⁶ Of course, *Coming Home* is not really about the "invasion" of Vietnam but about its aftermath, and thus it is situated in a tradition of films about veterans' adjustment to civilian life. Nevertheless, critics have faulted it for not focusing more on "the problems of returning veterans than on the clichéd love story."

Coming Home is about a disabled veteran, Luke (Jon Voigt), returning to civilian life and meeting a woman named Sally (Jane Fonda) who is married to a Marine serving in Vietnam and whose consciousness about the war is raised when she serves as a volunteer in a VA hospital. This consciousness receives a gigantic boost when Sally makes love with Luke, who brings her to climax orally, giving her her first orgasm. Critics have struggled to understand this scene in symbolic terms. Jason Katzman writes, "[*Coming Home*] uses the love story as a metaphor for the impotence of an entire country in understanding Vietnam. . . . Sally's ability to reach an orgasm with Luke where she could never before with her husband Bob (Bruce Dern), is one of the more widely discussed symbols."[7]

It is not clear to me what exactly the woman's orgasm is a symbol *of*; but it is clear that the event was less satisfying to some male critics than it was to *her*. Albert Auster and Leonard Quart express uneasiness about this plot element, but they do not really specify the source of their discomfort. "Unfortunately," they write, "Sally's transformation seems unconvincing and mechanical, especially in the emphasis it places on her achieving orgasm . . . while making love to Luke."[8]

Suppose, however, we take the orgasm to be an end in itself, rather than a symbol or a metaphor for something else. Suppose, in fact, that one of the problems with Vietnam War films in particular is their relentless exploitation of experiences and events for the sole significance that they have to the soldiers who fought rather than to the men's loved ones, allies, or enemies. I suspect that the critics' discomfort stems from the vividness with which the film demonstrates the point that men's losses may be a gain for women. Kaja Silverman has identified a similar theme in *The Best Years of Our Lives*, William Wyler's 1946 film about men returning from World War II and attempting to adjust to civilian life. Silverman heralds the film as a kind of feminist milestone in the history of what she calls "libidinal politics," since at this moment of "historical trauma," in which men came back from war mutilated both psychically and physically, non-phallic forms of male sexuality presumably emerged.[9]

One might argue that from the point of view of women on the home-front, *Coming Home* is even more important in this regard. Unlike *The Best Years of Our Lives*, in which the women must find satisfaction in an eroticized maternal relation to their men, *Coming Home*, based on a story by a woman, Nancy Dowd, was made at a time when feminists were vociferously proclaiming the myth of the vaginal orgasm and agitating for the requisite attention to be paid to the clitoris. Women were hardly passive beneficiaries of the historical vicissitudes of male sexuality (and male warmongering) but were actively demanding and sometimes winning their sexual rights. However attenuated the film's politics are in other respects (and these politics are indeed feeble in some regards),[10] it is important not to overlook the film's significant place in the ongoing struggles over sexual politics.

To see how far women have been forced to retreat from this position of sexual advantage, we need only briefly compare *Coming Home* to a film on the same subject made a dozen years later. In Oliver Stone's *Born on the Fourth of July* (1989), a fictionalized version of the story of Ron Kovic, a war hero, who is obsessed about his dysfunctional penis; the Kovic character reaches a low point when he goes to Mexico and spends days and nights whoring, gambling, and boozing with other disabled veterans. Going to bed with a whore, Kovic appears to bring her to climax through manual stimulation; but in contrast to *Coming Home,* which focuses on the woman's tears of joy and passion, in this film we see *the man* crying out of self-pity for his lost potency. (*In Country* also contains a scene in which a vet is impotent with the young heroine, who is kind and understanding. Nowhere is there a hint that his hard-on might not be the sine qua non of *her* sexual pleasure, for the question of her sexual pleasure is not even on the horizon.) In a brief subsequent scene in *Born on the Fourth of July,* the camera assumes the hero's point of view as he wheels himself around a whorehouse in which various Mexican women beckon him with lewd remarks. Racism and misogyny combine in a scene meant to demonstrate the depths of degradation to which the hero has sunk. Here we see an example of the commonplace phenomenon in Vietnam films in which exploited people (in this instance, the prostitutes) are further exploited by the films themselves for the symbolic value that they hold for the hero. Thus do the films perpetuate the social and cultural insensitivity that led to America's involvement in the war and the atrocities committed there.

The film *Casualties of War* (1989) presents an even more extreme example of this phenomenon. In this film, as Pat Aufderheide argues, the rape and murder of a Vietnamese peasant girl by American soldiers signify "the collapse of a moral framework for the men who kill her. The spectacular agony of her death is intended to stir not the audience's righteous anger at the grunts . . . but empathy for the ordinary fighting men who have been turned into beasts by their tour of duty."[11]

Equally extreme and still more bizarre is a scene in *Born on the Fourth of July* in which Kovic and another vet quarrel over which of them has killed more babies. The implication is that the superior person is the one who has killed the most babies since he has to carry a greater burden of guilt![12]

It is eminently clear from *Born on the Fourth of July* that historical trauma does not necessarily result in a progressive politics—"libidinal" or otherwise. Nor is the phallus necessarily relinquished by men who have suffered from such trauma. As Tony Williams has shrewdly observed, the trip to Mexico allows Kovic to "confront his dark side . . ., confess his sins to the family of the man he shot, and gain the phallus (if not the penis) by speaking at the 1984 Democratic convention before an audience mainly composed of silent . . . and admiring, autograph seeking women."[13] Earlier in the film, when Kovic is released from the hospital, he goes to see his girlfriend from high school days. She, however, is so caught up in antiwar activities that she is unable to connect with him. As Kovic wheels through her college campus, he declares his love for her, and the camera focuses on her walking beside him but looking away, in the direction of a group of activists about to hold a meeting. The proper role for a woman in the antiwar movement, the film makes clear, is that of silent supporter of male protestors, not independent actor. The feminism that grew up partly in response to this attitude is, needless to say, nowhere evident in the film.

In addition to *Coming Home,* one other film is noteworthy not only for its incorporation of a love story, but, most importantly, for the emphasis it places on the effects of the war and the war's aftermath on women's subjectivity. In *Jacknife* (1989) Ed Harris plays Dave, a pent-up alcoholic vet whose antisocial attitudes and behavior are ruining the life of his school-teacher sister (Kathy Baker) with whom he resides.[14] A friend called Megs, played by Robert De Niro, comes to visit and attempts to break Dave out of his shell, and in the process Megs falls in love with Dave's sister. Dave violently opposes the relationship and (often unconsciously) works to sabotage it. One night when his sister and Megs are at the prom (she is there as a chaperone, but clearly they are both attempting to capture something lost in their youth), Dave comes to the high school and smashes a glass trophy case. After he runs off, his sister attempts to go on with the evening as if nothing had happened, while Megs is understandably distracted and distraught. Astonishingly, the film does not demonize the woman for resenting the way her brother's trauma has circumscribed her life. Indeed, it shows that the brother needs to accept a certain amount of responsibility for casting a pall on her existence. At the end, we see him in group therapy coming to terms with the fact that he betrayed Megs one time in battle and coming to terms as well with what he sees as his own cowardice. It hardly seems necessary to point out how seldom issues relating to inglorious combat behavior get raised in Vietnam films.

Rick Berg, a combat veteran, has written in an influential essay, "Losing Vietnam," that "the vet can begin to overcome his alienation" only when he recognizes that Vietnam's "consequences range throughout a community."[15] While, for the most part, Berg's concern is with issues of class, *Jacknife* has the merit of focusing on the vet's recognition of the way that the war affects the relations between the sexes. *Jacknife* is certainly not without many of the problems characteristic of films about post-Vietnam life. It never, of course, even alludes to the feminism that arose from antiwar activity; on the contrary, the sister is cast in stereotypically spinsterish terms ("I know what I am," she says, not, however, voicing the dreaded term), and to her the goal of a life of her own is having a husband and family. The ending of the film is especially problematic in suggesting that somehow the union of Megs and Dave's sister will allow the two to capture the lost innocence of their high school days. Nevertheless, if the film accomplished nothing more than granting Baker the line, "Don't you want a point of view?" when her brother says he won't take her, a woman, fishing with his friend, it would have done more than almost any other Vietnam film in granting a woman an independent subjectivity and hinting at the possibility that she can be the maker and not just the bearer of meaning.[16]

The ending of *Jacknife*, with its nostalgic promise of a return to innocence, is characteristic of the genre. "Return" is a constant motif in Vietnam films, as many critics have noted. In the Rambo-type films there is one heroic man's return to Vietnam so that he can "win the war this time." For all its apparent liberalism, the same concern is detectable in *Born on the Fourth of July*. Ron's activism at the end of the film is *explicitly* associated with warfare: turned out of the Republican convention that the veterans have stormed, Ron uses militarist language in instructing his men to return and "take the hall." Not only is Kovic thus positioned as a victorious warrior, a man who wins the war against the war, but in taking the hall he reverses another ignominious defeat—that is, a wrestling match he lost in high school, to the tremendous disappointment of his mother and girlfriend.

The notion of return is also present in the constant process of "metaphorization" that occurs in these films, which as we have seen make everything and everyone (raped women, murdered babies, etc.) refer back to and stand in for the American soldier (or veteran) and his plight. Difference and otherness are recognized only to the extent that they are seen to signify something about the American male. Some feminists have identified metaphor, which reduces differences to versions of sameness, as basic to Western "phallocentric" thought; in making this point, they draw on the paradigmatic Freudian scenario whereby the male reads the female body in terms of his own standard—the penis, which the female body is judged to be lacking.[17] Now, given the crisis in America's "phallic" authority that opened up with the loss of the war, and given the occasionally literal severing of the penis from the phallus that symbolizes it, it is not surprising to see metaphorical operations such as those described move into high gear in representations of Vietnam. Thinking again of the coming in *Coming Home*, we can see why a feminist would appreciate the sex scene in that film, might prize its brief acknowledgment of feminine difference, and want to insist that sometimes a clitoris is just a clitoris, and a woman's orgasm simply that.

This preamble is designed to put in relief the achievement of Nancy Savoca's understated, small-budget film *Dogfight* (1991) since it is easily lost in the midst of the loud, frantic, spectacular representations with which we have been bombarded by so many male directors of Vietnam films. *Dogfight* is a film about a group of marines about to be shipped overseas (they do not know it, but they are destined for Vietnam). The group members set up a "dogfight," a dance, to which each of the marines is supposed to bring an ugly date. The man who finds the ugliest woman is the winner and receives a cash prize. One of the four men, Eddie (River Phoenix), is unsuccessful at convincing the women he encounters to go out with him; more or less giving up on the attempt, he goes into a diner and meets Rose (Lili Taylor, padded up a bit for the role), a waitress whose mother owns the diner and who, when he first sees her, is picking a folk tune on her guitar. He invites her to the dance, and excited about the prospect of going to a party and escaping for a time her humdrum life, she accepts. One of the interesting aspects of this setup is that the spectator is not sure whether or not Eddie really considers Rose to be "dogfight" material, a question that the film never clears up. During the evening Rose discovers the purpose of the dogfight, slugs Eddie, and leaves in a rage. Remorseful, Eddie goes to the apartment over the diner and gets Rose to agree to go out with him for the night. Scenes of their evening out together, which is sweetly romantic, despite their arguments (Rose is clearly a budding peacenik), are intercut with scenes of Eddie's three friends spending their last night in the States brawling with sailors, watching pornography in a theater, being fellated by a prostitute, who does each of them in turn in the theater, chewing gum in between bouts, and getting tattoos of bees on their arms (each has a last name beginning with "B") to mark their loyalty to one another. Eddie and Rose sleep together, and then he runs to meet his bus. After a very brief battle scene set in Vietnam, Eddie returns, wounded, to San Francisco, and the film very movingly presents us with the point of view of a man who sees an entirely different world from the one he left. Flower children fill the street, and one walks by and softly asks, "Hey, man, did you kill any babies over there?" For all the preoccupation of films like *Born on the Fourth of July* with soldiers' readjustment to civilian life, in my view no scene in any other movie captures more vividly their estrangement and confusion. Eddie

13 Last chance for romance: the pivotal prom scene in Oliver Stone's *Born on the Fourth of July* (Universal, 1989).

goes into the diner that Rose now runs and encounters a more mature woman; as they look at each other, they are at a loss for words, and in a mournful ending they embrace.

I want to argue against the grain of the voluminous criticism on Vietnam War movies and to propose that it is precisely because the film is a love story and gives us a woman's perspective on war and the warrior mentality that it is less compromising in its opposition to war than the films in that most paradoxical of genres, the antiwar war film. The antiwar sentiment is present not only in *Dogfight*'s narrative but is conveyed at the level of style: much of the film's subversiveness lies in the peacefulness and restraint of its pacing, rhythms, and soundtrack. There is a sweetness in the encounter between the boy and girl that is genuinely moving. But it must be said that this film is less sentimental than most Vietnam films, which Andrew Martin has convincingly shown to be, for the most part, male melodramas.[18] To measure the gap between the sentimentality of some of these films and *Dogfight*'s uncompromising view of the cruelties of which people are capable and which, after all, have some bearing on our desire or at least our willingness to make war, we need only compare one event that is featured prominently in Vietnam films with a very different one in Savoca's film. I refer to, respectively, the high school prom and the dogfight—the dogfight being an event for which it is difficult to muster up the same sort of nostalgia inspired by the prom.

The eponymous event of the dogfight may be seen as the antithesis of the prom scenes in both *Jacknife* and *Born on the Fourth of July,* both of which nourish us in the dangerous illusion of a time of lost innocence and unimpaired relations between the sexes. In *Born on the Fourth of July,* Ron Kovic, on the eve of his departure for military service, runs through a storm in his old clothes and arrives soaked at the prom to dance with his starry-eyed girlfriend. In *Dogfight,* of course, the "dance" is the cruel event of the dogfight itself (staged again on the eve of the men's departure for overseas, and as it turns out, for Vietnam). Rose finds out about the dogfight in a scene that takes place in the ladies' room. Standing in front of a mirror, one of the girls, Marcy, who has gone toothless to the affair and earned the prize for her date, tells Rose about the rules of the contest while putting her teeth back in. "The thing that gets me," says Marcy, "is how great they think they are. Did you ever see such a pack of pukes in your life?" Rose is, naturally, appalled and marches out to Eddie and punches him, confronting him directly with his cruelty and lack of feelings for others. In this regard, the script, written by a former marine, Bob Comfort, has the infinite merit of focusing on the anger and humiliation of the object rather than the sad plight of the subject of the cruelty.

Yet in creating Rose, who in the original script was supposed to be overweight, Comfort intended to make the woman serve as a "metaphor" for the marine, her unacceptable looks symbolizing his "outsider status."[19] Comfort has publicly expressed his unhappiness with Savoca's changes in his script, and one can only speculate that his feeling of dispossession—his sense of dis-Comfort, as it were—stemmed from Savoca's resistance to turning the heroine into a metaphor, a reflection of the hero. According to Savoca, in the original script Rose "was more of a catalyst for change," and, she says, "this bored me and Lili to tears." They resolved to make the character "someone in her own right." Importantly, however, the transformation in the female character does not occur at the *expense* of the male character (almost a primal fear of men when women make art); on the contrary, Savoca maintains, "as her IQ comes up, so does his. Because rather than reacting to a thing, he's reacting to a complicated person. Something happens between two people and not just between this guy and his revelation." Savoca continues:

We decided that the first thing to do was give her a passion—so that regardless of what she looks like there's something going on within herself. And that something is music. . . . He becomes attracted to her because she has a love in her that goes beyond their small world and the rigid narrow existence he's used to living—

and not because one day she takes out the ribbon in her hair and, oh my god, she's stunning.[20]

In view of the narcissistic self-referentiality of many male-directed Vietnam films, which repeatedly and utterly disqualify women as authorities in matters of war and peace (we recall Ron Kovic's girlfriend being judged harshly by the film *Born on the Fourth of July* for turning away from Kovic and toward the antiwar demonstrators), we can perhaps appreciate Savoca's audacity in having her heroine's aspirations and values point a way out of the trap in which the soldier finds himself. Indeed, we might say that, by extension, just as Eddie is required to treat Rose as a person with an independent subjectivity who has the potential for giving him a glimpse of more expansive horizons, so too does Savoca's encounter with Comfort's text strengthen it while respecting and underscoring its powerful indictment of a society that devours young working-class men and spits them out.

Rose is a young girl who dreams of changing the world, of possibly joining the civil rights movement in the South or engaging in some other form of social activism; she also aspires to be a folksinger, and during their date she argues with Eddie throughout the evening about the most effective means of changing the world. He, of course, has opted for guns; she for guitars. That the aspirations and values of the heroine are expressed in her love of folk music is particularly appropriate. Vietnam was, as more than one critic has noted, America's rock-and-roll war, and many of the films about the war are edited to the supercharged rhythms of the Rolling Stones and similar groups; even if the lyrics of many of these songs are intended to make an ironic or critical comment on the war, the music itself often serves to pump up the testosterone level, working viscerally against the antiwar sentiments supposedly being conveyed.[21] Such music, as David James has argued, is always at least ambivalent.[22] In this regard, then, we might compare the nihilistic song by the Stones, "Paint It Black," that ends Stanley Kubrick's antiwar, antimilitary film, *Full Metal Jacket* (1987), with the Malvina Reynolds' song that Rose haltingly sings sitting at a piano in a nearly deserted café as Eddie looks on: "The grass is gone / the boy disappears / and rain keeps falling like helpless tears."

Eddie and his friends do disappear, and only Eddie returns. When he comes back to San Francisco, the world has changed drastically, and he is viewed with disdain and suspicion by the flower children milling around in the street. Eddie, wearing his uniform, limps into a bar across the street from Rose's café. When the bartender sees Eddie's tattoo, he shows him his: a girl who jiggles and performs a "belly dance" on his protruding gut. Eddie asks about Rose, describing her as kind of chubby; the bartender responds, "She ain't no prize," and one of the men chimes in, "yeah, like you're really something, eh, Carl?" At one point the bartender asks Eddie if he served in Vietnam. When Eddie responds in the affirmative, the bartender, clearly at a loss for words, says, "Yeah, bummer," and walks away. "No charge on that," says the owner of the bar when Eddie gets a second drink. "Thank you," Eddie replies. "Thank *you*," says the man quietly, but he cannot look Eddie in the eye.

As I have said, these few moments convey the pathos and isolation of the returning veteran more eloquently than a thousand bombastic moments in an Oliver Stone movie. They strike a note of pure loss, prolonged through the final shots when Eddie goes across the street to see Rose, and she too seems not to know what to say or do. They embrace, though this is clearly not the embrace of two people destined to live happily ever after; it is an act of mutual consolation over all the sorrow and loss that has occurred in the intervening years, including the severing of their slender connection, which of course cannot ever be reforged. Warner Bros., encouraged by a preview audience's positive response to the first part of the movie ("Rah, rah, dogfight,"), exerted intense pressure on Savoca to change the ending to make it more upbeat. Finally, an exasperated Savoca asked sardonically, "Do you want us to change the ending so we win the war?"—apparently not realizing that that was *exactly* what Hollywood wanted and what, over the last decade or so, it has gotten from movie directors all too happy to oblige.

Dogfight, in a certain sense, may be seen as the second in a two-part series that Savoca filmed on relations between the sexes. Her brilliant film debut was a movie that gained a cult following among women. Drawing on some techniques of documentary, her first film, *True Love* (1989), is a dystopian "wedding comedy" that focuses on the rituals leading up to the big day. Donna and Mikey, the engaged couple, battle over whether or not Donna will get to go out with Mikey after his bachelor party. Throughout the film, Savoca cuts between the group of guys (Mikey and his friends) and the group of girls (Donna and hers), showing us two worlds that are so separate that their inhabitants might as well live on separate planets. "Sometimes she says things to me and I don't understand what the fuck she's talkin' about," says Mikey at one point. As the boys party on throughout the night, getting drunker and drunker (while the camera gets wilder and wilder, positioned, for example, as the ready bank teller

14 The morning before in Nancy Savoca's dystopian wedding comedy, *True Love* (United Artists, 1989).

machine and the jukebox, capturing Mikey in unflattering close-up as he leans down and pushes the buttons), the girls just hang out, waiting, talking of marriage, home decorating, etc. Although at one point they go to watch male strippers, it is clear they do so defiantly, in order to prove their ability to have the same kind of fun as the boys—thereby, of course, proving the opposite. The climax of the movie occurs during the wedding itself when Mikey proposes to his friends that they go out for a while on the wedding night. He begs Donna to let him go with the boys for just an hour or two, and she runs crying into the ladies' room, where she is followed first by her friends and then by Mikey. (Savoca adds a wonderful touch when she has a mother drag her protesting little son into the ladies' room; glancing at Mikey, she says, "See, this boy is in here, too.") The film ends inconclusively, though one senses that the two will go on with married life, living it as happily, or rather as unhappily, as most.

True Love documents better than any film I have ever seen the asymmetries of life as it is generally lived by the two sexes in modern America. It submits to a trenchant analysis the relations among men that are glorified and idealized by the overwhelming majority of Hollywood films. *Dogfight* continues in this vein, alternating scenes of the developing relation between Eddie and Rose with scenes of Eddie's friends out on the town boozing and whoring. From wedding to war and home again, Savoca's films, taken together, cover the same territory as Michael Cimino's lengthy and controversial Vietnam film *The Deer Hunter,* and they may be read as a rewriting of that film in feminist terms.

Roughly the first third of *The Deer Hunter* depicts a wedding that takes place in a highly sex-segregated working-class community of Russian-Americans, just as *True Love* is situated in a working-class Italian-American section of the Bronx. In an excellent analysis of *The Deer Hunter,* Susan Jeffords has shown that the wedding sets up the basic conflict of the film, which is most clearly played out by the character of Nicky, who is killed off by the film precisely because his loyalties are divided between the world associated with women (sex, marriage, domesticity) and his affiliation with men (specifically one man, Michael, played by Robert De Niro). Jeffords writes, "One must either live all of the points of the code ['discipline, endurance, purity'] or not attempt it at all . . ., but not attempt and fail; one must fulfill either the masculine or the feminine, but not both."[23] Of course, these two options are not equally valued by the film (for the character who chooses the feminine comes home a paraplegic); rather, its primary emotional investment is in the relations among men, in particular, the relations between Nick and Michael, whose friendship is highly idealized by Cimino. Jeffords goes so far as to claim that "Vietnam is . . . not the subject of *The Deer Hunter* but merely the occasion for announcing the primacy of the bonds between men."[24]

While many Vietnam films, most notably perhaps *Full Metal Jacket,* have exposed the misogynistic aspects of military life, none of them focuses as unwaveringly as *Dogfight* on the more unappealing aspects of the male bonding that is part and parcel of the misogyny—the dirty jokes, the lies about sexual prowess, the animal behavior and brawling, the humiliation of those further down the pecking order, and so on. Soldiers who fought in Vietnam were, after all, just boys out of high school (a fact we are inclined to forget when a thirty-something Robert De Niro is the star). *Dogfight* goes very far indeed in contravening one of the most basic assumptions of Hollywood war films in regard to women and male bonding, suggesting *not* that men must give up ties to women and families in order to survive, but that the unthinking loyalty to the all-male group (marines vs. "squid shit," for example), an ideal promoted by military life and by much of our civilian culture as well, is what threatens their survival. The point may seem obvious, but it is never made in Hollywood films.

To emphasize the irrationality of these bonds, the film includes a discussion at the dogfight in which the boys explain to the girls how it is that their surnames all begin with the same initial. One of the four explains that they had to line up by alphabetical order when they were in infantry training. Rose conversationally concludes, "So you got to be friends by standing in line?" There is an awkward pause, and then Marcy, the toothless girl, guffaws loudly. The guys call themselves the four B's, and the film makes much of the ritual in which they get themselves tattoed with bees on their arms—all except Eddie, who is with Rose. When Eddie rejoins the group the morning after sleeping with Rose, he tells Burzon that he has learned from Rose that Burzon "fixed" the dogfight. Burzon, in turn, tells Eddie that when he was getting his tattoo he saw him with Rose— and not with the gorgeous officer's wife Eddie has lied about. In a rare moment of honesty between men, Eddie asks his friend Burzon, "How'd we get to be so full of shit like this?—such idiots?" All these lies are "bullshit," Eddie says. Burzon replies:

Let me tell you something about bullshit. It's everywhere. You hit me with a little, and I buy it. I hit you with a little, you buy it. That doesn't make us idiots. It's what makes us buddies. We buy what the Corps hands out, and that's what makes us Marines. And the Corps' buying all the bullshit from President Kennedy and President Kennedy's buying all the bullshit from everybody in the U.S. of Fuckin' A and that's what makes us Americans.

"It's still bullshit," Eddie insists. "Right, and we're in it up to our goddam lips, Buddy. . . . I don't know if I'm making sense, but [here he rolls up his sleeve to show the bee tattoos], this makes a hell of a lot of sense to me. There's no bullshit in this." At this point, one of the four "bees" farts loudly; the guys all start laughing and joking again about the officer's wife. In a chilling gesture Eddie tears up the address that Rose has given him and throws the pieces out the window of the bus, where they are scattered to the wind. The act of renouncing association with women would help in the logic of most war films to secure the man's safety, which as I noted at the outset is endangered whenever men keep mementos of their attachments at home.

In *Dogfight* the outcome is very different. The film devotes about one minute of screen time to depicting the men in combat, thereby avoiding the contradictions involved when films "rely on combat sequences for their antiwar message." In this brief scene we see the four young men sitting around playing cards, and one of them brings up a joke, "What did the ghost say to the bee? 'Boo, bee.' " As two of the four "bees" laugh at a dumb pun that condenses various themes of the film—male bonding (as in the four bees), the degradation of women (the crude reference to "booby" as part of female anatomy), and death (the ghost)—a mortar round falls into their midst and apparently kills Eddie's friends. It might be said that, at the fantasmatic level, Eddie is allowed to survive *because* his loyalties were at least temporarily divided between the male group and a woman—the very reason, in Jeffords's argument, that Nick in *The Deer Hunter* must be killed off.

In Savoca's *True Love,* the heroine, Donna, resents the all-male group but recognizes its primacy. The night before the wedding, Donna takes Mikey aside and cuts both their hands to intermingle their blood—like "blood brothers," she explains. The acknowledgment of the primacy of male bonds, along with the yearning to become a member of the privileged male group, is characteristic of representations of Vietnam created by women. In her discussion of Bobbie Ann Mason's *In Country,* for instance, Jeffords shows how the heroine longs to have the same kind of understanding of war that men have and how finally the novel confirms "collectivity as a function of the masculine bond."[25] When Sam/Samantha, who is named after her father, goes to the Vietnam War Memorial, she is able to touch her own name and symbolically become part of the collectivity from which she has felt excluded. In *Dogfight,* however, the woman stands for a higher form of collectivity—higher than the military swarm, exemplified by the four bees—and her vision of a better world achievable through artistic endeavor and political activism (*not* through any essentialized categories such as feminine nurturance) is presented by the film as admirable, if vague and a bit naïvely idealistic.

On the film's horizon is a faint glimpse, barely discernible, of another kind of collectivity. Taking place on the eve of a feminist revolt that would gain momentum in the seventies, *Dogfight* reminds us, most particularly in the casting of feminist folksinger Holly Near as the mother of a heroine who believes in the power of music to change the world, of a time when *women* would bond together in lesbian separatist spaces, preeminent of which was the woman's music festival. Additionally, in locating Rose and her mother in a place called "Rose's Coffee Shop," which is handed down from mother to daughter, beginning with Rose's grandmother, the film privileges matrilineality and presages the alternative economies that feminists would be attempting to devise.

The fullness of that story, though, is left to another time, another film. For Savoca and her husband, screenwriter Richard Guay, with whom she works closely, the principal concern is the relation, or more accurately nonrelation, between the sexes.[26] One finds a fairly persistent pessimism in Savoca's work about the possibilities for meaningful union between male and female. But the connection between Rose and Eddie when it

15 A tender moment in Nancy Savoca's *Dogfight* (Warner Bros., 1991).

does occur is luminescent. The physical part of the relationship begins after Rose has sung for Eddie in the café. He takes her to a musical arcade where they put money in all the machines and dance to a cacophony of music box tunes. Then, as the music winds down, it is as if the raucous soundtrack of every Vietnam film ever made were being stilled, the rhythm cranked down, and a temporary truce called in the hostilities between men and women. In place of the noisy soundtrack, there is the sound of two people awkwardly embracing, fumbling to get their arms right, breathing unevenly in excitement and surprise at the intense pleasures of newly awakening sexuality.

When Eddie goes to Rose's bedroom, they begin kissing as we hear a Malvina Reynolds song playing on Rose's phonograph; then Rose goes into her closet "to change." Eddie whisks out of his clothes and stops when he remembers to go through his wallet to find a condom. He slips it under the bib of Rose's teddy bear and quickly gets into the bed. Rose comes out wearing a long flannel nightgown, and Eddie is dazzled: "You look good, you look real good." The process in which Rose goes from (possibly) being perceived as dogfight material to being looked upon as a vision of loveliness is complete and entirely believable, without the film's ever stopping to make a point of it. As if responding to the commentators who have criticized Vietnam representations for seldom acknowledging that warriors are just boys, the film in this scene touchingly evokes precisely the liminal space between childhood and adulthood so crucial to the future of humanity (its end—in war; its beginning—in sex).

Because the movie is about teenagers, Warner Bros. thought it should be marketed as a teen comedy—the ghetto to which so many women directors are relegated in Hollywood. Such a confusion might seem laughable on the face of it, but the fact that the movie is as much about a teenage *girl* as it is about boys makes it especially vulnerable to being judged trivial. Nor is *Dogfight* alone in being patronized because of its protagonist. In an article entitled "Men, Women, and Vietnam," Milton J. Bates writes of Bobbie Ann Mason's novel *In Country:* "Mason, having elected to tell her story from a teenage girl's point of view, cannot realistically venture a more mature critique of the War or sexual roles. Why inflict such a handicap on one's narrative?"[27] One might as well say that *Huckleberry Finn* is handicapped by having a young boy as its protagonist and that consequently Mark Twain cannot venture a mature critique of American society or slavery. I am arguing, however, that having a teenage girl as (co)protagonist enables Savoca a unique vantage point from which to advance an important critique of the war mentality and especially of war narratives—a critique, in part, of their exclusive emphasis *on the white male soldier's point of view.*

During a discussion of the sex scene in class, one of my students remarked that this was the first time in the movies she had seen a man ask a woman if it was okay to proceed in his sexual "advances." It is in such small details that the subversive nature of women's popular cinema often lies. This detail might be dismissed as another instance in which "the invasion is rewritten as erotic melodrama." But I would counter that in being anti-invasion on a very minute scale, the film points to the existence of other subjectivities and other desires besides those of the white male hero. In so saying, I do not, I hope, commit the same error I lamented in mainstream representations of Vietnam and implicitly offer Rose as a metaphor for Asia. I do, however, mean to suggest that in the film one man takes a crucial *step* toward recognizing otherness in the variety of forms it takes; the movie thus prepares the ground for the emergence of an antiwar sensibility, even though it does not go the entire distance.

Dogfight itself respects the integrity and independence of its viewers, and it never bludgeons them with a moral; rather than beginning by condemning the men for their barbarous treatment of women, for example, the film depicts the staging of the dogfight in such a way that the point of the men's search for ugly women, which may initially strike us as somewhat amusing, only gradually becomes clear. We tend to be identified with Eddie until Rose learns about the dogfight, and then we find ourselves implicated in his lack of sensitivity and his propensity for cruelty. This development, along with the film's somber finale, angered audiences who wanted the comfort of a happy and predictable ending: boy loses war but at least wins girl (the boo-bee prize).

The preview audience's anger at the film's ending is certainly understandable. The war itself seemed to many Americans to have had the wrong ending, and rather than mourning its loss they have fallen victim to a widespread melancholia in which they cling to the lost object rather than let it go.[28] According to Sigmund Freud, when people are unable to mourn a loss and put it behind them, they internalize it; preserving it within themselves, they become inconsolable. Turning inward, they appear narcissistic. This malady is what Freud calls "melancholia."[29] Having denied loss, denied the fact of having lost the war, America internalized the war and cannot seem to move beyond the question of what this loss has meant to itself, much less what it has meant to others who also were affected. *Dogfight* points to the necessity of moving beyond narcissism and coming to terms with the losses that incurred in being vanquished—the necessity, then, of mourning.

And in mourning, who better than women to lead the way? In her important study, *The Gendering of Melancholia,* Juliana Schiesari, following and reassessing Freud, discusses mourning as a social ritual that has generally been performed by women. Melancholia, by contrast, is a category by which a solitary male elevates and glorifies his losses (a comrade; a lover; a war) into "signifiers of cultural superiority."[30] A literary critic, Schiesari is speaking primarily of poets and other creative writers who through masterful elegiac displays put their own exquisite sensibilities forward for us to admire, offering themselves as objects to be pitied for the losses they have sustained, and—in some cases—hoping to acquire immortality through having created for posterity a work that supposedly acknowledges death. For our purposes, Oliver Stone, who cannot seem to stop making movies about Vietnam, might be seen as an example of the inconsolable melancholiac. A quintessentially melancholic scene would be the one in *Born on the Fourth of July* in which we are asked to focus on *Kovic's* pain and guilt when he has to talk to the parents of a dead buddy. Schiesari concludes that mourning is a social ritual which "accommodates the imagination to reality," while melancholia accommodates reality to the imagination. In her capacity as mourner, too, then, woman stands for the collectivity over male individualism.

Doing the work of mourning, accommodating the imagination to reality, *Dogfight* confronts us with both the reality of our losses and, concomitantly, with the real—as opposed to figural—status of woman. The struggle to establish this reality by asserting a female vision and female authority was waged at several levels of the film's production and reception. First, in order to make the female character of equal importance to the male, Savoca had to take on scriptwriter Comfort, who was displeased at some of her revisions. Second, many of the changes were decided on during the filming and were worked out collectively between cast and crew, most particularly between Savoca and Taylor. Finally, after preview audiences responded unfavorably to the film's ending, Savoca refused to comply with Warners' attempts to get her to change it. She conceded that Warners had the authority to do what they liked with the film, but she demanded that her name be taken off the project if the ending was reshot. Warner Bros. executives called River Phoenix and Lili Taylor, but both refused to reshoot the ending without Savoca. Savoca believes that neither she nor Taylor counted for much with the studio, but she thinks that Phoenix's refusal to reshoot forced Warners' decision to stop pursuing the issue (whereupon, according to Savoca, the project was dumped "into the toilet"). I like to think that Phoenix's role taught him something about resisting the orders of a male power structure and the desirability of sometimes deferring to the vision of a woman—in this case, Nancy Savoca, who elicited from him one of the finest performances of his tragically short life. Fittingly, Phoenix ended up being eloquently memorialized in a work that asserts the legitimacy of female authority.

Notes

1 I want to thank Jon Lewis for his very helpful editing. I am indebted to the National Endowment for the Humanities, which provided a fellowship that enabled me to interview Nancy Savoca and to research and write this essay.
2 Gilbert Adair, *Hollywood's Vietnam: From "The Green Berets" to "Full Metal Jacket"* (London: Heinemann, 1989), p. 159.
3 Ibid., p. 169.
4 See my chapter, "A Father Is Being Beaten," in *Feminism Without Women* (New York: Routledge, 1991), pp. 61–75.
5 David James, "Rock and Roll in Representations of the Invasion of Vietnam," *Representations* 29 (Winter, 1990): 91.
6 See John Hellman, *American Myth and the Legacy of Vietnam* (New York: Columbia University Press, 1986).
7 Jason Katzman, "From Outcast to Cliché: How Film Shaped, Warped and

Developed the Image of the Vietnam Veteran," *Journal of American Culture* 16, no. 1 (1993): 12.

8 Albert Auster and Leonard Quart, *How the War Was Remembered: Hollywood and Vietnam* (New York: Praeger, 1988).

9 Kaja Silverman, *Male Subjectivity at the Margins* (New York: Routledge, 1992), pp. 52–121.

10 As David James reminds us, though, *Coming Home*'s "unequivocal assertion that the invasion is *wrong* distinguishes it from all other films made in Hollywood." See James, "Rock and Roll," p. 90.

11 Pat Aufderhide, "Good Soldiers," in *Seeing Through the Movies,* ed. Mark Crispin Miller (New York: Pantheon, 1990), p. 88.

12 One might be tempted to read the scene as satiric; yet the entire film presents its hero and its subject with such hysterical immediacy and apparent overidentification that it lacks the distance necessary for a satiric commentary.

13 Tony Williams, "Narrative Patterns and Mythic Trajectories in Mid-1980s Vietnam Movies," in *Inventing Vietnam: The War in Film and Television,* ed. Michael Anderegg (Philadelphia: Temple University Press, 1991), p. 129.

14 This film also has been criticized for involving a "love story," which, writes Katzman, "digressed from the film's more interesting element, which was Dave's healing process." "From Outcast to Cliché," p. 21.

15 Rick Berg, "Losing Vietnam: Covering the War in an Age of Technology," in *From Hanoi to Hollywood: The Vietnam War in American Film,* ed. Linda Dittmar and Gene Michaud (New Brunswick, N.J.: Rutgers University Press, 1990), p. 65.

16 These are Laura Mulvey's terms in "Visual Pleasure and Narrative Cinema," in *Movies and Methods,* vol. 2, ed. Bill Nichols (Berkeley: University of California Press, 1985), p. 305.

17 See ibid.

18 Andrew Martin, *Reception of War: Vietnam and American Culture* (Norman: University of Oklahoma Press, 1993).

19 Quoted in Julie Lew, "After *True Love* There Comes the *Dogfight*," *New York Times,* July 22, 1990, p. H20.

20 All quotations are from an interview I conducted with Nancy Savoca on November 30, 1994.

21 In addition to James's "Rock and Roll," see David James, "The Vietnam War and American Music," in *The Vietnam War and American Culture,* ed. John Carlos Rowe and Rick Berg (New York: Columbia University Press, 1991), and Douglas W. Reitinger, "Paint It Black: Rock Music and Vietnam War Film," *Journal of American Culture* 15, no. 3 (1992): 53–59.

22 James, "Rock and Roll."

23 Susan Jeffords, *The Remasculinization of America: Gender and the Vietnam War* (Bloomington: Indiana University Press, 1989), p. 95.

24 Ibid., p. 99.

25 Ibid., p. 62.

26 Savoca's third film moves into different terrain, with mixed results. Savoca tries her hand at magical realism in *Household Saints* (1993), a story of three generations of Italian women. The granddaughter, played by Taylor, has visions of Jesus and ultimately dies. The film leaves an explanation for the visions (whether they result from madness or mysticism) unresolved.

27 Milton J. Bates, "Men, Women and Vietnam," in *America Rediscovered: Critical Essays on Literature and Film of the Vietnam War,* ed. Owen Gilman, Jr., and Lorrie Smith (New York: Garland, 1990), p. 29.

28 J. Hoberman speaks of the war's "bummer of a finale that's left us with a compulsion to remake, if not history, then at least the movie." See Hoberman, "America Dearest," *American Film* 13 (May 1988): 41.

29 Sigmund Freud, "Mourning and Melancholia," in *The Standard Edition of the Complete Psychological Works of Sigmund Freud,* vol. 17, trans. James Strachey (London: Hogarth, 1974).

30 Juliana Schiesari, *The Gendering of Melancholia: Feminism, Psychoanalysis, and the Symbolics of Loss in Renaissance Literature* (Ithaca, N.Y.: Cornell University Press, 1992), p. 62.

From Pillar to Postmodern:

Race, Class, and Gender in the Male Rampage Film

Fred Pfeil

Lethal Weapon (1987), *Die Hard* (1988), *Lethal Weapon 2* (1989), *Die Hard 2: Die Harder* (1990), all four of these top-ten–grossing films produced or coproduced by Joel Silver for Warner Bros. and Twentieth Century-Fox follow the same basic narrative formula. In each film a white male protagonist, portrayed by an actor of proven sex appeal, triumphs over an evil conspiracy of monstrous proportions by eschewing the support and regulation of inept and/or craven law enforcement institutions, ignoring established procedures, and running "wild" instead, albeit with the aid of a more domesticated semi-bystanding sidekick. What I want to explore in this essay, then, are the specific ways in which that formula is worked from film to film; for these films' peculiarly overdetermined, multivalent "dreamwork of the social"[1] has much to tell us about the irresolutions, anxieties, and contradictions of contemporary straight white masculinity and patriarchal power.

We might begin our look at the unsettled and unsettling features of these four "male rampage" films by noting those formal elements within them that are, when we take them out and look at them, surprisingly postmodernist. True, the straightforward plots of *Lethal Weapon* and *Die Hard* and *Die Hard 2* have little in common with the loopy surprises and multiple thematics of a *Brazil* (1986) or *My Beautiful Laundrette* (1986); their formula is simply to enact the amusing yet brutal struggle of a white male protagonist trying to get his anguished head straight while defeating a criminal conspiracy of monstrous proportions. But on the more formal levels of cinematic style, narrative structure, and constitutive space-time, our fast-paced smash-em-ups have a lot more in common with recognized postmodern art films like, say, *Diva* (1981), than one might think:

Stylistically, the *Lethal Weapon* and *Die Hard* films project a particular "urban look," their palettes shifting between bleached or even washed-out pastels and dark metallic blues and grays, with an occasional burst of apocalyptically lurid orange, managing, as a *Variety* reviewer noted, "to be gritty and glamorous at the same time."[2]

In terms of their characteristic mise-en-scène, leaving aside, for the moment, the domestic space of home in the *Lethal Weapon* films, they contrast two different spaces: the spanking-new high-rise office building, penthouse suite, or airport terminal versus the functioning viscera, oily and steaming, beneath the polished façades and abandoned industrial landscapes now grown gothic and obscure. The freighter's massive hold at the end of *Lethal Weapon 2,* the ventilation ducts and elevator shafts of the Nakatomi office tower in *Die Hard,* both are surprisingly close kin to the cool postindustrial ruins of *Diva.* Similarly, spatial relations in all five films are perplexing. Neither the cop heroes of the *Die Hard* and *Lethal Weapon* films nor the Jules and Gorodish of *Diva* seem to go anywhere, or, alternately, in the *Die Hard* films we could say that John McClane always has to go somewhere, but, as with Fredric Jameson's famous postmodern subject lost in L.A.'s Bonaventure Hotel, we never really know where he is / we are. The action is simply taking place *here*—and *here*—and *here*—in spaces whose distance from one another is not mappable as distance so much as it is measurable in differences of attitude and intensity.

Narratively, too, an older model of plot development, moving from an initial, relatively tranquil state through destabilization and development to a new and more fully resolved stasis, is largely superseded in all five films by the amnesiac succession of self-contained bits and spectacular bursts. A difference does remain between the *jouissance* offered us in *Diva*—between, say, the delightfully twisted bits of aphasic narrative trash that Gorodish's Vietnamese lover tells Jules as he bleeds away in the video arcade telephone booth—and the brutal, industrialized thrills of the Hollywood films, whose narrative logic and rhythm are so eloquently described by rampage hero and auteur Sylvester Stallone: "It's chop-chop-chop-chop, chop-chop-chop-chop. It's almost like a diced salad. You have to keep it going."[3] Still, such a salad is substantively different from the meat and potatoes of *The Maltese Falcon* or *High Noon.*

We will return to that difference, which for now I evoke merely as one more instance of these films' discontinuity with classical Hollywood cinema, a formal aspect of their postmodernity. Yet the mention of Stallone reminds us of how much the *Lethal Weapon* and *Die Hard* films have in common with other Hollywood productions past and present. For starters, one only has to think of the central icon of both sets of films, that is, the taut, torn, upper torso of the white star brandishing his lethal weapons (giant pistol, machine gun, deadly martial-arts-skilled hands) in perfect readiness for whatever comes next, despite the wounds, burn marks, and other signs of mortification marring the muscle-rippled skin we can see through the rags of his clothes. The locus classicus of this image, of course, is that of Rambo himself [4] —yet less the Rambo of the blockbuster hit *First Blood II* (1985), the one who fights and wins a second Vietnam War all by himself in the terrible year of Reagan's reelection, than the one who, in 1982's *First Blood,* finds himself humiliated and hunted by an obsessed local sheriff upon whose small mountain town Rambo, goaded beyond endurance, ultimately unleashes the sad and terrible violence he must otherwise, as a post-traumatic-stress-syndrome-ridden Vietnam vet, carry always within himself.

Thus, the importance of *First Blood* as precursor of the films we will be examining: as the movie that introduced the image and theme of the wounded warrior alone in the wilds, deserted and even abused by the authorities who ought to have appreciated and supported him, and through that figuration offered U.S. audiences a new synthesis of what Robert Ray has called the "left and right cycle" films of the 1970s. According to Ray, the mass audience for Hollywood product in the 1970s was offered a choice between two kinds of antiestablishment film: a "left" version, in which the protagonist uncovers an evil conspiracy of power elites and is usually defeated and killed before he can publicize or contest it in any effective way (*Chinatown,* 1974; *The Parallax View,* 1974); and a "right" version, in which the established authorities are so corrupt or impotent that they leave the hero no choice but to wage his own war against the scum who threaten him, his family, and All That Is Decent from below (*Dirty Harry,* 1972; *Walking Tall,* 1973).[5] If so, *Rambo: First Blood* was one of the first movies of the 1980s to dream these two sides or cycles together and thus to offer us the sight of a downscale, deauthorized figure going native (Stallone as canny proto-Indian "savage" in the long central section of the film, tracking and killing his human and nonhuman quarry without resort to rifle or gun) against those irrational authorities who oppose, misunderstand, and abuse him again and again.

This new synthesis presents us, in turn, with more than we have time to follow out in anything like full historical detail. Just for starters, there is the Vietnam trope itself, used here, as in *The Deer Hunter* (1978) and elsewhere, as a perfectly ambivalent emblem of a certain kind of sadomasochistic fantasy experience: that is, signifying not so much that the U.S.-Vietnam War was either evil (left populist) or incompetently and faintheartedly waged (right populist), as that it was *both* a place where white guys learned to practice a lot of especially mean and violent techniques—to "go wild," in short (Rambo, we learn, picked up his forest survival skills "over there")—*and* where they suffered horrible physical and psychological trauma.[6] There is the "wild man," the white man gone native / wild, which, thanks to Ronald Takaki, Richard Drinnon, and Herman Melville, we could track back to the mythic history of John Moredock, the white man who lives to kill the Indians who massacred his family and who lives like an Indian to kill them.[7] And Rambo's misunderstood, abused, and embattled goodness contains a trace of yet another old—and quintessentially American—popular narrative of white masculinity: that of the heroically strong and honest worker who fights the greedy, corrupt, and powerful for the benefit of all, and whose concealed elite or aristocratic origins, Mike Denning has argued in his study of nineteenth-century dime novels, signal his allegorical status as a figure of the rightful American republic in whose special virtues we (white males) all presumably may share.[8]

The function of the privileged origins in our dime-novel narrative would, then, in the *Rambo, Lethal Weapon,* and *Die Hard* films be nicely taken up by the star system itself. Mel Gibson and Bruce Willis, the then-newly established white stars of the *Lethal Weapon* and *Die Hard* films, both offer themselves as figures of the common (white) workingman and allow us (white men) to be flattered by that image with which we identify. Yet, especially in view of the blockbuster audiences hauled in by our four films, it is also worth pointing out that the stardom Gibson and Willis bring to these films is to some extent a cross-over phenomenon, both of them having some measure of proven box-office appeal to women—and perhaps to upscale, professional women at that: Gibson as the heartthrob and / or beefcake of *The Year of Living Dangerously* (1982) and of a half-dozen other Australian (ergo, in this country, somewhat classy) films, and Willis as costar with Cybill Shepherd in the popular television series *Moonlighting,* much-favored site of "TV slumming" for thirty-something white professionals in the mid- to later 1980s. Moreover, as we shall see, their definition in these male rampage films is precisely, indeed crucially, that of *wild yet sensitive* (deeply caring yet killing) guys—which mutation, moreover, is braided with that other "nonracist" liberal trope, running back from *I Spy* and the interracial buddy film all the way to *The Adventures of Huckleberry Finn,* that makes the white guy and his black pal both supposedly equal brothers-in-arms, yet domesticates the black man while keeping the white man out there, wild, on the edge.

But with these observations we already have begun to change the direction of our gaze from the vertical toward the horizontal, from genealogy to affiliation. Given the erosion of generic boundaries in contemporary film and TV production, we can see bits of our four films scattered all about the terrain of mainstream Hollywood film in the 1980s. Does the interracial partnership of Riggs (Gibson) and Murtaugh (Danny Glover) in the *Lethal Weapon* films owe more to the foulmouthed madcap duo of Eddie Murphy and Nick Nolte in *48 HRS.* (1982) or to the impulsive, anti-authoritarian Richard Gere and the rigid disciplinarian Louis Gossett in *An Officer and a Gentleman* (1982)—or, indeed, is it some recombinant blend of the two? For that matter, is the partnership of urban professional FBI man Sidney Poitier and Northwest mountain man Tom Berenger in *Shoot to Kill* (1988) a relatively inert variant of the complementary roles played by Glover and Gibson, and is the scene in which, wisecracking steadily, Poitier and Berenger try to survive in a snow cave cheek-to-cheek on top of one another a knockoff of Glover and Gibson's homosocial play? Or how about those closing moments in both the initial *Die Hard* and *Lethal Weapon* films when the monstrously strong, skilled, and monomaniacal double / Other that we thought our hero had vanquished stages one final, terrible, impossible reappearance: are they—aren't they?— spinoffs of the thing's last attack on the Sigourney Weaver character in the original *Alien* (1979) or of the robokiller's last attempt to destroy *The Terminator's* (1984) female protagonist?

Such lines of influence run out virtually to infinity—and, indeed, could be made to run backward in time as well as sideways (e.g., *Die Hard* as bastard offspring of an unholy union between Chuck Norris action films and the disaster flicks—*The Towering Inferno, Airport*—of the 1970s). The point of this hasty sketch of our films' antecedents and traveling companions has simply been, first of all, to situate them, albeit roughly, with regard to industrial-popular cultural production in this country in general, and film production in particular; and, secondly, to suggest something of just how highly condensed and overdetermined these action films really are.

Finally, though, alongside these aesthetic mutations and generic syntheses, we also should keep in mind another, equally widespread but far less fun kind of shift: the long changeover, very much still in progress, from "Fordism" to "post-Fordism" that is the economic context in which these films and their national reception take place. In the United States, in particular, in the deindustrializing 1980s, this "squeeze in the middle of the labor market" was accompanied by a small increase in the number of skilled technical or professional positions and a veritable explosion of low-paying jobs in service industries and small-scale production sites, with people of color and women especially targeted for such low-wage employment, if indeed they are "lucky" enough to find work at all.[9] A deepening gulf in the division of labor, then, *within* what had been called the working class, and doubly so: that is, *both* in terms of the distance between those with enough luck and training to reach the lower strata of the professional-managerial class and those pumping gas, flipping rat-burgers, or sewing up slacks in a sweatshop at piecework rates down at the bottom, *and* in terms of the faultlines of gender, race, and ethnicity, simultaneously deepening and crosshatching that larger divide. All of this is not only the backdrop against which the newest crop of mainstream rampagers must be seen, but in very real ways it is their subject matter as well; for out of just such distances, disappearances, and convergences, and within an ever more generalized postmodern culture, the new subjects of domestic post-Fordism construct and receive various narrative versions of their own uneasy dreams.

But now, let us home in on the *Die Hard / Lethal Weapon* films themselves, starting where they do: with two convergent yet distinct problems per film, one private or personal and the other public, even international, and the resolution of these problems through the violent action of a white male protagonist—Mel Gibson as Riggs, Bruce Willis as McClane—working outside the system and aided by a much tamer sidekick who happens to be black.[10] In the *Lethal Weapon* films, Danny Glover portrays middle-aged family man Roger Murtaugh, the cautious yet nurturant soul who seeks in vain to curb Martin Riggs's daredevil tendencies. In *Die Hard,* Reginald Veljohnson gives us plump, good-natured Al Powell, the cop with a pregnant wife waiting for him back home, who functions via walkie-talkie as our hero John McClane's only tie to the outside world. Moreover, the stable satisfactions of Al and Rog's domestic lives stand in stark contrast to those of Riggs and McClane—McClane confused about his professional wife (Bonnie Bedelia), a rising executive in a Japanese multinational; Riggs trapped in a shabby trailer with a cheap TV set, a few photographs of his dead wife, and a pistol to suck on experimentally when the pain is at its worst.

Failing or defunct marriages thus constitute the personal problems of the *Die Hard* and *Lethal Weapon* films; criminal conspiracies by mercenaries, drug lords, evil South African government ministers, and European technovillains constitute the public concerns.

Yet these conspiracies themselves take two different forms, each with its own type of archvillain, and the two sets of films present us, neatly enough, with two examples of each. In the first *Lethal Weapon* film and *Die Hard 2*, the bad guys are either American mercenaries or paramilitaries who, usually with some foreign ties or foreign help, are either running a heroin network out of Laos back to the States (*Lethal Weapon*) or attempting to spring a Latin American druglord-general in the process of his extradition to the States (*Die Hard 2*). In the original *Die Hard* and in *Lethal Weapon 2*, on the other hand, the evil ones are less military and more socially upscale in their general profile. Arjen Rudd, the wicked minister of diplomatic affairs for the government of South Africa, and his gray-suited henchmen are running a triangular trade in drugs, American money, and South African Krugerrand gold (the precise logic of which, by the way, is never all that clear—but what the hell) in *Lethal Weapon 2*. In *Die Hard* a mostly European group of skilled technicians, headed by suave, impeccably tailored ex-terrorist Hans Gruber (Alan Rickman), take over the spanking new office tower in which the Japanese corporation for which McClane's wife works has its American HQ, passing themselves off as real terrorists to buy the time they need to computer-hack their way into the vault in which their real goal, some $640 million in negotiable bearer bonds, is stashed.

It goes without saying that our heroes and their buddies ultimately triumph over both these flavors of dastardliness, but not thanks to normal law enforcement procedures or officials, whose guidance and aid invariably prove to be irrelevant, ineffectual, or worse. In the first *Die Hard* film, the official efforts of the LAPD, presided over by a chronically irascible buffoon of a lieutenant, are both belated and easily crushed; and the subsequent involvement of the FBI nearly proves fatal to both the corporate hostages and McClane himself. Likewise, in *Lethal Weapon 2*, when the South African conspirators begin an assassination campaign against the police department in retaliation for the harassment they have received, the department appears powerless to predict or prevent the attacks, especially given the evil minister's invocation of diplomatic immunity as his ultimate line of defense. And in *Die Hard 2*, we are in for yet worse; though McClane does receive some grudging help from the local airport police at Dulles Airport who dismissed him at first, the crack antiterrorist military team sent in to defeat the paramilitary plotters and recapture the Latin American general turn out to be in league with their presumed antagonists. Only in the first *Lethal Weapon* are enforcement officials other than our heroes and their nonwhite companions depicted as

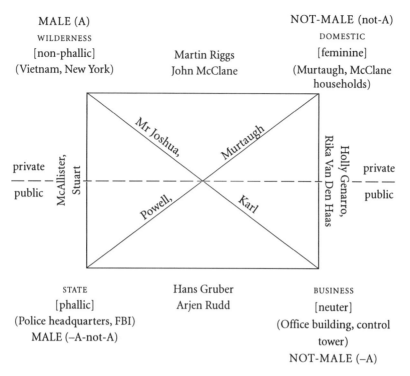

MALE (A)   NOT-MALE (not-A)
WILDERNESS   DOMESTIC
[non-phallic]   Martin Riggs   [feminine]
(Vietnam, New York) John McClane   (Murtaugh, McClane
                                                 households)

private   private
——————   ——————
public   public

STATE   Hans Gruber   BUSINESS
[phallic]   Arjen Rudd   [neuter]
(Police headquarters, FBI)   (Office building, control
MALE (–A-not-A)   tower)
                                                 NOT-MALE (–A)

Figure 1

anything but explicitly disabled, incompetent, or corrupt; yet even in this
film, clear (albeit usually comic) signs are thrown off concerning the
deauthorized slackness of the police department from which both Riggs
and Murtaugh dissociate themselves once the bad guys make off with
Murtaugh's daughter Rianne.

From all of this plot material, we can extract four basic nodes of opposition and difference from which these films generate their dramatic energies. I place them in figure 1 in the familiar form of a Greimasian rectangle
in which position (A) is the nameable yet never represented space of "the
Wild"—the place the white protagonist has come from before he lands in
L.A., be it Vietnam (in *Lethal Weapon*) or, in *Die Hard,* that other well-
known symbol of multiracial rot and riot, New York City. The Other, or
(not-A) of this space would then clearly be "Domesticity," the site of heterosexual, family life. The Opposite (−A) in these films must be the place
of Business or professional-technical work. And finally, filling the place of
(−not-A), functioning as the inverse of the nurturant realm of hearth and
home, we find the official space of law, authority, and order—that is,
the State.

Set up in this way, these positions permit some further subdivisions along two additional axes: one between private and public (or, perhaps more accurately, between individual and collective), given here as a dotted horizontal line bisecting our rectangle; and the other between gendered lines of force, here given as our two uprights. The lefthand "male" vertical connects what, loosely drawing upon Lacan and his feminist appropriators, I call the "Phallic" State with what I will argue is a non-Phallic "male-rampage" Wilderness; the righthand vertical—not, significantly enough, so much "female" as *not-male*," lines up wife and mother in the "patriarchal-Female" Family Home with the "non- or post-patriarchal Neuter" professional in her / his Office or (in *Die Hard 2*) Control Tower.

This figure, complex as it may be, is a road map merely, not the ground itself. The real significance of these films lies in the complex shifts, shadings, exchanges, tensions that play upon these categories—in the relationships, for example, between the political / historical / newsworthy materials that viewers are given as background to the public plots and the simple manichean bad guy / good guy dualities they disturb. Take, for starters, the explanation that (typically) flies by in the original *Lethal Weapon* of how the evil paramilitaries of "Shadow Company," the Special Forces unit now turned vicious heroin cartel, got their start working with the CIA in Vietnam. It is simply sound business practice in the realm of mass cultural production to scramble your left-wing ideologemes (runaway CIA and paramilitary elements engaging in renegade criminal activities) together with some from the right. But it is equally the case that behind the official oppositions and antagonisms a kind of relay system of partial incorporations is also at work, through whose operations the evil Other is not only resisted, but partially, covertly taken in. In *Lethal Weapon*, to continue our example, if you recall that Riggs, the Mel Gibson protagonist, was also Special Forces in the war and, indeed, bears the same tattoo upon his arm as the ruthless Mr. Joshua (Gary Busey), that relay system in this case looks something like: Viet Cong → Shadow Company → Riggs, where the relationship between each term and the next in the series is one of both opposition *and* identity. (As, by the way, the predominance of specifically "Asian" martial arts techniques in the final mano-a-mano between Mr. Joshua and Riggs also suggests.)

Something of the same type of antagonistic linkage is set up between the extradited Latin American general of *Die Hard 2*, the renegade paramilitary units, in and outside the government, who are trying to spring him loose, and our hero, John McClane. Here again, the political explanation we are given is both fleeting and steeply ambivalent. General Esperanza, we are told, is a Central American officer whose flagrant human rights abuses finally resulted in his fall from power (shades of the brutal autarchy of the U.S.-backed Guatemalan and Salvadoran armed forces, ergo a "left" element, even if, of course, no such officers in real life to date have been punished or removed for their crimes); *and,* we learn, the general then turned to cocaine smuggling (shades of Panamanian leader Manuel Noriega, a "right" element) after he was deposed. This burst of ideologically Janus-faced info, moreover, is given in the form of a TV news blip whose continuous sound and visual image also function as the hinge of a straight editing cut from McClane, idly gazing at a screen inset in the central information booth in the main lobby at Dulles, to the paramilitary archvillain we soon will come to know as Colonel Stuart, staring at the screen with cold contempt while he concludes his T'ai Chi ritual, his musclebound nakedness enclosed by a motel room so stripped of all adornment and so drab olive-green as to suggest some preternatural military space. The point here is not only the effective subordination of the scrambled ideological message to the exigencies of plot, but the blurring of that scramble with another similar linking of opposites, that is, of McClane with Stuart, insofar as on some level, despite the ominous music that strikes up around Stuart, we recognize or suspect a certain kinship or brotherhood here—a sense that once McClane's topcoat and scarf are off, once he too gets stripped down for action, he's going to prove every bit as tough an hombre as this other guy. Later on, the film itself will spell out explicitly the implications of this covert affiliation. When McClane says to hard-bitten Major Grant, the head of the antiterrorist team sent in to locate and defeat Stuart's unit, "Guess I was wrong about you—you're not such an asshole after all," Grant replies, "No, you're right. I'm just *your* kind of asshole." And later still, when Grant is revealed to be one of the bad guys, he and McClane punctuate their final confrontation out on the wing of a taxiing 747 with the following exchange:

Grant: Too bad, buddy; I kind of liked you.
McClane: Too bad, buddy. I got enough friends.

Such kinship in antagonism, simultaneously asserted and refused, flickers at the core of all four films, given their definition of heroes and villains alike as men highly skilled in pain- and death-dealing techniques, perfectly willing to commit virtually any act of outrageous violence to persons or property in pursuit of their ends. Thus, the secret dream sense and magical power of that ostensibly gratuitous moment in the first *Die Hard* film, just at the end of archvillain Hans Gruber's victory speech to McClane,

16 Under siege: straight, white, and male in *Die Hard 2* (Twentieth Century–Fox, 1990).

delivered in the mistaken belief that McClane is finally disarmed and destructible, and McClane's last-second turning of the tables. The elegant, cosmopolitan pseudo-terrorist Gruber says, "What was it you said to me before? 'Yippi-ky-aye, motherfucker' "—throwing McClane's defiance, and the film's most famous line, back in his face, and laughing contemptuously. Then—and here the moment changes, the underside comes bubbling up—that laughter is joined, first by the big, square-shouldered thug who is the only member of Gruber's gang left, and then by McClane himself, as his wife Holly, still literally in Gruber's clutches, stares around in shocked bewilderment: for that laughter of Gruber and the thug is no longer contemptuous, no more than McClane's seems merely strategic, but *exuberant;* for just that one scant second all three men, all three outlaws, compose a community no woman can enter and share a joy no woman can know.

The current of pleasure and affiliation that radiates from this covert brotherhood of rampagers thus runs across and against these films' structured oppositions—and not merely from the bad guys around to the good. According to our grid, the African American figures of Al Powell in *Die Hard* and Roger Murtaugh in especially the first *Lethal Weapon* film cross in opposition the line connecting frenzied wildness and professional competence that defines the pointedly Aryan figures of Karl (Alexander

Godunov) and Mr. Joshua. Powell and Murtaugh—mildly incompetent cops who play by the rules and unproblematically love their wives and families and who, it is worth noting, hardly threaten white audiences at all—are the perfect antitheses to their blond, blue-eyed white Others' combinations of steely, square-jawed efficiency and perfect murderousness. Yet, for both these good guys, as for the white villains and heroes, where professional concerns give way to personal ones, the brotherhood of the wild begins. Shadow Company's kidnapping of Rog's daughter Rianne in *Lethal Weapon* seals his determination to go on the warpath after them. The shift is nicely signaled by the movement between two scenes and shots: the first, a tableau, in the russet Murtaugh home, of Rog being held and comforted on the staircase by his wife, standing a step above him and holding his head against her chest, while in the foreground, filling up most of the screen, Riggs's handsome psycho-loner lethal-weapon face and head, stiff with grim resolve, appear in close-up; the second, a two-shot of Roger and Riggs preparing for battle in the lurid orange light (whose ostensible source is, ironically, the lights of the family Christmas tree), two figures of equal strength and value now, Roger acknowledging with pressed lips and curt nods his acceptance of Riggs's rampager's code: "We do this my way—we shoot and shoot to kill. We're gonna get bloody on this one, Roger."

And so, of course, they do, up to and including that climactic moment when the vanquished Aryan warrior, having risen or reappeared from the ashes of his defeat, is dispatched for good by the black sidekick. Only *Die Hard 2,* which departs from its companion films in this regard by more or less "randomizing" race across its spectrum of male actants, fails to provide us with some such moment, which the original *Lethal Weapon* and *Die Hard* films deliver in all its bristlingly multivalent glory.[11] When in *Die Hard* Al Powell lifts his weapon to blow away wild-eyed, blood-spattered Karl when, miraculously undead, the latter shows up on the steps of the ravaged office tower from which McClane and Holly at long last are free, we know he has thus healed himself (with, one feels, the help of McClane's good example) of the sickly hue of irresolution that has rendered him impotent as a street officer ever since he mistakenly shot and killed a thirteen-year-old. Likewise, albeit less explicitly, when despite the terrible clobbering that Mr. Joshua of *Lethal Weapon* has just taken at the punching, chopping hands, high-kicking feet, and hard-squeezing thighs of Riggs, that bad guy too manages to slip yet one more gun into his wicked hands and lift his head to aim and shoot one last time; though here both Riggs and Murtaugh manage to whip theirs out, and plug the Evil One together simultaneously, it is Murtaugh's shoot-to-kill reflex action that signifies most, given his frequent criticisms earlier in the film of Riggs's propensity to kill whatever he draws his gun on.

Yet this literally invigorating tale of cross-racial influence—in which, one cannot help but notice, the black man seems to receive from the white man's hands not only the capacity for effective violence, but something very like virility itself, both gifts the white male imaginary has often enough feared it might lack in comparison to blacks who might possess them in excess—is only half a story that also includes the equally proto-sexual healing of the white man by the black. Or, more accurately, *gendered* healing, *feminine* healing, insofar as Powell and Murtaugh manage to curb McClane's and Riggs's worst, most excessively male tendencies toward self-destructive behavior by bringing them back to the pleasures of the hearth. Murtaugh lures Riggs in from the suicidal shoals by bringing him literally into the family, an action whose intended cure we realize is complete when, at the film's conclusion, Riggs hands the daughter whose life he and her father have saved the silver bullet (shades of the Lone Ranger!) he had once meant to use on himself. In *Lethal Weapon 2*, moreover, while working in their kitchen on a meal for the Murtaughs, Riggs can now tell Rog's wife, Trish, at last what happened on the day his wife was killed. And the self-recriminations that crowd Riggs's account of that fateful day on which he was too bound up with work to meet his doomed wife on time find their counterpart in McClane's confession of sins to Al over the walkie-talkie, at the Gethsemane-like moment in *Die Hard* when, hiding in a men's room, blood oozing from the glass in his feet, he begins to feel that perhaps he will not make it out alive after all. In this down mood, McClane calls for his friend; Al replies, "I'm here, John"; and McClane asks the only person in the world to whom he is both literally and affectionally connected to transmit the following rambling, tearful message to his wife:

tell her that—uhm—tell her it took me a while to figure out what a jerk I've been—but that when things started to pan out for her I should have been more support—and I should have been behind her more—tell her that she's the best thing that ever happened to a bum like me. She's heard me say I love you a thousand times, she's never heard me say I'm sorry—I want you to tell her that, Al. I want you to tell her that John said that he was sorry.

The manifest motivation for this flurry of sensitive insights derives from the hoary narrative trope of the need to come clean in the face of impending death. But its latent suggestion is the sense, as in the *Lethal Weapon*

17 Black and white together in *Die Hard with a Vengeance* (Twentieth Century–Fox, 1995).

films, that only thanks to the presence and influence of the nurturant, supportive, domestic black buddy can the white hero at last let himself be straightforwardly soft and sensitive too.

To sum up the last several paragraphs, we might say that one primary aspect of these films' ideological / entertainment project is the construction of a dizzyingly double-sided hot spot at the heart of the action, where some of the bad guys as well as the good, and black guys as well as white, both care and kill—kill, indeed, *because* they care. Even Karl of *Die Hard,* vile assassin that he is, knowing the pain of losing a brother, hisses when he at last gets his hands on McClane, "Ve're both professionals, but zis iss *personal.*" Our black sidekicks' execution of the Aryan bad guys functions not only as a clear sign of their enhanced potency, but as a sacramental emblem of their purgation of the inward equivalent of such testosterone-crazed thugs from our white protagonists' souls. Nor, in addition to this psychotherapeutic component, does this zone where quasi-emasculated blacks and thuggish whites trade domestication and *cojones* lack its own eros as well. McClane and wife Holly may have just embraced, following the death of Gruber, in the light provided by the flaming building they are still in; but once they walk outside together, the movie's largest gush of romantic violins is reserved for that drawn-out moment when the two men, Al and John, at last come face to face, approach one another, and fall into an embrace (followed, in its stead, by Al's prompt dispatch of

Karl). Likewise, in the first *Lethal Weapon,* where Riggs's and Murtaugh's joint execution of Mr. Joshua is capped by Riggs's backward collapse into Murtaugh's waiting arms: "I've got you, partner," Murtaugh says, and we go to dissolve, for the climax is complete. At the end of the sequel, the same maternal-erotic charge is pushed beyond the point of parody. *Lethal Weapon 2*'s final image, before the helo-shot lifts us up and away, is of Riggs sprawled across Murtaugh's lap in a perfect Pietà, trading joke lines (despite the clip of bullets just emptied in Riggs by Afrikaaner archvillain Arjen Rudd before—you guessed it—Murtaugh blows him away) whose main butt is the sensitive, health-conscious, "postfeminist" guy ("Hey Rog—anybody ever tell you, you really are a beautiful guy?" followed by hilarious snorts and brays).

Hence, the truth of the revelation laid by one of producer Joel Silver's friends on the *Vanity Fair* writer profiling Silver: the friend who exclaimed after watching the widely promoted scene from *Lethal Weapon 2* in which Riggs shares with his friend Murtaugh the risk of death from a gravity bomb wired to the toilet, " 'This movie's about *love.*' "[12]

But now let us move on and turn our attention to those actants and sites that such erotic and comradely rites and festivals either leave on the sidelines or destroy: women and their supposed "natural habitat"—the family home. Our four films are, as we have seen, manifestly supportive of both family and home; yet the underlying dream logic of the *Lethal Weapon* films in particular suggests a somewhat different attitude. Something of that attitude is implied, for example, in the very scene that we have just mentioned, and it surfaces as soon as we allow our minds to play with the montage logic of its climax. Riggs and Murtaugh, hands firmly clasped, rising swiftly then toppling, embracing, in the bathtub, underneath the black bomb cloth—and we cut to the front lawn to see part of the Murtaugh house's second story blow up. Likewise, with the Murtaugh family station wagon in the same film, which we see getting more and more battered with each action sequence, beginning with the opening chase scene—and even with the otherwise rather curious degree to which the full comic potential of this bit of shtick is left unrealized, given the tepid and occasional nature of Murtaugh's protests at this abuse.[13] And while we are talking about cars, homes, and families, how about that moment near the end of the first *Lethal Weapon,* just after the frenzied Mr. Joshua blasts his way into the Murtaugh home, when suddenly an empty squad car careens through the front picture window, sent by our rampaging, no-holds-barred partners Murtaugh and Riggs?

Again, the manifest content here would have it that this action is a ploy to distract Mr. Joshua so that our boys can get the drop on him. Yet it seems at least as significant that our squad-car-gone-wild delivers a spectacular trump to the action on whose heels it lands: Mr. Joshua's comic-demonic drilling of the TV set on which *A Christmas Carol*'s Scrooge has just inquired of a passerby what day it is. "Christmas Day!" the maddened Mr. Joshua shrieks, pumping lead into the exploding set next to the bedecked Christmas tree—and then, on top of that comes the squad car, smashing through the walls. It is Christmas, the time, symbolically, mythologically, of maximally happy domesticity, in three out of four of these films (*Lethal Weapon 2* being the exception), all three of whose endings prominently feature an ironic play between the visions of domestic coziness conjured up by familiar renditions of canonical Christmas songs on the soundtrack and the disordered or ravaged landscape presented to our eyes. According to coproducer Silver's friend in the *Vanity Fair* puff piece, Silver's response to the line on love quoted above is, "Why do you think I play so much Christmas music in these films?"[14] But the visual contexts in which those corny old renditions occur—the wrenched, flaming rubble of the office tower at the close of *Die Hard* and the swirling tumult around the jet on the still-disabled airstrip of Dulles at the end of *Die Hard 2* (both shown against the sound of Vaughn Monroe purling, "Oh, the weather outside is frightful / But the fire is so delightful / So as long as you love me so / Let it snow, let it snow, let it snow!"). Even the close of the first *Lethal Weapon*, as Riggs enters the Murtaugh home together with his dog, who immediately, though offscreen, takes off after the cat, so that over the strains of "I'll Be Home for Christmas" we hear the sounds of smashing objects, hissing snarls, and barks—all invite us to enjoy a far more aggressive, even sadistic, and certainly antidomestic holiday than Silver's response implies.

If all four films, and the *Lethal Weapon* duo in particular, thus wring the full ambiguity out of the old domestic injunction to "keep the home fires burning," their depiction and deployment of women is also ambivalent at best. We have described how these films translate racially coded actants into gender-coded ones. What bears emphasizing now is not only the culturally androgynous and complete killing-caring guys who result from this translation, but the perfect adequacy of this all-male couple and, accordingly, the relative superfluity of all those around who remain merely biologically female. We might arrive at this same point slightly differently by noting that the one secondary woman character in these movies who comes closest to embodying the old nuclear-patriarchal ideal—Roger Murtaugh's wife, Trish—has, as her single character note, only her partial divergence from that very role: the running joke is what a terrible, incompetent cook she is after all. On the other hand, the definition of Holly Gennaro and Rika van den Haas, the official love interests of the first two *Die Hard* films and *Lethal Weapon 2*, respectively, presents the male protagonists of those films with a problem to resolve: both are business-suited, dress-for-success professionals pursuing careers that take them away from home.

So, in the first *Die Hard,* following the brisk exposition of his uncertain marital situation that McClane gives to the young but sympathetic (i.e., in these films, black) chauffeur who, courtesy of the corporation Holly works for, has picked him up on his arrival in Los Angeles, we watch McClane's frown as he looks in vain on the video directory screen in the office tower's lobby for the listing under McClane, then his features crumple when he finds Holly under her maiden name, Gennaro—all, moreover, while he is still reacting to the guard-receptionist's mocking / menacing claim for the machine he is using ("if you have to take a leak, it'll even help you find your zipper"). Next, upon his arrival upstairs, where a posh yet casual corporate Christmas party is under way, McClane is immediately cruised by an attractive female employee and kissed on the cheek by a male one. "California," he mutters with sour humor, thus suturing geographic region and multinational enterprise by finding them complicit in the same unseemly behavior. Such lax, unwholesome goings-on may be found even in his wife's own office, where, as it happens, Ellis, the glib, sleazy yuppie executive who hit on Holly just a few minutes earlier is presently stuffing his nose with coke. Mr. Tagaki, the company president who shows McClane into the office, merely issues a hushed warning to his mega-deal-making hot-shot employee: "Mr. McClane," he says, "is a *policeman,*" and leaves it at that.

Such is the amoral but sumptuous world for which Holly McClane née Gennaro has broken up her home with husband John. Yet it is clear from her breathy close-up when they too are at last together that he still excites her—not to mention that later moment when, as one of the hostages, she looks on at Karl's fit of rage, and sighs joyfully in the knowledge that her man is still alive and on the job: "Only John," she exhales, "can drive somebody that crazy." Moreover, despite all these hints (or, just as accurately, thanks to them), we are not invited to vote openly against this professional woman's pursuit of a career. For one thing, McClane does not seem to know or credit what we know: that she is still a fully domesticated, family woman underneath, who rejects Ellis's invitation to "mulled wine, a nice aged brie" and a fire in favor of her traditional Christmas vision of "families, stockings, chestnuts, Rudolf and Frosty," even as she strides down the corridor to look at some spreadsheets. Moreover, in the couple's first and only scene together before the pseudo-terrorists burst in, we are given one moment whose intent is clearly to position us on Holly's side—the moment when McClane, washing up in Holly's own corporate washroom, converts the loving warmth in her eyes to blazing anger with a tirade, whose immaturity even he will condemn a minute later, on the perfidy of her deceitful use of her unmarried name.

Yet in other equally obvious ways, the film actively subverts its own lip service to this bourgeois feminist ideology and the right it asserts for all women fairly deemed qualified to have their own high-salaried jobs in tall buildings. For not only does the film amply demonstrate the impotence of all these corporate professionals in the face of real danger—in which a fast-talking schmoozer like Ellis, for example, in his greasily unscrupulous attempts to "negotiate" a way out, simply gets himself blown away, and from which—wouldn't you know it?—the suit types can only finally escape thanks to the efforts of a low-class, wisecracking yet inarticulate ordinary guy. The film's script, indeed, even arranges things so that when the evil Gruber, shot by McClane, falls backward through the office tower window, he nonetheless manages to drag Holly, his final hostage, with him to the edge of death by holding on to the strap of the Rolex—the watch that Ellis pointed out to McClane earlier, back when the party was still on, as the Nakatomi Corporation's token of appreciation for a mega-deal well made—so that McClane must undo the strap of this trinket-turned-handcuff to free her from the evil that is literally attached to it, and thereby save her life.

The case of Rika van den Haas in *Lethal Weapon 2* is both similar to and far cruder than that of Holly Gennaro in *Die Hard 1* and *Die Hard 2* (in the second of which, still a businesswoman, she must be rescued again, this time from a circling jet), so we need not linger on it long. Suffice it to say, first, that Rika, Riggs's only (briefly) living love interest, is also a business-suited employee of a major firm, albeit one headed by jowlishly sinister, pale-eyed Arjen Rudd, and dealing drugs, currency, and gold under its cover as an embassy arm of the South African government. Secondly, beneath her gray suits it soon turns out Rika has all the giggling sexual good health of a Playboy Bunny, and so is immediately susceptible to Riggs's rather fumbling yet domineering moves. Third, on the other side of their one lovemaking bout, she first turns immediately domestic (as evidenced by her rescue of Riggs's dog from the bad guys, her immediate rejection of her job, and her prompt invitation to Riggs to move in with her) and then gets iced herself. Her death, in turn, will once again trigger Riggs's trauma and rage over his wife's death, all the more so since it turns out that the same wicked South Africans were to blame in both cases—and so, with him and Murtaugh, we are soon off on the rampage again. But meanwhile, the film's soft-porn characterization and perfunctory treatment of Rika merely underlines the double-bind brutality of all four films' traffic in women—dismissed or disappeared insofar as they settle for being love toys and / or domestic servants, they must have their noses rubbed in their lack of real capacity and power if they do not.

Finally, a slight yet revealing sub-motif of the *Lethal Weapon* films takes us to an unexpected stake within all four films' depictions and deployments of the gender divide. I refer to the female police psychiatrist whom we meet in the first *Lethal Weapon* film on her way down a corridor in police department headquarters, badgering the male, middle-aged lieutenant who oversees Murtaugh and Riggs on the subject of Riggs's suicidal tendencies and resultant unfitness for duty. The lieutenant wants to dismiss her diagnosis and recommendation as just a "bunch of psych bullshit," but she so relentlessly dogs his trail that he has to duck into the nearest men's room to escape her. Subsequently, in *Lethal Weapon 2*, she reappears in the same setting just after Riggs has dislocated his shoulder, then just as painfully put it back in, in order to escape unaided from a straitjacket and so win a bet with his fellow cops. With a sadly disapproving shake of the head, she tells the still-panting Riggs that the door to her office is always open. But Riggs's sarcastic reply is swift and devastating: "I think we ought to keep this on a professional level, don't you, Doctor?" And once again the female shrink is humiliated and cast aside by the invocation of sexual difference, this time accompanied by the sneering laughter of the crowd on the screen as well as of the audience looking on.

Now, obviously, the pleasure that the films invite us to share at these moments is a sexist one—but not exclusively so. It is equally the pleasure of watching *a professional* get her comeuppance from a regular working guy, in organic relationship to other working folks, men and women alike, gathered round—and all the more so, insofar as what deflates her claims and brings her down is the fact of her gendered embodiment itself. Correlative, I have already hinted that no small part of the appeal of the *Die Hard* films for mass audiences of nonprofessional people (not to mention those professional folks with an anticorporate, antiprofessional populist cast) might well lie in the fact that in both cases McClane literally brings his corporate professional wife down from the heights to earth. In this sense, in these ways, just as these films' black-white racial code transmits messages that are as much about gender as they are about race, the woman-man code fuels a reactionary sexual politics with the high octane of antiprofessional, anticorporate resentment.

But this angry fuel of envious desire and impotent hatred is sparked for us in a wide range of actions strewn throughout these films. In *Die Hard* alone, one could compile a list of all such actions, ordered in terms of what we might call their "degree of disavowability." The shockingly sudden, perfunctory, and, in terms of narrative time, *early* murder of Mr. Tagaki, head of Nakatomi-U.S. operations, by the suavely elegant Gruber would probably top this list. Tagaki seemed to be a nice enough guy when he was showing McClane around; McClane seemed to like him; plus when the bad guys take over the building and Gruber stalks the crowd of corporate execs for him, Mr. Tagaki faces up to his nemesis with courageous dignity. On the other hand, though, he *did* make that smooth little crack when he first met McClane at the Christmas party about the Japanese being "flexible—Pearl Harbor didn't work out, so we got you with tape decks"; he did try to slide the fact of Ellis's coke-sniffing under the rug; he *is* a ruling-class capitalist manager of enormous power, wealth, and smugness, *and* a foreign one at that. So let's kill him anyway; or no, let's *have* him killed, but have Gruber do it; we'll even have McClane himself there, looking on from a hidden place, so that when Tagaki gets his, right in the center of the forehead, so close the force of the shot knocks him straight back off his chair, we immediately cut to McClane's shocked and horrified reaction. That way we can enjoy the death at the same time we enjoy our disavowal of that enjoyment in the form of a savory horror at the villainy of our new archvillain and his crew.

The subsequent execution of yuppie cokehead Ellis provides us with another example of hostility simultaneously discharged and disavowed, of the mechanism which so clearly resembles the "double / Other-ing" whose spider web of likenesses / antagonisms spun from black sidekicks to white wild-guy protagonists to Aryan nasties we traced out a while ago. Yet in *Die Hard* the relation of the mobile functionality of Gruber and his bad guys to both their assigned iconic value on the one hand, and our audience pleasures on the other, soon becomes an even richer and more slippery subject. For starters, as the film's voice of (delegitimated) authority, LAPD Deputy Chief Duane Robinson is wont to point out to McClane's outside friend Al Powell, both the pseudo-terrorists *and* McClane are responsible for the destruction of the brand-new shiny office tower itself, whose wrenched and shattered form is, as we have noted, surrounded by smoldering heaps of ruin as the Christmas music strikes up at the film's close. It is with places of business as with homes in these films, as, according to the American army officer's notorious comment almost thirty years ago, it used to be with Vietnamese villages: one has to destroy them, apparently, in order to save them—and one does so virtually side by side with those villains from whom one is saving the particular space or site.

But all that is still really *only* for starters, since what truly complicates *Die Hard* and its proferred pleasures is the degree to which the Gruber gang's "double-Othering" is itself doubled. Or, to speak more plainly, what complicates is the extent to which they function as doubles for the corporate managers and professionals they have taken hostage in the film, *as well as* for the rampaging McClane (whose opposition to them becomes, accordingly, more resonantly multivalent still for those of us looking on). Clearly enough, the entire group of bad guys are specialist-technicians, each and all, in their respective fields: electronics, explosives, and, perhaps most of all, computer-hacking, performed here by a brainy, witty, and rather likable black techno-twit named Theo, who not only keyboards himself past a battery of key-code gates protecting the boodle in the corporate vault, but presides over the destruction of the LAPD's crack antiterrorist force and its most sophisticated weaponry with the wholesome élan of a brainy teen at video games. As a specialized armored car is explosively struck by a rocket whose launch signal he has given, Theo shrieks in self-mocking delight, "Oh, my god, the quarterback is *toast!*" Except for Theo, moreover, who sports a rather collegial sweater, the group's sartorial taste runs to stylishly loose-fitting turtlenecks and baggy trousers, albeit in somber hues of gray and black—really a kind of uniform of off-duty elegance for the casual yet industrious professional-managerial man of the 1980s, with a dash of European flair.

The group's leader, Hans Gruber, is, on the other hand, in another class altogether, if not quite a class of his own. As the head of his own international group of experts and specialists, he is befittingly garbed in a quietly expensive, custom-tailored gray wool suit. Indeed, on the way up to the chief executive's office and boardroom with Tagaki, Gruber, in between humming the "Deutschland über Alles" theme he has picked up from the Haydn quartet performed at the Christmas party when he and his boys first busted in, makes a point of telling Tagaki that he recognizes the Saville Row tailor of his suit, and in fact owns two of the same man's suits himself: "Rumor has it," he murmurs with a disarming insider's charm, "Arafat buys his there." Moreover, between the time of Tagaki's execution and the herding of the hostages up to the roof, Gruber presides over the hostage-employees of Nakatomi with a blend of casual grace and perfect firmness that is itself the mark of one of the manner born. Even the subtle relays of the film's soundtrack music, diegetic and not, from the Bach and Haydn that the string quartet is playing at the party, through Gruber's humming to the strains of Beethoven's "Ode to Joy" that pour forth as the vault finally opens up, underline the extent to which Gruber and his band are continuous with their ruling-class and / or corporate professional hostages: all those, that is, with the cultural capital to listen to classical music, like it, and know it as their own.[15] And that continuity, in turn, makes possible yet another double pleasure for the viewing audience. Besides getting off on (while disavowing, if need be) the rush of seeing yuppie professionals terrorized and office towers smashed to bits, we get to see surrogate yuppie professionals and one incredibly smooth ruling-class creep eat lead and die.

Again, as in the case of women, when we turn to the equivalent characters in the *Lethal Weapon* films, we find a bolder, cruder version of the same, this time in the person of the Afrikaner diplomat and his well-coifed thugs in their uniformly formal silver-gray suits and ties. Yet Arjen Rudd and the bad boys of his South African syndicate remind us in turn of another aspect of all these proto-corporate figures of evil we have left unexplored until now, that is, their use of *politics as camouflage*. Gruber and his gang pose as political terrorists to lure the authorities into a series of helpful false moves; Rudd protects himself and his men with the invocation of "diplomatic immunity." In *Lethal Weapon 2*, even Riggs and Murtaugh get into the act, as Murtaugh fakes a political incident downstairs at the embassy then whips up the enthusiasm of the antiapartheid lobby outside so that Riggs can slip into the building all the way to Rudd's inner office and tweak the devil's beard in his own den. What are we to make of this consistent—and, judging from the films' receptions, apparently consensual if not downright pleasurable—depiction of international politics as a ruse and a shuck?

On one level, the answer to this question coincides with another we have already given, on the overriding importance of the *personal* motive in our heroes' struggle against the evil people and powers they face. In *Lethal Weapon 2*, just before Riggs takes off in his truck to pull down singlehandedly the multimillion-dollar stilt house Rudd and his cronies live in, thereby handing us another jolt of left-populist eat-the-rich joy, he tells Murtaugh, "I'm not a cop tonight, Rog. It's personal." And in the film's final confrontation, Rog, having taken the cue, will complete the sentence that Arjen Rudd, having just gunned down Riggs, has begun—

"Diplomatic immunity"—with "has just been revoked," and a bullet of his own. In the same way, an exchange between Tagaki and Gruber in *Die Hard* (Tagaki: "What kind of terrorists are you?" / Gruber: "Who said we were terrorists?") receives its rhyme in another that follows shortly between McClane and the first of Gruber's gang that he kills (Bad guy: "You won't hurt me 'cause you're a policeman; there are rules for policemen." / McClane [as he starts the fight that ends in his opponent's death]: "Yeah? That's what my captain keeps telling me."). Through such rhymes and echoes still another link in the slippery chain linking Brothers to Others is forged in the form of a tacit agreement between heroes and villains that state and legal institutions finally have nothing to do with what is really going on.

Clearly, this anti-institutionalist, antipolitical theme offers its own populist pleasures; there are satisfactions for both left- and right-populists, after all, in depictions suggesting that underneath the level of politics and statecraft which the media always babble about, there is only naked greed, and, finally, beyond all talk of government and law, it is up to us—and primarily us white men—to clean up the muck. Yet these films also hint at other, darker sentiments and forebodings concerning their own anti-institutional cast. We have only to recall the apparent helplessness of the LAPD in *Lethal Weapon 2* in the face of the offensive staged against their members by the South African cartel, or the spectacularly, disastrously ineffective shows of force mounted against the pseudo-terrorists of *Die Hard* by both the LAPD and FBI, to sense that there is anxiety as well as delight in the distrust of authority that all four films in some measure tap and display.

When we begin to look in these films for evidence of what precisely is amiss in the house of law and authority, we come up with an interestingly similar pattern of details. I am thinking here once again of that scene from *Lethal Weapon* between the female psychiatrist and the male superior officer in which the lieutenant must duck into the men's room to get away, and of the other scenes and details that surround it: of the shot of the female police officer conducting a blue-uniformed chorus of "Silent Night" with her nightstick that opens this whole sequence introducing us to police headquarters, and of her reappearance in a sort of Christmas mambo line of herself and her fellows just after the scene with the shrink—a laughing, singing collective whose ludic image provides the spatial suture between that scene and the next one in Murtaugh's office, between Rog and another, unnamed plainclothes cop delivering a jeeringly ironic rap on what a sensitive guy he is. Then, in *Lethal Weapon 2*, there is the scene in which Riggs and Murtaugh are taken off the case involving the South African cartel and given state's witness Leo Getz to protect instead. These orders, which we pretty much already know they will largely ignore, all the more since Riggs even now is flaunting his refusal to respect the No Smoking sign that their superior officer, the same one badgered by the shrink in the original film, has posted in his office.

The amused contempt we are invited to feel for this officer in the *Lethal Weapon* films is thus close to that we come to have, with Powell and McClane, for blustering, ineffectual Deputy Chief Duane Robinson in the first *Die Hard* film, and even for the parodically hard-bitten FBI officers who ignore his presence and dismiss his authority as soon as they enter in turn—and for the same general reason. For soon enough in *Die Hard* we will discover what the *Lethal Weapon* films have had, more insistently and explicitly, to tell us: that the underside of even the FBI guys' rulebound power trip (in a voice drenched in sarcasm and disgust, Powell tells Mc-Clane, "They got the Universal Terrorist Playbook, and they're running it step by step") is the outright *childishness* of white agent Johnson's creepy glee as their helicopters swing in toward the office tower roof on a disastrously ill-conceived mission of first picking up the hostages and then blowing the "terrorists" away. (White Johnson [following war-whoop]: "Just like Saigon, huh, Slick?" / Black Johnson [grinning with amused contempt]: "I was in junior high, dickhead.")

But there is an old joke from the U.S.-Vietnam War itself that sums up what all these comic-disastrous examples of weakness in the site of institutional Power have in common. "What's the difference between the U.S. Marine Corps and the Boy Scouts?" the riddle goes, and the answer is, "The Boy Scouts have adult leadership." Or, more precisely, for our purposes at least, *male* adult leadership. To bring it back to our films, it is as if both the municipal and federal police departments were modeled on the pattern of Rog Murtaugh's family, in which the father—Rog himself, that is—is clearly loved yet equally clearly too much of an incompetent child himself, for all his bluster, to be seriously obeyed. In short, the State is weakened by both its *domestication* and by the *subversion of patriarchal-male authority* that contemporary domestication brings in its wake.[16]

If any further proof be needed that such a hollowing-out of Phallic authority is not only to be enjoyed as crypto-domestic sitcom but felt as a nagging problem as well, let us close our trip around the map of these films with the briefest of glances at those genuinely patriarchal-Phallic figures of fully empowered male authority, Shadow Company leader Peter McAllister (in *Lethal Weapon*) and Colonel Stuart (in *Die Hard 2*)—or, more precisely, of the common iconography of their deaths. For if both these uniformly rigid, disciplined and disciplinary (ex)-military men must be finally defeated by our wild-guy heroes, it will not be before they have given us ample proof of their ability to command exceptional obedience from their men (think of Mr. Joshua's roasting his own flesh at McAllister's command in *Lethal Weapon*) and to hand out punishments and executions with relentless swiftness to their enemies (the murder of Hunzacker in *Lethal Weapon* the moment he has spilled the Shadow Company beans, or, more impressively still, the catastrophic air disaster that Stuart coolly stage-manages as object lesson in *Die Hard 2*)—proof, that is, of the very ability to lay down the Law, be obeyed, and be effective that is woefully absent from the sites of legitimate state authority in these films. Nor, it is equally worth noting, will either of these evil yet authorized archvillains be dispatched without literally trailing clouds of glory as he goes. McAllister's escape car, loaded with explosives and grenades, carries him off to hell in a giant fireball; Stuart vanishes in an even more spectacular explosion, literally on his ascent in the 747 jet that bursts in the night sky more gloriously than the most gorgeous fireworks. Is it too much to suggest that such grandiose demises give evidence of yet one more ambivalence that they invite their audiences to indulge—toward the figure of a Patriarch who deserves to be honored in the very act by which He is canceled or refused, in the form of the death / disincarnation of a god?

Conclusion: Bodies and Buildings, Open and Shut

diehard, *adj.—stubborn in resistance; unwilling to give in.* n.—*a stubborn or resistant person; especially, an extreme conservative.* Webster's New World Dictionary, *College Edition* (1966)

It's the only thing I was ever good at. Martin Riggs, in *Lethal Weapon* (1987)

White, working, men. We have seen how these films define each of these crucial terms of race, class, and gender by means of a relatively complex yet highly specific network of contrasts, codes, and correspondences; and we have studied how that network, in turn, is offered to the audience as an equally complex, multiply gratifying pleasure machine. But now, when we stand back from the network / machine we have articulated, what seems most deserving of contemplation is the indisputable ubiquity, and therefore the apparent centrality, of *gender* in all these operations of knowledge and naming and narrative pleasure: gender, that is, as these films' fundamental medium of exchange, even when the nominal point of the transaction is good guys versus bad guys, not to mention race and class.

It is not hard to discern behind this general deployment of gender the shaping influence of shifts in the dominant sex / gender system, including those of which the ideologies and movements of feminism itself are either cause, effect, or both. The increased presence of women in professional positions in the national work force and the erosion and decline of phallic-patriarchal power in both the national state and the family home—these events and processes are clearly registered by our four films as faits accom-plis that must be acknowledged and accepted, even as the anxieties and resentments they have provoked within the dominant male culture are rechanneled and released in a variety of directions and ways. To name only the most prominent of those we have seen: the projection of gender code and homosocial bonding across racial lines; the marginalization, humiliation, and rescue of (or revenge for) the *incapacitated* woman; and the ambivalent pleasures of "keeping the home fires burning." Yet gender is even more deeply and consequentially present in these films than we have so far admitted, even as, at a certain level, what we have been calling gender begins to turn into—or to become, *inextricably*—something else. To find this level, though, we need to move away from narrative analysis per se and back to those elements of dramatic rhythm and mise-en-scène that we touched on as aspects of these films' strictly *formal* postmodernity. For now, on the other side of our prolonged look at the network of relations and pleasures provided by our films' plots, it is possible to say a bit more about just what and how much those forms might signify.

Thinking back to the unsublimated energies of these films' dramatic forms, of their ready and indeed almost regular ("chop-chop-chop") grati-fications of the desires they excite for graphic outbursts of violent action, we may now discern, despite and even through their industrialized reg-ularities, a deep and virtually "organic" connection between this formal aspect and our films' more explicit thematics of postpatriarchal male "wildness"—a breakdown and rearranging of the Oedipal patterns of classical emplotment in such ways that the slow-cooked and massively overdetermined climactic shootout is now dispersed into the rhizomatic form of one affiliated bit before and after another, each typically carrying its own relatively self-contained buzz or jolt. Yet it also seems that at precisely this deepest—or at any rate this least coded and most directly felt—level of experience and pleasure, the invocation of gender alone as an explanatory category becomes inadequate; or, to put it differently, that from here onward and inward, gender must be thought of in conjunction with the traditionally Marxist concept of mode of production, even as the meanings and energies called up by the gender concept transform it irrevocably.

For who, after all, by the late 1980s could fail to see the extent to which the present transformations of our national capitalist system (in conjunction with and athwart those of other national systems and of multinational capital) are, subjectively and objectively, gender-coded and delineated through and through? Postmodern feminist theorist Donna Haraway, in fact, has gone so far as to speak of the displacement and destruction of the old, semiorganized, male-dominant working class as, precisely, a "feminization" of work:

Work is being redefined as both literally female and feminized, whether performed by men or women. To be feminized means to be made extremely vulnerable; able to be disassembled, reassembled, exploited as a reserve labor force; seen less as workers than as servers; subjected to time arrangements on and off the paid job that make a mockery of a limited work day; leading an existence that always borders on being obscene, out of place, and reducible to sex.[17]

What, we referred to, quoting Marxist geographer Edward Soja, as the "squeeze in the middle of the labor market" is here described as a new, gendered kinship stretching across the gulf left by the disappearance of the First World proletariat. For Haraway, post-Fordism and postfeminism are indissolubly interconnected, and the literal and symbolic feminization of both the professional-managerial layers and the subproletariat relates to "the paradoxical intensification and erosion of gender itself" as cause and effect.[18]

In the light of these formulations, the rhythms of excitation and satisfaction in these films—as well as those of their even more obnoxious cousins, for example, *Rambo II* and *Top Gun* (1986)—appear as expressions of this "paradoxical intensification and erosion," asserting male violence and / or death-trip spectacle again and again, even as their own speeded-up processes of gratification undermine any claim to male authority. Riggs's boastful yet plaintive disclosure of his talent as crack-shot assassin in Vietnam—"the only thing I was ever good at"—thus rhymes with "die-hard" John McClane's wisecracking New York street savvy. Both evoke, with various admixtures of pride, embarrassment, and wistfulness, the resonance of hard-won skills discounted or dismissed in the new late-capitalist Processed World of L.A. within films that go on to reassert the ongoing value, even necessity, of such skills and savvy in the struggle against international criminal-commercial enterprise. Yet at the same time, such old-fashioned skills serve as trigger and alibi for the desublimated, spectacularized pleasure of free-floating aggression that these same films so casually and frequently let loose.

I am trying here to speak of the ultimate ground of these films, the reactive core of their overall imaginary, as a place where fantasies of class- and gender-based resistance to the advent of a postfeminist / post-Fordist world keep turning over, queasily, deliriously, into accommodations. "Who are you, then?" Euro-villain Gruber demands of our hero over the walkie-talkie; and John McClane, skipping nimbly from room to room across a corporate landscape that includes a stone statue of an Asian god framed by bonsai, says, "The fly in the ointment. The monkey in the wrench. The pain in the ass." And so he is, of course, in one sense; but in a less plot-bound and verbal and more visual-iconographic sense, as guerrilla warrior in a new, wild, foreign space, he fits right in. Likewise, our heroes' abilities to move behind and through the skin of these new surfaces to the mechanisms and generators that run them—the beltways of Dulles Airport, the heating ducts of the Nakatomi tower, even those bizarre industrial back rooms of the heavy-metal nightclub in *Lethal Weapon* with their array of hooks, pulleys, and chains, from which Riggs and Murtaugh must make their escape—suggests not only their ability to get *behind* the surfaces of the new space, and to strike back at those surfaces from those forgotten and repressed undersides, but also their uncanny familiarity with this whole new world, their weirdly intimate knowledge of their way around the very spaces they do so much—to our delight—to destroy.

These buildings / bodies, moreover, which literally in-corporate Fordist old and post-Fordist new, these sites or spaces both ruined and saved: do they not rhyme in turn, or even coincide, with the bodies of our oh-so-desirable heroes themselves, simultaneously displayed as beefcake and mortified as beef? In *Lethal Weapon 2,* recall how we first see Riggs dislocate his shoulder and smash it back into place on a bet, then see him do the same again when, tied and tossed into the water to drown, he comes face-to-face with the drowned body of his new lover, Rika. In that second smashing of his shoulder to put it right, then, and in the howl he gives out, we feel and hear an amazing mélange of significance in which our dripping, sexy hero (at the moment of maximum pain and undress in these films, our heroes often come wet) demonstrates all at once his wildness and his sensitivity, his vulnerability to pain and his incredible ability to take it. And those torn but still beautifully exposed slick-muscled bodies, those bulging arms and firm springy pecs, how do we distinguish between their fierce (re)assertion of gendered difference and their submission to the camera—surely every bit as abjectly / potently complete as any submissions by Rita Hayworth or Marilyn Monroe, in their own ways—as objects of its gaze and our own? What, likewise, is the boundary line between the diehard assertion of rugged white male individualism and its simultaneous feminization and spectacularization?

I confess that I do not know the answers to these questions; indeed, I suspect that our four films and their protagonists have been constructed in just such a way as to make them unanswerable. The wild, violent, mortified, desirable white male body at the center of these films can hardly be taken as an icon of progressive movement, or even of its unambiguous possibility; yet the permeability that comes along with that body's persistence, the sensitivity that accompanies its violence, the affiliations on the underside of its mechanisms of rejection and disavowal—all at least suggest something is loose and shaking in the old centered, hard-shell masculinity they simultaneously defend and subvert. Our films depict a specifically white / male / hetero / American capitalist dreamscape, international and / or multinational at the top and multiracial at the bottom, in which the interracial is eroticized even as a sharp power line is reasserted between masculine and feminine, in which, indeed, all the old lines of force and division between races, classes, and genders are both transgressed and redrawn. If the results of all these constructions and operations are scarcely to be extolled as examples of radical or liberatory cultural production (and who would have ever thought they could be, given the economics and social relations of blockbuster filmmaking?), they nonetheless suggest a version of white straight masculinity subject to new and vertiginous psychosocial mobility, displaced and mutating under the pressures of a systemwide restructuring and flux.

Coda, October 1996: Lost in Space

Since 1990, when I first composed the argument rehearsed above, so many more films have appeared in which the representation of a threatened or unstable white straight masculinity is crucially at stake that it seems pointless merely to take this discussion of the *Die Hard* and *Lethal Weapon* films any farther and impossible to cover the overall field of play. What I offer here instead by way of an update is something closer to a sample than an overview, a sounding rather than a map—and thus an invitation or provocation to more substantial work by others along the same lines.

To begin this sketch, however, I first need to introduce a slightly different and more global way of looking at the capitalist restructuring in which we are all still enmeshed and a third set of terms to set alongside those proposed by Soja and Haraway. In *The Long Twentieth Century,* Giovanni Arrighi's magisterial structuralist history of the capitalist system, Arrighi proposes that hegemonic capitalist power has until recently always been achieved and maintained via particular, historically specific combinations of two types of actors in pursuit of two logically opposed yet practically compatible "logics of power." One is *territorial,* having to do with the political control of space; the other is *functional,* concerned with what, quoting Ruggie, he sometimes calls the economic "space-of-flows" of raw materials (including, of course, labor power itself), products, and capital.[19] Thus, U.S. hegemony in the postwar period up until the early 1970s, for example, is explained as a complex alliance between political actors constructing, by means of the legitimating terms and discourses of the Cold War, a military-industrial condominium of interests and a stunted form of the Keynesian welfare state while prosecuting a neoimperial foreign policy, and their economic partners, the giant, centralized, nationally based corporations, ever extending and deepening their control over the national and international space-of-flows.

All well and good, you might be saying, but what do such terms and analyses have to do with the renegotiation of white straight masculinity in contemporary American popular culture? They are linked, I would say, only if and insofar as you will allow the following claim: that once we enter the era of putatively democratic statehood in the West, each and every hegemonic regime of capitalism / territorialism that Arrighi studies must inevitably propose its own version of the *normative gendered subject* as part and parcel of its legitimating work. This would be my general claim; and my particular historical claim would be that in the era of mass production, mass culture, and mass consumption, crossed and combined with the political hegemony of the American state—that is, the era just described, and still just past—that normative, gendered subject was conceived to be an economically productive, reliably employed protector-provider, Oedipally hard-shelled (tough and inexpressive on the outside, but tender within) and, of course, staunchly heterosexual white American man.

If this edition of officially approved white straight masculinity is specifically appropriate to an alliance of economic and political interests that has now unraveled and begun to turn into something else, it makes sense to try and site the deconstruction/reconstruction efforts of contemporary American film around questions of white straight masculinity in relation to what Arrighi sees as a truly original potential within the world capitalist system today. "By about 1970," he writes,

when the crisis of U.S. hegemony as embodied in the Cold War world order began, transnational corporations had developed into a world-scale system of production, exchange, and accumulation, which was subject to no state authority and had the power to subject to its own "laws" each and every member of the inter-state system, the United States included. . . . The emergence of this free enterprise system—free, that is, from the constraints imposed on world-scale processes of capital accumulation by the territorial exclusiveness of states . . . marks a new turning point . . . and may well have initiated the withering away of the modern inter-state system as the primary locus of world power.[20]

My thesis, then, is that this unprecedented, ongoing shift within the world capitalist system away from territoriality and the power logic of the "space-of-places" toward a new form of global capitalism under the sole directorship, as it were, of a deterritorialized "space-of-flows" managed by transnational corporations is bound to have provoked in and of itself a corresponding crisis in our collective definitions and representations of white straight masculinity—a crisis that the *Die Hard* and *Lethal Weapon* films seek to manage, mine, and recontain. But what evidence can we find in contemporary film that this latter crisis is still under way in the 1990s; what new or old, or new *and* old, definitions and images of de-ranged or re-arranged white straight masculinity are being proposed in such films today?

Or—to specify these questions still further in the light of Arrighi's terms—how is the deterritorialization of global capital and its ongoing decoupling from U.S. political hegemony registered in the images of white straight masculinity that we have seen on the screen in recent years? Here is one very provisional quick sketch of an answer: one sampling of the range of a broad spectrum of contemporary films in which the deterritorialization and dis-placement of white straight masculinity itself is variously contested, fretted over, and/or offered up for our enjoyment, sometimes all at the same time, using four films that at first glance seem hardly related at all: a primitivist swashbuckler, *Rob Roy* (1995); an action-adventure film, *Die Hard with a Vengeance* (1995), offered with a tip of the hat to the preceding analysis; a critically acclaimed, narratively adven-

18 Three men posing: Quentin Tarantino's *Pulp Fiction* (Miramax Films, 1994).

turous gorefest, *Pulp Fiction* (1994); and a monumentalizing epic of the American adventure in outer space, *Apollo 13* (1995).

In this coda to an already lengthy essay, I can of course provide no full reading of any one of these films, much less all four of them together. But happily, it requires no more than a first pass over the most salient features of each to descry the features that matter for my argument. Take *Rob Roy*, for starters—and take the start of *Rob Roy*, which opens with a scrolling historical headnote to inform us that we are now at the beginning of the 1700s, when "the centuries-old Clan system was slowly being extinguished" by the "greed of great Noblemen," and to announce, "This story symbolizes the attempt of the individual to withstand those processes and, even in defeat, retain respect and honor." Fade-out message; fade-in on Liam Neeson as that great clan leader Rob Roy.

The film's title character is indeed the most transparent and uninflected figure of an Oedipal-Fordist masculinity under assault from all sides to be found in these four films. While defending his turf against the incursions and depredations of the new mercenary aristocracy from the south, he also endorses a curiously modern individualist view of male power and integrity. When one of his followers asks, "Will McGregors ever be kings again?" Rob Roy replies, "All men with honor are kings." And likewise women, we learn from our hero at a later moment, "are the heart of honor; we [men] cherish and protect it in them." It would, in effect, be hard to find a more succinct formulation of the "doctrine of two spheres" that served as the middle-class norm of gender relations during the period of Fordist monopoly capitalism now past. The name for the threat that endangers Rob Roy and his clan's way of life may be land tenantry and the commutation of the old feudal ties of obligation and loyalty into the new coldly exploitative hierarchies of a precapitalist economy brought in from England and centered on cash; but the gender identities and relations modeled by Rob Roy and wife Mary are anachronistically Oedipal-patriarchal in nature. Even the dual catastrophe with which they are visited by the English villains of the film makes this point, as Rob Roy's home is, in effect, foreclosed on for debt and Mary is raped—so that the sexual violation rhymes with and redoubles the economic one.

However grim the long-term prospects may be for the "individual"—that is, the subaltern white male—seeking to keep hold of his "respect and honor" in the face of a new system's tendencies to wrench them from him, Rob Roy in the end, of course, will win the conflict that we see in the film. But before moving to our next film, we need to take note of *Rob Roy's* equally evocative spatial politics, in which the evil protocapitalist powers are identified with England, and the good localist resistance to them is, as in *Braveheart* (1995), linked with Scotland. In *Rob Roy*, then, the reassertion of an Oedipal-patriarchal masculinity goes hand in hand with a national allegory in which England functions as the nation-state name for the abstracting, deterritorializing power of capitalist relations of exchange and exploitation, and Scotland as the nation-state name for virtuous territoriality, for the stability of local placement, the security of bounded habitation, and the strong sense of identity-in-place that it confers.

With such emphatic linkages in mind between a given mode of masculinity and an equally value-laden spatial allegory in which the nation-state operates as the mediating or hinge term between abstracting knowledges and practices of deterritorialization (i.e., the capitalist space-of-flows) on the one hand, and the particular knowledges and identities that derive from local placement on the other, we can now take a glance at *Die Hard with a Vengeance*, the next film along the continuum I am trying to sketch. At first sight, this most recent addition to the *Die Hard* series would seem to propose much the same definition of normative white straight masculinity as *Rob Roy*, both in itself and in relation to the film's own political allegory of space. After all, its plot is single-mindedly devoted to our hero John McClane's pursuit of the terrorist gang that has stolen the nation's entire gold reserve held in the Federal Reserve Bank of New York. So we have McClane and Zeus (Samuel L. Jackson), his black sidekick in this Joel

Silver production, two native New Yorkers fighting to regain control of their home territory and recover the nation's wealth from the Euro-internationalist thugs who have ripped it off. But *Die Hard with a Vengeance*, to an even more heightened degree than the films we have discussed, exhibits the same queasy and delicious ambivalence toward its official terms and values—and not least toward its politics of space. After all, the newest contribution to the *Die Hard* series follows in the steps of its distinguished forebears in regularly offering us the pleasure of *deranging* the territory of the local, of disordering it and blowing it up, from its department stores to (almost) its neighborhood schools. For another, that local habitation itself, the home of our hero, is depicted as an *already* disorganized, chaotic state, a site whose messiness, clutter, and anarchy is the spatial counterpart of our hero's own emotional and physical disarray at the opening of the film, when he is scooped up, drunken and disheveled and disgraced, from the streets and put back into action, to undergo the "die-hard" process of pummeling and mortification yet again.

If, then, we are to see Bruce Willis's John McClane as the figure of an older Fordist patriarchal-Oedipal, hard-shell / soft-center masculinity under siege, and New York City as the site of the hero's local habitus that must be protected at all costs, we also must acknowledge the degree to which *Die Hard with a Vengeance* invites us to view and enjoy the decadence and degradation of both that masculinity and its local home base. Similarly, the linkage of that habitation and masculinity with the nation-state is an uneasy and troubled one, at least iconically, insofar as the nation in this film gets represented mainly in the form of a great, uniform stack of interchangeable bars of gold, that medium of exchange, which, however tenuously, is still taken as the basis for all other mediums of exchange—the United States as no more and no less than its gold reserves. If the gleaming abstraction of such a representation of the nation stands in stark contrast to the funky disorder of the film's territory—the streets of New York—it is, I would argue, all the more connotatively, if covertly, akin to the world of the terrorists, whose mastermind, as in earlier films, only pretends to be prosecuting a political and / or personal agenda as cover for his sheer, simple greed, and whose technocratic wizardry and dry unflappable wit alike are offered for our enjoyment as the counterpart to McClane's funky, homegrown aggressiveness and corny jokes.

Finally, of course, our mastermind Simon (Jeremy Irons) and his gang have been defeated by McClane and his black sidekick, as he and his kind have been before. Yet it also seems notable that this time the finale itself occurs outside and beyond the space of both nation *and* home base, via a transnational climax that brings us to an utterly abstract late-capitalist landscape of transit, shipping, and exchange outside Toronto. In placing its finale in an international air transport no-place within a country that is merely adjacent to the United States, the film all but severs the ties it officially valorizes and endorses between the national and the local; the die-hard working guy is putatively at the center of both, as the old satisfactions inherent in what Ruggie and Arrighi call the "space of places" give way to the new excitement and romance of the transnational, late capitalist space of flows, and an older, eroded Oedipal masculinity of controlled release metamorphoses almost entirely into a new post-Fordist, post-Oedipal model, composed of equal parts of high-tech game-playing (Simon's specialty) and unsublimated, wild rampaging (McClane's), both together and each alike virtually untethered to any space of place.

To turn, then, from the most recent *Die Hard* film to *Pulp Fiction*, the crossover cult film success of 1994, is to note immediately the full emergence of an unabashedly pre-Oedipal white straight masculinity in a landscape from which the nation as a mediating category between the space of flows and the space of places has disappeared. What is left behind in *Pulp Fiction* is less a sense of an unfolding drama in a single fixed place than a spasmodically violent skid across various generically rendered sites—a bar, an apartment, a coffee shop-restaurant, a motel room—in which action may or may not violently erupt out of and into their paradoxically localized placelessness at any moment. Likewise, the film's spatial interrelations of these placeless localities are as radically contingent and problematically tangential as the relationships of the film's several diachronically out-of-synch plots are to one another, like a set of epicyclical motions around a once-fixed center that has long since fallen away.

Through the work of Roland Barthes and others, we are by now familiar with the notion that there is at least a deeply rooted and organic affiliation between an older kind of story, with its accumulation of unspent dramatic or suspenseful elements throughout the narrative's so-called "rising action" into a force that is discharged most completely at the story's climax, and the preferred rhythms of saving and spending, of repression and release, inscribed into the operations of Oedipal masculinity. And the three films we have been sampling in this update seem to bear out this hypothesis with an almost schematic precision. *Rob Roy*, in its unabashedly nostalgic celebration of an endangered Oedipal protective-provider masculinity like that of its protagonist, builds outrage upon outrage and trial upon trial, until the final duel between our legitimate clansman and husband Rob Roy and the foppish bastard and begetter of bastards, the evil Archibald (Tim Roth), in which victory is at long last secured. *Die Hard with a Vengeance,* with its only putatively "die-hard" Oedipal-industrial-era protagonist, "pretends" like its predecessors to follow a similar structure, while in fact it offers us an altogether different economy of pleasure, in which the giddying blur of the high-speed chase and / or the gratifyingly spectacular release of aggressive impulse occurs at regularly recurring intervals throughout the film. And *Pulp Fiction* gives up all pretense of offering us either such an authorized developed narrative or such a self-restrained masculinity. Its displaced narrative structure, spatialized and detotalized, are matched by the devolution of the white male protagonist himself into an only peripherally related assortment of white boys (Travolta's Vince, Willis's Butch, Tim Roth's Pumpkin) cruising around their landscapes of interchangeable parts splattered here and there with the messes they make, and laced with their hilariously obscene yet insipid dialogue, itself the verbal equivalent of the gelatinous world they drift across to pull their jobs or make their drug connections— protagonists so deauthorized, it must be said, that a couple of them work for a *black* man, even as they each and all hit new highs—or would it be lows?—in the area of poor impulse control.[21]

Viewed in this way, however, our three films would seem to describe or suggest a too-simple and straightforward narrative of cultural change, in which rear-guard actions in defense of a once organic interrelationship between nationality, territoriality, and authorized Fordist-Oedipal man and the whole economy of pleasure based upon it, fall away in the face of a new postnational landscape of giddy de-rangements, zigzagged across by so many deauthorized, pre-Oedipal / post-Fordist, havoc-wreaking guys offering us a new set of kicks. But of course such cultural and ideological shifts are never all that smoothly linear and uncontested. So, if only to acknowledge a more crowded and various picture than we can draw here, let us consider the relationships proposed, masculinity modeled, and pleasures offered in a fourth and final film, *Apollo 13,* a movie that, perhaps not incidentally for our argument, grossed in its first theatrical release close to three times what *Pulp Fiction* took in.[22]

Most of what I want to note here is fairly obvious—and arguably all the more significant precisely for that reason. For starters, of course, by dint of its subject matter alone, director Ron Howard's representation of the near-catastrophe of the Apollo 13 spaceflight operates to recuperate and revalorize the category of the nation, whose value and resonance *Die Hard with a Vengeance* so eroded and *Pulp Fiction* so thoroughly dismissed. Yet if *Apollo 13*'s men are, like Rob Roy, national subjects, unlike Rob Roy they are, first of all, obviously enough, both *plural* and curiously *placeless* protagonists. As characters, of course, they are minimally distinguished from one another by their individual foibles and preexisting star texts—Tom Hanks the cigar-smoking family man, Kevin Bacon the swinging single, Ed Harris the Camel-smoking, flat-topped head of ground control, etc. But such characterizations, and indeed the plot itself, are entirely subordinated to operations of what C. L. R. James described as "abstract intellect, abstract technology . . . serving no purpose but the abstract purpose," in a film whose climactic drama is a technical exercise of literal spin control and whose climactic instant, pumped up with great gusts of stirring, monumentalizing music, is notably enacted as a thrilling change of numbers on a Lunar Excursion Module's LED.[23]

In this respect, then, we could say that *Apollo 13*'s reinvigoration of white straight masculinity and the nation-state itself are effected by means of a deterritorialization completely distinct from the others that we have here described. This reduction of masculinity to technical functionality, and of the nation-state that is figured in this film as the collective agent that moves forward and back as an abstractly quantifiable and malleable void called "outer space," restores the potency of both by rendering them as devotees of the same relentlessly abstracting force that in late capitalism reduces all places and all lives into so many quantifiable units and all problems into so many technical calculations in its quest for power through and over the space of flows. Thus, for me at least, the peculiar resonance in a film I found otherwise as creepy as it was dull, of that impossible shot / reverse shot that occurs just before the film's climax when Jim Lovell gazes apprehensively and longingly out the window of the lunar module at planet Earth to which he well might never be able to return, and his wife stares anxiously and lovingly "back" from the kitchen window of their home in Texas—for this simultaneously eviscerated yet empowered masculinity is indeed, both figuratively and literally, quite frighteningly lost in space.

As, in a different sense, this coda is lost as well. For there is no real conclusion to the exercise that it has been about. My only purpose in conducting this analysis, in fact, has been to suggest how unsettled the very concept and normative definition of white straight masculinity remains within American popular culture in this protracted moment of perhaps unprecedented structural change within what nonetheless remains a deeply racist, sexist, and exploitative world system: how unsettled and, perhaps as well, how unsatisfactory. But if we want better models and representations of what white straight men ought to be, we not only will have to remake them ourselves, outside Hollywood, but we will have to restructure and collectively remake a great many other things—institutions, representations, and relationships—in both our imaginary worlds and our real lives as well.

Notes

An earlier version of this essay was published in Fred Pfeil, *White Guys: Studies in Postmodern Difference* (London: Verso Press, 1995). It appears in this volume by permission of Verso Press.

1 The phrase comes from Michael Denning, *Mechanic Accents* (London: Verso, 1987).

2 Quoted from the review by "Jagr.," *Variety*, March 4, 1987, p. 24.

3 Quoted in Pat H. Broeske's critical summary of Stallone's *Rambo: First Blood Part II, Magill's Cinema Annual, 1986,* ed. Frank N. Magill (Englewood Cliffs, N.J.: Salem Press), p. 323.

4 I can imagine some argument on this point from others who would insist that the real progenitor of Riggs and McClane is not Stallone / Rambo but Chuck Norris (in, for example, *Invasion U.S.A.,* 1985, or *Delta Force,* 1986)—and, I suppose, before him, it would be Charles Bronson and Bruce Lee. Of course, these figures, and the brutally violent, explicitly reactionary and / or jingoistic cinematic subgenre in which they appear, have their role in the genealogy of the films we are studying here. However, I would maintain that Rambo is nonetheless the most immediate and direct forebear of Riggs and McClane and their respective films, insofar as the characters they portray and the bodies they display are defined by their sensitivity, suffering, and mortification, as well and as much as by their lethal strength and competence.

5 Robert B. Ray, *A Certain Tendency of the Hollywood Cinema, 1930–1980* (Princeton, N.J.: Princeton University Press, 1985).

6 Of course, I do not mean to deny the validity of this account of the U.S.-Vietnam War insofar as it corresponds to the experience of any number of veterans still suffering from the effects of their participation in it. What is at stake here, rather, is Hollywood's projection of this traumatized individual perspective as "the Truth" of the war. Such a "therapeuticization" of our country's genocidal and imperialist aggression is most flagrantly ideological precisely in its offer to sidestep larger questions of policy, outcome, and responsibility. Yet, as will soon become clear, my concern here is less with the political utility of this narrative for dominant elites or within mainstream culture than with the gender-specific and race-specific fears and desires that it runs on and expresses. For a provocative, insightful (if occasionally monological) analysis of the Vietnam narrative(s), which links up such desires and anxieties to a project of ideological retrenchment, see Susan Jeffords, *The Remasculinization of America: Gender and the Vietnam War* (Bloomington: Indiana University Press, 1989).

7 For profiles of Moredock and analyses of the symptomaticity and spread of his behavior, see Ronald Takaki, *Iron Cages: Race and Culture in Nineteenth-Century America* (New York: Knopf, 1979); Richard Drinnon, *Facing West: The Metaphysics of Indian-Hating and Empire-Building* (Minneapolis: University of Minnesota Press, 1980), and Herman Melville, *The Confidence-Man: His Masquerade*, chaps. 25–27 (first published 1857). More generally, the standard work on the historical construction of American masculinity as a synthesis or negotiation between violent "savagery" and "civilized" behavior and belonging is Richard Slotkin, *Regeneration Through Violence: The Myth of the American Frontier, 1600–1860* (Middletown, Conn.: Wesleyan University Press, 1973).

8 Denning, *Mechanic Accents*.

9 The quoted phrase is taken from Edward Soja, *Postmodern Geographies* (New York: Verso, 1989), p. 187; but see also *The State of Working America* (Washington, D.C.: Economic Policy Institute, 1990), and the guided tour of post-Fordism offered in *Socialist Review* 21 (January-March 1991).

10 I use the term "black" here and throughout the analysis that follows rather than the politically preferable "African American" insofar as the films themselves construct this axis of racial difference as precisely as a "black/white" affair.

11 In *Die Hard 2*, Al Powell makes only a token appearance, while McClane encounters other blacks (or should we now, in this context, say "African Americans") in the persons of the paramilitary thug he fights but fails to capture in the first action scene of the film; Barnes (Art Evans), as a sympathetic and helpful control tower technician; and the good-evil character of Major Grant (John Amos). I owe to my statistics-hip partner Ann Augustine the wonderful insight that "if you want to discount a variable, randomize it"—along with many other insights and editing suggestions along the way.

12 Jesse Kornbluth, "*Die Hard* Blowhard," *Vanity Fair*, August 1990, p. 66.

13 This may be the place to note the shift in the characterization of Roger Murtaugh himself in the follow-up film—or rather, the sense in which that shift picks Murtaugh up as he is at the end of the first *Lethal Weapon* film, a more or less unhesitating partner and accomplice of wildman Riggs. In structural terms, then, the position Murtaugh occupies in the first *Lethal Weapon*, is given to a new sidekick, Leo Getz (Joe Pesci), the money launderer turned prosecution witness who is put under their supposed protection and who is pointedly depicted as being even more hysterically inept and comically domestic (he dons an apron and wields a vacuum cleaner in a vain attempt to clean up Riggs's bachelor-pad trailer) than Murtaugh ever was.

14 Kornbluth, "*Die Hard* Blowhard," p. 66.

15 There is more to this musical signifier and its meanings than I have space to develop here. For one thing, the classical music whose presence is introduced at the Nakatomi Christmas party strikingly contrasts to the Christmas rap music that Argyle the chauffeur has just been playing for McClane on his way in from the airport, prompting McClane to ask, "Aintcha got any Christmas music?" The suggestion here, then, would be that neither Argyle's nor the corporate folks' music is genuine (white) people's Christmas music: *that* is up to McClane to produce or make possible himself, as, by the end of the movie, he does. Yet, on the other hand, when his victory and the happy end of the film finally prompt such tunes to purl out on the closing soundtrack, the images of ruin they play against give them a rather ironic lilt.

There is also the interesting and no doubt significant sense in which it is important that the classical music is identifiable with "Europeanness"—which in turn attaches itself easily enough to Gruber and his cronies but has interesting implications for the ideological resonance of Japanese corporate figures Tagaki / Nakatomi. But I am sure that most readers would agree that the present reading is sufficiently detailed and lengthy that I need not delve further into these matters.

16 Here again, there are some curious echoes of the Rambo films, especially of *First Blood*, in which the local sheriff (Brian Dennehy) behaves as a hysterically punitive yet ultimately impotent paternal / institutional power, in contrast to the authorized *and* ultimately nurturant figure of Rambo's former commanding officer (Richard Crenna) from—where else?—his Special Forces days back in Vietnam. Yet, as we now know from *Rambo II,* this officer will ultimately be shown to be less than fully supportive and empowered, leaving our hero to suffer and rampage through the wilds all on his own. For an interesting feminist-psychoanalytic speculation on all that may be represented and at stake here, see Jessica Benjamin, "The Oedipal Riddle: Authority, Autonomy, and the New Narcissism," in *The Problem of Authority in America,* ed. John Diggins and Mark Kann (Philadelphia: Temple University Press, 1981).

17 "A Manifesto for Cyborgs: Science, Technology, and Socialist Feminism in the 1980s," first published in *Socialist Review* 80 (1985); here quoted from *Feminism / Postmodernism,* ed. Linda Nicholson (New York: Routledge, 1990), p. 208.

18 *Feminism / Postmodernism,* ed. Nicholson, p. 209.

19 Giovanni Arrighi, *The Long Twentieth Century: Money, Power and the Origins of Our Times* (New York: Verso, 1994), pp. 33, 80.

20 Ibid., p. 74.

21 For a far superior and more comprehensive reading of *Pulp Fiction,* see Sharon Willis's brilliant essay, "The Fathers Watch the Boys' Room: Characters in Quentin Tarantino's Films," in *Camera Obscura* 32 (September-January 1993-94): 40–73—an essay that I would quibble with only insofar as it too quickly and imprecisely describes the white straight masculinity modeled in

Tarantino's films as a new type of Oedipality, despite the overwhelming evidence that Willis herself cites and explores from within those films (e.g., their emphasis on the messy splatter of bleeding bodies, or on excremental humor and the scene of the bathroom) that the pleasures, anxieties, and preoccupations of that masculinity are preeminently *pre-Oedipal.*

22 The figures, according to *Magill's Cinema Annual* for 1994 and 1995, are $63 million for *Pulp Fiction* and $172 million for *Apollo 13.*

23 Quoted from James's *Renegades, Mariners, and Castaways,* from Paul Buhle, *C. L. R. James: The Artist as Revolutionary* (New York: Verso, 1988), p. 106.

Your Self Storage: Female Investigation and

Male Performativity in the Woman's Psychothriller

Sabrina Barton

Let's begin with an image from *Pacific Heights* (1990; directed by John Schlesinger). One hand shielding her eyes from the bright Los Angeles sun, a woman wearing blue jeans and a blazer watches from shore as a man and a woman drink champagne and cavort on a small yacht, unaware that they are being watched. The woman who watches, Patty Palmer (Melanie Griffith), has followed the man on the boat, Carter Hayes (Michael Keaton), from San Francisco where he destroyed her just-bought Pacific Heights Victorian home and perhaps her relationship with her live-in boyfriend. Unhelped by the law, Patty has set out on her own to seek clues, answers, restitution, revenge. She watches as this scheming but psychotic imposter works his wiles on another victim.

Women's psychothrillers[1] unabashedly employ the larger psychothriller genre's disturbing convention of depicting women who are terrorized and attacked. However, they do so from a different perspective. The story is told predominantly from the point of view of its female protagonist who pursues, even as she is pursued by, her antagonist. In granting a more central and more active investigating role to the woman in jeopardy, these films significantly depart from most movie psychothrillers (in which women function almost solely as victimized objects).

The image of Patty voyeuristically exerting her gaze upon the unwitting villain is a useful starting point for undertaking a feminist analysis of the woman's psychothriller. As feminist film theory's single most influential article, Laura Mulvey's "Visual Pleasure and Narrative Cinema" (1975), has argued, classical Hollywood cinema—exemplified by such Hitchcock suspense films as *Vertigo* (1958), *Rear Window* (1954), and *Marnie* (1964)—tends to equate the male with an active gaze that is associated with knowledge, power, and agency, while the female tends to be placed in a sexualized, objectified, "to-be-looked-at" position.[2] The image from *Pacific Heights* with which I began reverses this male-gaze / female-object gender dynamic and signals Patty's narrative transformation from victim to investigator.

My interest, however, is not in gaze / object relations per se, but in how these and other conventionally gendered traits, actions, and associations can, potentially, be redistributed among the characters of a given film in such ways that cinema's traditional gender roles and gender meanings are revised. Since the early 1970s, feminist film theory has been grappling with (among other things) the issue of how femaleness and maleness, femininity and masculinity, are represented in narrative film, and how spectators identify (or do not identify) with such representations. In *Pacific Heights*, Patty's femininity is not of the voyeured or fetishized sort theorized by Mulvey. In contrast, for example, to the meticulously made-up and elegantly costumed Tippi Hedren (an icon of blonde 1950s-style femininity) in Hitchcock's *The Birds* (1963) and *Marnie*, Patty is dressed and coiffed with deliberate casualness. This female protagonist is neither coded for visual pleasure nor portrayed as an enigma whose feminine performance mandates male investigation. Rather, Patty's subjectivity is anchored internally, in her perceptions and actions, in what the movie represents as a stable, core self. She is visually and narratively coded, I wish to argue, as "real" rather than as "performative."

I am using the term "performative" in relation to gender as described by theorist Judith Butler who argues that, however natural one's femininity or masculinity may feel or appear, gender identity is, in fact, the constituted *effect* of repeated poses, gestures, behaviors, positionings, and articulations. Butler's influential argument builds on the poststructuralist-psychoanalytic claim that "femininity" and "masculinity" are not inherent and not necessarily tied to anatomy; sexual identity is psychically and linguistically based, and thus it is far more mobile and multiple than that.[3] Although some people might choose to foreground their own gender performativity (drag queens are an obvious example), socially acceptable gender identities tend to be presented as naturally emanating from an essential core self—a core self that, Butler would insist, is an illusory effect, a performance.

However illusory such a "core self" may be, my impression from teaching film courses to undergraduates and from reading popular press film reviews (as well as from my own viewing experiences) is that many audience members seek out and value characters who are represented as possessing coherent core selves.[4] For in spite of the performative dimension of gender that Butler discusses, Hollywood cinema most often emphasizes and highlights the performativity of only certain categories of characters, while coding others as authentic. Through a filmic foregrounding of femininity in terms of body parts, costume, makeup, masquerade, seduction, and makeover plots, women far more often than men have been explicitly aligned with the performative. This is significant from a feminist perspective because performativity tends to be negatively coded as duplicitous, fragmented, and unstable, whereas the depiction of the "real" is coded as strong, unified, and stable—the more desirable identity category.

Let me clarify by way of an example. In *Marnie,* Hedren plays a character who performs her femininity, performs her identity. Marnie constructs a series of false selves out of stories and suitcases; hard-luck tales and false identification papers; changing wardrobes, makeup, and hair dyes. By contrast, the film's hero, played by Sean Connery, emerges as self-evidently stable (largely through his role as investigator of the mysterious and unstable Marnie, a woman who, in addition to her masquerading of multiple false selves, fundamentally does not herself know who she "truly" is). In this manner, feminine identity comes to be defined as performative, while masculine identity is associated with authenticity. In *Pacific Heights,* as in other women's psychothrillers, the opposite dynamic is at work. It is the male who is duplicitous and unstable—a scam artist whose identity is constituted from carefully coordinated images of socially successful masculinity. When Patty looks at (and sees through) Carter's manipulative performance, she not only defines herself against the traditional to-be-looked-at version of femininity exemplified by Hedren, but, in exposing masculine masquerade, she also lays claim to a "real" self.

The image from *Pacific Heights* with which I began is especially resonant for feminist film analysis and film history. Patty is played by Melanie Griffith, the real-life daughter of Tippi Hedren, who, interestingly enough, is cast as the older woman on the yacht whom Carter is drawing into his snares. In this instance, Griffith has followed her mother's footsteps into the suspense genre shaped by Hitchcock.[5] But, unlike her mother, the blonde daughter of the Hitchcock blonde watches from a critical, investigative—one might even say feminist—distance at the scenario of male manipulation and exploitation that she seeks to understand and from which she seeks to extricate herself. Thus, in this brief, poignant pairing of mother and daughter, *Pacific Heights* juxtaposes two generations of women as well as two generations of the psychological suspense film genre, charting a shift in the representation of gendered subjectivity.

This essay will examine four exemplary women's psychothrillers: *The Silence of the Lambs* (1991), *The Stepfather* (1987), *Sleeping with the Enemy* (1991), and *Pacific Heights*. Each of these four movies reveals the complex interplay of gendered traits and actions at stake in violent encounters between female protagonists and male antagonists. These women's psychothrillers thereby make available for analysis the implications of somewhat different arrangements of the same basic competitive economy. That is, in all four movies, a woman manages to achieve an identity that the film codes as "real," an identity centered in a coherent interior self. She achieves this "real" identity at least in part through her action of resisting and, ultimately, investigating a psycho-male antagonist's ostentatiously "performative" identity, an identity made up of formulaic poses and manipulated surfaces.

Your Self Storage: *The Silence of the Lambs*

From the unmasking of masculine masquerade there emerges, if not a feminist role model, at least a strong woman character. (Here I deliberately use what may sound like dated, 1970s feminist terms such as "strong woman" and "role model.") The goal of this essay is to recover a means of speaking affirmatively about cinematic representations of empowered women and to revalue the structures of viewer identification mobilized by such representations. An important motivation has been my own pleasure in watching the strong female protagonist negotiate the horrors of the woman's psychothriller, a pleasure that underscores this critical project.

Consider the "realness" of Clarice Starling's subjectivity in *The Silence of the Lambs* (directed by Jonathan Demme), a subjectivity defined by depth rather than external props and bodily surfaces. Foster's characteristic blondeness is darkened, and she is dressed conservatively in ways that do not emphasize the body (work clothes or sweats hardly coded for Hollywood-style visual pleasure). Additional formal techniques—above all the cinematography's use of point-of-view shots—work to code the female protagonist's subjectivity as stable and active. The look and who deploys it is a powerful cinematic means to empower a character in mainstream narrative cinema: the camera looks *with* certain characters, not merely *at* them as visual objects.[6] Point-of-view shots are repeatedly granted to Clarice as she investigates the long series of enclosed spaces in this film, beginning with FBI supervisor Jack Crawford's (Scott Glenn's) office and Hannibal Lecter's (Anthony Hopkins's) cell, and concluding with Buffalo Bill's (Ted Levine's) mazelike basement.

In another important strategy for establishing depth of self in its female protagonist, the film explores Clarice's own interior space of subjectivity. Two memory flashbacks to childhood, as well as the "therapy" sessions with Hannibal Lecter (in which she candidly answers the imprisoned killer's questions about herself in exchange for information about the serial killer at large), serve to take the audience into Clarice's family history, psychic structures, feelings, and motivations. Ultimately, we are brought to see her commitment to saving women as entirely coherent with her own self-rescue from childhood trauma—watching the scream-ing lambs being taken to slaughter, the death of her father—a coherence that serves to code her character as one in possession of a core self.

The female protagonist's access to a more substantial, interior-coded identity is implicit in the remark that Hannibal offers to her early in the film: "Look deep within yourself, Clarice." Clarice knows that the line is, as she puts it, "too hokey" for this genius-psychiatrist. It must be a clue, and it must point to something beyond an apparently trite call for intro-spection. As things turn out, it does and it doesn't. Clarice eventually tracks down evidence on the serial killer at a company in Baltimore called "Your Self Storage." There she investigates a space filled with a dark clutter of things that help reveal the secret self and psycho-past of (we later learn) Buffalo Bill. The psycho-killer's self is routed through a confused mass of cultural artifacts: mannequins, flags, jewelry, weaponry, cos-tumes, makeup, the "stored" head of a dead male lover. Penetrating this storage space filled with personal items is an important stage in Clarice's attempt to understand why Buffalo Bill kills and then skins his female victims. As she later figures out, he wishes to stitch for himself an external "suit" made from pieces of different women's skin; putting on this grue-some outfit will let him performatively inhabit the femininity that ob-sesses him. Along with forwarding Clarice's investigation of Bill's perfor-mativity, however, Hannibal's pun—"Look deep within yourself"—also resonates with Clarice's growing ability to enter and explore her own self storage, which takes the internal form of memories and dreams, and thereby to find and become not just a self, but *herself*.

Because I went to see *Silence* with a gay male friend, my first reaction to the film was painfully fraught. We both heard the spectators in the row

19 Jonathan Demme's *Silence of the Lambs* (Orion Pictures, 1991). a. The female investigator on her own, in the dark. Photo by Ken Regan. b. The women of the FBI. Photo by Michael Ginsburg.

behind us mocking Buffalo Bill's cuddling of his white poodle "Precious," tittering "yuck" at the sight of Bill's nipple ring, and rooting for the demise of the fag. Even so, I loved the female protagonist and therefore "liked" the movie that brought her to me.

I later found my own conflicted viewing experience replicated in two distinct but overlapping critical debates that arose around *Silence,* both of which concerned the gender and sexual politics inherent in embracing the character of Clarice Starling. First, what did it mean for spectators to enjoy the movie and its female protagonist in view of the rampant homophobia expressed through the Buffalo Bill character? And, second, given all of the sophisticated critical work that has been done, by feminist and queer theorists among others, in deconstructing simplistic notions of identity and identification, was it theoretically and politically naïve, even retrogressive, for a feminist film critic to claim Clarice as (to use another old-fashioned term) a "positive image"?

The media's encounter with *Silence* divided along gendered lines, expressed most dramatically in the *Village Voice*'s March 1991 forum, "Writers on the Lamb: Sorting Out the Sexual Politics of a Controversial Film."[7] In that forum, those male contributors writing from a gay-identified political position condemned the film for adding yet another "fag-as-psycho-killer"[8] to the popular imaginary, while female contributors writing from a feminist-identified political position supported the film for its empowering representation of Clarice Starling. Amy Taubin, for example, wrote in support of the film because it portrayed "a woman solving the perverse riddles of patriarchy—all by herself."

Elsewhere, filmmaker Maria Maggenti offered the following formulation to three gay male anti-*Silence* copanelists at Outwrite (the San Francisco lesbian and gay writers' conference). "You boys just don't get it . . . there is more than one way to look at a movie and women see something entirely different here than you do."[9]

What to make of this division between the "girls" and the "boys"? One way to understand, and perhaps move beyond, this critical split—feminist women vs. gay men—would be to think of these two critical sides not so much as having read "entirely different" films but as having identified with opposed characters, and as having read those characters more or less in isolation from the film's overall system of representations. Let me be clear: I think that responses based on how a film portrays a single character are important. Indeed, it is my argument that academic critics and theorists need to develop a more nuanced understanding of these deeply passionate responses. To do so, however, I believe that we need to grasp with more precision how films deploy, code, and gender characters as parts of a representational, psychological, and ideological system of desirable "realness" and undesirable, dangerous "performativity."

Within a particular film, certain key characters "energize" certain psychical experiences and social concepts and ideas. The circulation and distribution of these elements is often gender-based. In women's psychothrillers, because performativity is associated with the psychotic male antagonist, a female protagonist can emerge who is coded as having an intact, stable identity. Thus, a positive female image will come at a cost: in *Silence*, Buffalo Bill—the not exactly gay (or so claimed Jonathan Demme), not quite a transvestite, not yet a transsexual, feather-boa- and makeup-wearing serial killer—bears the burden of (classically feminized) performativity so that Clarice can occupy the traditionally male-identified role of investigator. In the economy of selves that the movie sets up, if one marginalized group is allowed to find a subjectivity that feels stable and authentic, another must take its place at the negative pole of performativity.[10]

In light of *The Silence of the Lambs'* scapegoating of a gay character, one can understand the ambivalence registered in B. Ruby Rich's wry conclusion after describing in the *Village Voice* forum her own warm response to Starling: "guess I'm just a girl." A critic with overlapping queer and feminist sympathies, Rich is not only forced to choose sides (identifying with the feminist girls), but to enter into a critical discourse that, as her use of the term "girl" hints, some might perceive as juvenile. Rich also happens to be that rare film critic who writes for both academic and popular publications, and hence she is especially well-positioned to recognize that for a sophisticated feminist film critic these days to claim a conscious identification with a positive image, she risks being construed as naïve or even ignorant.

I understand her consternation. While critical theory ought not simply to line up with an untheorized, naturalized audience response, I do think it can do better at working with (not against) audience members' powerful investment in positive identity.[11] In a study of *Silence's* reception by a range of newspaper and other periodical reviewers, Janet Staiger found that "women—both straight and lesbian—uniformly defended [Jodie Foster] and the movie as a positive, powerful representation of a female."[12] Staiger's study helps to render visible the second critical division that the movie provoked, which was quieter but nonetheless significant. Consider the distance between the "I love Clarice" response that Staiger found on the part of women writing in the popular press and, for example, academic critic Judith Halberstam's Butler-influenced analysis of the performative character of Buffalo Bill. Halberstam concludes that (what she terms) Buffalo Bill's gender trouble "challenges" the "heterosexist and misogynist constructions of humanness, the naturalness, the interiority of gender"; "he rips gender apart and remakes it as a suit or a costume."[13] I do not contest that this character indeed exemplifies performative subjectivity. But performativity is not in and of itself a subversive challenge.

The Silence of the Lambs channels "negative" performativity into what gay male critics aptly view as little more than the familiar homophobic stereotype of an effeminate psycho-killer. Moreover, it would not be going too far to say that the film associates performative versions of femininity with death. The opposition of realness and performativity as a matter of life and death for women is dramatized when Clarice visits and investigates the bedroom of a past victim, Frederika Bimmel. Most obviously, Buffalo Bill has reduced Frederika to sheer performativity by using her dead and flayed body as part of the female-suit costume he is constructing. But Frederika also has been reduced to the accoutrements of cultural femininity she has left behind: posters of glamorous women, framed family photos, romance novels, china animals, frills and chintz, and, most poignantly, a ballerina jewelry box. In this young woman's version of self storage—the ballerina figurine a perfect emblem of performed femininity—Clarice finds concealed a handful of Polaroids of Frederika, too large (for conventional beauty) in her cotton briefs, assuming a series a culturally scripted poses intended to signify feminine eroticism. In opposition to this painful instance of woman-as-failed-image—woman as purely the static materiality of mise-en-scène—Clarice is defined by her investigative, analytic gaze. That gaze is inscribed by the active and mobile camera as it cuts to a point-of-view shot and pans around the spaces and surfaces of Frederika's bedroom.

The dangerous linkage of feminine performativity and death is precisely what Senator Ruth Martin, the mother of Buffalo Bill's final victim, Catherine Martin, attempts to forestall. In an effort to save her daughter, Senator Martin goes on television to plead with the serial killer for the return of her child. Her strategy is to show family snapshots of Catherine, share anecdotes, and describe what her daughter must be thinking and feeling. Clarice watches the broadcast transfixed, then explains to the friend standing next to her that "if he sees Catherine as a person and not just as an object, it's harder to tear her up." Paradoxically, the senator's powerful performance of Catherine through images and words can, at least potentially, produce a "realness-effect" that may save her life. As this example suggests, identity may at some level always involve a performance that relies on images, props, and narrative, but that does not mean that an individual must be reducible to that performance. Just such a threatening reduction is implicit in Hannibal's menacing compliment to Senator Martin. "Nice suit," he tells her. The comment links him to Buffalo Bill: both are serial killers who reduce the complexly human to its theatrical materials.[14]

We seem to have arrived at an impasse between many viewers' commonplace wish to feel whole, to feel real, and academic criticism's sophisticated, poststructuralist critique of identity undone. How, then, can academic feminist film criticism go about reclaiming such filmic versions of female "realness" as potentially empowering without repeating the mistakes of the past? The 1970s' critical approach that emphasized positive "images of women" and surveyed and criticized negative stereotypes,[15] alas, has some significant blind spots. Early feminist film criticism failed to acknowledge, for instance, that what counts as "positive" or "negative" is highly complex, situation-specific, and not self-evident. Not only are different spectators (of different genders, races, classes, sexualities, ages, and so forth) likely to read images differently, but it is also simplistic to posit one-to-one correlations between images and reality, or to assume one-to-one relations between, say, women viewers and women characters. After all, female spectators might just as well be identifying with images of maleness, or vice versa.[16]

Beginning in the mid-1970s and early 1980s, psychoanalytic theory, especially as reconceptualized by Jacques Lacan, seemed to many feminist film critics to offer a much richer way to think about gender identity and representation. Lacan's work emphasized the place, not of biology, but of language.[17] Butler's work on performativity, which is informed by certain key aspects of Lacan's work (although critical of certain others), also defines "maleness" and "femaleness" as symbolic positions that give meaning to the self, but which can never be taken for granted as fixed and unitary.

Psychoanalytic theory, especially the notion of a destabilizing division between our conscious and unconscious lives, has been especially useful in critical analyses of the horror / slasher / psychothriller genres in which the "normal"—located in identities or social institutions—is violently disrupted by unconscious fears and desires. Starting with critic Robin Wood's groundbreaking 1979 article, "An Introduction to the American Horror Film," a series of psychoanalytically based, politically progressive readings of the genre have focused on and valued the ways in which figures of monstrosity transgress and subvert the norms, boundaries, and normative identity positions of patriarchal culture.[18] Rhona Berenstein continues this tradition in her 1996 *Attack of the Leading Ladies: Gender, Sexuality, and Spectatorship in Classic Horror Cinema*, and she adds a significant focus on the concepts of fantasy and performativity. Berenstein argues that horror cinema invites audiences to "identify against themselves" and enter into "multiple" and "transgressive identifications and desires" that disrupt their "day-to-day" gender and sexual positions.[19] Although Berenstein (unlike Wood) does not claim political progressiveness for the horror genre, she suggests, as does a great deal of critical work in recent years, that filmic representations of gender identity are most challenging to the status quo and hence, by implication, most subversive and liberatory when identity is exposed as performative.

In focusing on the feminist politics of the (more or less) intact identity of the strong female protagonist of the woman's psychothriller, I hope to offer an alternative to the critical emphasis on monstrosity or performativity as subversion.[20] For it seems to me that such criticism has come to rely too readily on the equation: *performative = transgressive = politically progressive.* Given such a formula, it is hard to theorize, value, or even talk about our investments in representations of unified and stable selves (as if these, by definition, were retrograde).

In fact, spectatorial investments in such real and stable selves can be profitably illuminated by returning to psychoanalytic theory. Despite the overwhelming attention received by the destabilizing aspects of Butler's theory, her work explores not only how identity is undone or troubled, but how identity is, over and over again, remade, "constantly marshaled, consolidated, retrenched" as well as "contested, and, on occasion, made to give way."[21] I would stress a similar point about Lacan's theory of subjectivity, that, as he argues, we are defined by a *lack-in-being,* by an incompleteness that attends ceaseless efforts to obtain a meaningful self through identifications with others. Even if such identifications—with film characters or stars, for example—are rooted in the misrecognizing of a pleasingly unified external image as one's self, such are the seeming completions of identity that make some film viewers feel *real.* Cinema offers a resource of gendered images in which we as spectators endlessly try (through fantasy and indentification) to find or complete our selves.[22]

Domestic Makeovers: *The Stepfather* and *Sleeping with the Enemy*

In *The Stepfather* (directed by Joseph Ruben) it is the sixteen-year-old stepdaughter, Stephanie (Jill Schoelen), who discovers that the role-model husband and father whom her widowed mother has married is precisely that: a man performing a scripted role, a constructed model. In their first scene together, Jerry Blake (Terry O'Quinn) comes home with a new puppy for Stephanie, prompting his wife, Susan (Shelley Hack), to declare of her husband of one year, "you're perfect." Stephanie is less convinced; she has been uncomfortable and suspicious from the start, resenting Jerry's efforts to substitute for and *act* like her dead father. Indeed, Stephanie is uneasily aware of the performative dimension of her stepfather's behavior. Talking to her best friend on the phone after a family supper

20 Just like the families on TV: daughter, mother, and lunatic in *The Stepfather* (ITC Productions, 1986).

during which she has confessed to being expelled from school, she remarks: "It's freaky, the way he looked at me like he wanted to erase me off the face of the earth. . . . He has this whole fantasy thing, like we should be like the families on TV and grin and laugh and be having fewer cavities all the time—I swear to god, it's like having Ward Cleaver for a dad." (Ward Cleaver, indeed.)

As with *Pacific Heights* and *The Silence of the Lambs*, the female protagonist of this film is coded as *real* through her successful investigations into male performativity. The relation between the investigating woman and the performative man is illustrated by a pair of sequences that occur roughly an hour into the film. In one sequence, we watch the stepfather manufacturing a new identity for himself in the bathroom of a ferry boat as part of his elaborate preparations to, once again, murder his family, transform his identity, and relocate. Then, in the sequence that immediately follows, we see the stepdaughter searching for clues to Jerry's murderous masquerade at an unoccupied house. The adjacent sequences employ a range of formal elements that help to set up an opposition between these two differently coded subject-positions—investigated (male) vs. investigating (female).

Compared to the preceding "Jerry sequence," for instance, the subsequent "Stephanie sequence" employs relatively longer camera distances, longer takes, and more camera mobility. As Stephanie pulls herself through a window of the locked house, the camera tracks back, thus aligning camera movement with the entering female protagonist. Stephanie constantly looks around as she walks, emphasizing her investigating impulse. We cut to a point-of-view shot as she fixes her look on what she intuitively senses is the exact location of a murder committed by her stepfather. These techniques of cinematography have the effect of producing Stephanie as an active, intact, and autonomous character.

By contrast, the Jerry sequence uses more close-ups and extreme close-ups on Jerry's own face. The scene is structured with faster cuts and less camera movement. These techniques all function to foreground and fragment his bodily surfaces.[23] Perhaps most significant, the camera closes in tight on Jerry as he manipulates his contact lenses and glasses. These devices are coded as props—costume rather than corrective lenses. They grant him no clear vision or perspective. Similarly, the camera is inches from Jerry's face as he applies glue to his upper lip and then presses on a fake mustache. During this process of disguise we regularly cut to the briefcase where he stores his "selves." Throughout the movie, mise-en-scène frequently associates Jerry with suitcases, briefcases, and garment bags, all of them props that allude to his externally packaged identity. In *The Stepfather* it is clearly the performative Jerry who holds the greatest visual interest for the spectator, the character who, like Buffalo Bill, plays dress-up. Visual cues continually emphasize the performative aspect of Jerry's pieced together, constructed image of respectable, middle-aged masculinity, thus reiterating his identity transformation.

In what may appear a paradox, however, the performative psycho-male of *The Stepfather* is obsessively invested in cultural images of seamlessly "real" family life. He believes that the perfect makeover of self and family will finally yield a reality in line with the utterly fixed vision that he carries within himself. As Stephanie's "Ward Cleaver" comment recognizes, the stepfather's murderous performativity is all in the service of his attempts to inhabit a particular scenario that for him fits the measure of a particular preconceived Reality with a capital "R." In a startling image from the film's opening scene, a wall adorned with smiling family photographs is spattered and stained with blood—the picture-perfect wife and children punished for falling short of the ideal image. We watch the stepfather in an upstairs bathroom as he washes the blood from his body, changes his appearance, and prepares to start over with a new, carefully selected, camera-ready family. This male desperately, violently wants to be what he conceives as the real thing—the powerful, patriarchal head of a stable household. (To make even clearer that Jerry's notion of "real" family life is based in outdated television ideals of the patriarchal family—that is, that it derives from a reiterated cultural performance—the stepfather justifies his refusal to let Stephanie go to boarding school, which would break up the family, by intoning with an eerie smile: "Father knows best.") At a telling moment in Jerry's growing "disappointment" with his new family, we see him walking through his new neighborhood, stopping as he spots a little girl running from a happy Mom at the threshold of their home to greet Dad as he arrives home from work. The stepfather looks at this rosy image of family life with a yearning akin to that which Buffalo Bill might bring to a suitable female, whose skin he can imagine himself inhabiting because it has the right look. The visuals and narrative shape of the domestic scene evoke the fantasy of real families in which Jerry desires to find himself.

Just as Clarice finds deeper dimensions of her own "self" through coming to understand exactly why Buffalo Bill is so violently invested in manipulating surfaces, so Stephanie achieves more stability and coherence as a character through learning the truth of what is going on around her. The movie regularly cuts to shots of Stephanie as she reacts to or simply observes peculiarities in her stepfather's behavior, peculiarities that go unnoticed by others. When Jerry organizes a big neighborhood picnic to celebrate his new friends, his new family, and his new life, Stephanie hovers on the outskirts of the gathering, watching his smooth performance of a proud family man. She can barely bring herself to pose in the family photo that Jerry engineers.

We can better appreciate the significant link between female investigation and a woman's coding as a strong and active character in the woman's psychothriller by briefly comparing this group of films with a key precursor, the "woman's gothic"—for example, *Rebecca* (1940), *Suspicion* (1941), *Gaslight* (1944), or *The Two Mrs. Carrolls* (1947)—a subset of 1940s' women's films.[24] The plots of these films revolve around a woman who discovers that she has unwittingly stepped into a dire situation by marrying a man whom she comes to suspect may be a villain. Critics Mary Ann Doane and Diane Waldman[25] have discussed the effect of contingently granting a female protagonist such an active, investigatory look and making her the locus from which the story is told. Both critics conclude that giving a woman character such narrative agency creates a crisis that can only be remedied once the female protagonist's authority is undermined, dismantled, pathologized (even when her suspicions are confirmed to be true): "One can readily trace," writes Doane, "in the women's films of the 1940s, recurrent suggestions of deficiency, inadequacy, and failure. . . ."[26] Or as Waldman comments, "these films struggle with their representation of women," and, ultimately, many of them "invalidate feminine perception, establishing a polarity between masculinity, objectivity and truth on the one hand, and femininity, subjectivity, and false judgment on the other."[27]

By contrast to the 1940s' gothics, in the woman's psychothriller of the late 1980s and 1990s the female protagonist's investigation may be challenged or uneven, but on the whole it is not pathologized or deficient. Indeed, as a rule the opposite trajectory is at work as, from the very beginning, the female protagonist's suspicions are confirmed for the audience and thus her point of view validated and aligned with objectivity, truth, and a grittily realistic understanding of the world. When she sees through the lie of masculine masquerade, that discovery, while shocking and ultimately life-threatening, helps to dislodge her from the conventionally defined storybook role of lucky daughter / wife of some perfect man, a narrative in which she does not, in the end, fit.

How do the two types of subjective "realness" that I have discussed in relation to *The Stepfather* differ from one another? The male antagonist compulsively seeks to insert himself into external scenarios that to him embody a culturally endorsed version of the real thing—"real masculinity"; "real family life"; a "real marriage"; a "real home." By contrast, the female protagonist seeks to claim a personal and interior realness, one engaged with the vicissitudes and imperfections of her own individual history (in Stephanie's case, the loss of her father, her troubles at school, her therapy sessions). What counts for Stephanie is the self inside. By contrast, the stepfather is interested in a surface simulacrum of real family relations. He wants certain roles to be played by picture-perfect actors inhabiting a seamlessly designed set. Indeed, the disjunction between these two sorts of investments, in particular the female protagonist's inability, unwillingness, and ultimately her refusal to occupy her assigned role in the man's patriarchal scenarios of Reality, emphasizes the specificity, individuality, and ultimately the realness, of her true self. In showing us the violent excisions and cruel compressions necessary to sustain the male antagonist's vision of real family life (cut to the measure of 1950s' domestic ideology), the movies ultimately encourage the audience to see the space or disjunction between that scenario and the female protagonist's inner self as precisely the space that testifies to her realness.[28]

Sleeping with the Enemy best illustrates the space between these two sorts of investments in realness. The film opens with a married couple, Laura and Martin Burney (Julia Roberts, Patrick Bergin), summering in their spectacular modernist Cape Cod home—all white walls, black leather, and gleaming reflective surfaces—located right on the ocean. We quickly discover that Martin is a maniacally controlling, psychotically perfectionistic husband who masterminds with a terrifying precision every surface of both the home and Laura's image and behavior—her clothes, sexuality, daily activities, social calendar, housekeeping, dinner menus.

A tremendous source of anxiety for Laura is maintaining the household's perfectly ordered arrangements of gourmet canned items. Every time that Laura opens a kitchen cabinet she fearfully encounters this performative male's external self storage. To disarrange the punctilious mise-en-scène of home or self is to risk being brutally beaten by its "director" and "designer," Martin, whose identity seems to depend on each detail occupying its stage-managed place. "Is everything here as it should be?" he asks with menacing composure upon finding that the bathroom handtowels have become misaligned.

Laura must meticulously inhabit Martin's stage set—home life as theatrical display is underlined by the house's huge picture windows—as well as follow the costuming and stage directions that he indicates for her, all of which contribute to maintaining his image of true conjugal bliss. In the film's second scene, Martin appears behind Laura in a mirror and, while complimenting her outfit, muses "I wouldn't have thought of it." In the following shot, Laura's appearance has been made over in accordance with Martin's "thought": she now wears black instead of white, the hair is down instead of up, and even the earrings have been changed. After producing another perfect dinner one evening, Laura asks if she might work full-time at the local library, whereupon Martin asks if it is possible that she no longer cares about their home. In spite of the very 1980s' visuals of their lavish yuppie lifestyle, Martin is invested in scenarios of "genuine" marriage and "real" home that perhaps existed only in 1950s' pop culture, but his sense of coherent identity—and Laura's life—depend on the perfect performance of these scenarios.

From the start, however, the film provides visual cues that alert us to the fact that Laura's "real" self exists both outside of and beneath these frozen settings and scripts. The camera setups in front of Laura enable us to catch expressions of fear and depression as Martin comes up behind her (a recurrent pattern), such as when he returns with the (predictably selected) red roses and red lingerie after a particularly violent assault. The film's constant attention to her changing facial expressions and furtive glances speaks of the self behind the surfaces that Martin so compulsively manages. Shots of Laura gazing out the windows of their house emphasize the existence of a walled-up self. And her love of books codes the character as possessing a vivid interior life, one for which Martin has no feeling.

Although Laura does not literally investigate her male antagonist—she leaves him to start over in Cedar Falls, Iowa—*Sleeping with the Enemy*'s highly compressed version of "female investigation" can be located in two areas: the film's "pre-plot" backstory in which, after their honeymoon, Laura discovers that she has married a monster (as in the woman's gothic), and in her hard-won education in the techniques and intricacies of performance, learned at the feet (literally) of the master, knowledge that Laura has drawn on to free herself from him and to survive on her own.

In *Sleeping with the Enemy* the assertion of the female protagonist's personal and interior *realness* turns out paradoxically to entail her developing her own access and relation to performative identity. Approximately fifteen minutes into the film, Laura stages her own drowning at sea. After the funeral scene, the sudden emergence of Laura's voice on the soundtrack confirms for the audience that this story is actively being told from her point of view: "'That was the night that I died and someone else was saved. Someone who was afraid of water but learned to swim. Someone who knew there would be one moment when he wouldn't be watching. . . .'" The voice-over technique at once announces the rescue of a female self—its escape from an oppressive patriarchal gaze and narrative—and aligns that rescue with a woman's ability to tell her own story. The effect of Laura's voice-over is reinforced when the image track briefly shows us Laura at her secret swimming lessons, a feature of the story wholly outside Martin's "plotting" of their life together. Although initially dissociated from the "I" of her self—"someone else was saved" she tells us—the process of the female protagonist's journey (metaphorized by her subsequent bus ride) is toward a space in which she can reclaim her *real* identity.

Laura masterminds her own performative "makeover" to effect her escape. She cuts her hair and dons a wig; grabs a readied bag with costume (jeans and sneakers) and cash; flushes her wedding band in the toilet; covers her tracks; and catches a Greyhound bus to Cedar Falls where she renames herself Sara Waters, rents a house, wears simple cotton dresses, and finds a job working in a library. Finally laying claim to her own self storage, Laura not only paints and decorates her house to her own taste, but she happily jumbles the contents of her own kitchen cabinets. In contrast to Martin, Laura has a more flexible and heterogeneous relation to the performance dimensions of *identity*.

Laura's new life includes a romantic relationship with a more benign performative male, a local drama teacher. He takes her to the school auditorium and sits in the control booth working special effects around her; later he helps her play dress-up in the costume department. These sequences show the female protagonist repeating a scene of trauma— herself as visual object of a male-orchestrated makeover—but with a nonthreatening, indeed remarkably bland male (whom she nonetheless approaches with great caution). Here again, Laura is represented as mastering performativity in the form of an option and tool for accessing her own desires.[29] Her access to performativity is proven when, disguised as a man, she goes to visit her mother in a nursing home and crosses paths with Martin, who is there posing as a police detective as he tries to track down his escaped wife. If these two figures, both enacting assumed identities, seem momentarily like duplicate images of performativity—her brown wig and thin brown mustache establish a certain resemblance with Martin (and coincidentally with Jerry, the stepfather)—that mirroring only serves to foreground once again the distinction between a character defined wholly by seamless surfaces (Martin, Jerry) and a complex ("real") subject who encompasses both surface and depth. The point is voiced most explicitly by Laura's blind mother who looks past her daughter's disguise: "Honey, you're going to be fine. Inside you always were. There's nothing that Martin or any man can do or say to take that away: you have your self."

The final "showdown" sequence of *Sleeping with the Enemy* makes clear the differing relations to realness and performativity that the movie assigns to each of the characters. When Martin finally finds Laura's new home and breaks into it, his one goal is to re-create the same scenarios he forced his wife to enact on Cape Cod. He reorders into perfect symmetry her handtowels and canned goods in an unnerving violation of her space, and he announces his return with Berlioz's *Symphonie Fantastique* (his it's-time-for-sex music) on the CD player. For Martin, the proper setting and music, speeches, and gestures are all that it takes to realize the perfect conjugal reunion.

The state of Laura's interiority holds no significance for him. In response to Martin's typically theatrical bid for repossession, "we are one— we will always be one," a comment that would freeze his vision of true love into a rigidly unchanging eternity, Laura agonizes for several painful moments . . . then shoots him. Throughout this closing sequence, as in the movie's opening sequences, the disjunction between Martin's scenarios and Laura's inner feelings testifies to the specificity of her inner self.[30]

Power Tools: *Pacific Heights*

Pacific Heights begins with a woman who has mastered a number of "power tools." Patty and her romantic partner, Drake Goodman (Matthew Modine), share a yuppie dream: they renovate and inhabit a spectacular Victorian home whose rental units they hope will cover the mortgage. An early sequence begins with close-up shots of power tools (an electric nail gun, an electric drill), shots which then reveal that it is Patty who is doing this conventionally male-coded labor.[31] (We also see her working on the plumbing and showing Drake the proper way to paint a wall.) *Pacific Heights* displaces Patty's performativity by focusing on the gorgeous surfaces not of a woman but of a house (this film's object of visual pleasure), depicting a process of domestic repair and beautification in which the female protagonist plays an active role.

When psycho-tenant Carter Hayes moves in with the secret intention of appropriating the property of this naïve young couple, he directs his attacks on the "body" of the house and, more significantly, on Drake. What I find especially interesting about this women's psychothriller is its triangulation of the performativity / realness economy. Carter works his wiles on another *man*. He uses props, prescribed lines, and false identities to manipulate Drake into a self-destructive performance of macho masculinity. Because the performative psycho-male antagonist stages his scenarios around another man, the female protagonist is able to see and critique their effects. In this regard, Patty's expertise with power tools (we see her using them two more times in the movie) signifies her resistance to becoming herself a tool of their male performativity. As a result of Carter's destructive performances on and around the house—he shifts many of its physical surface and internal borders around, just as he remodels Drake's subjectivity—the patriarchal scaffolding that supports such idealized domestic scenarios is rendered visible.

Where Patty is skeptical of Carter, Drake is his dupe. Carter gets Drake to rent him the apartment in the first place by performing an image of patriarchal masculinity that Drake cannot resist—that of the smooth-talking, Porsche-driving, $100-bill-wielding professional male. Once ensconced inside the house, the new tenant brilliantly stages scenarios that lure Drake into the role of manly protector of his family and property. For example, after Patty has a miscarriage, Carter calls the police *before* he heads upstairs with the flowers and scripted lines (about the cruelty of nature) that are calculated to elicit Drake's violent reaction. (As Carter had planned, Drake gets arrested for his macho display.) Again and again, Drake's only recourse seems to be to lapse into an aggressive, protective masculinity that lurks just beneath the ostensibly egalitarian 1990s' new masculinity that he appeared to inhabit when the movie began. Drake utters such classic lines to Patty as: "I'm on top of this, don't worry about it"; "I'm not going to lose you, I'm not going to lose our child, and I'm not going to lose our house"; and, "I got us into this, I'm going to get us out of it—end of discussion."

His made-over masculinity prompts Patty to cry out: "What is happening to you?!" What she has discovered is the conventional maleness inside Drake's self-storage. His interiority turns out to be full of clichéd masculine performances that emerge once Carter pushes the right buttons. This version of maleness is represented as disempowering. By the end of the movie, Drake has been reduced to a helpless invalid lying under a quilt on a couch ingesting a series of flickering television images of performative masculinity (martial arts movies, westerns, music videos—even the home shopping channel features a western theme). Drake's performances of "man of the house" disqualify him from narrative agency and action. As their lawyer says angrily, Drake's been played "like a piano concerto."

As is typical of the woman's psychothriller, out of the association of masculinity with performativity emerges the figure of the empowered female investigator. Confronted with the wreckage of her boyfriend and her house, Patty gets out the power tools and goes to work. Amid the rubble she discovers a clue—a snapshot with "James Danforth" written on the back. It gets her out of the house, literally and figuratively. Tracking down Carter/James to a Marriott hotel in Century City, Patty gains access to his suite (by performing the role of his wife for a chambermaid). A quick set of eyeline-matched cuts aligns the investigating woman with the camera as she spots what is the signature prop of many of the performative male antagonists in the woman's psychothriller genre: a set of briefcases. This is the motif of carry-on luggage in which these male antagonists store their multiple selves. Going through his things, Patty finds the materials that Carter uses to construct his false identities and to manipulate other people—a social directory, newspaper clippings, financial profiles. A pile of passport-sized photos of himself, together with a family album of photos and clippings, further evoke the performative male's exteriorized self-storage. Patty also uncovers Carter's latest self-construction; she finds identity papers (passport, Social Security card, driver's license, credit cards) all in the (now performative) name of "Drake Goodman" and featuring Carter's photograph.

This hotel room scene, in which Patty turns the tables and invades her victimizer's personal space, culminates her movie-long investigation of how the system of money and power actually work. From naïve ignorance about capital (she is clueless about how to apply for a mortgage at the film's opening), Patty becomes a canny manipulator of the system. She concocts and directs her own script. After canceling "Carter's / Drake's" credit cards, closing the checking account, and reporting the travelers checks she finds as stolen, Patty (posing as "Mrs. Goodman") orders an elaborate dinner party for fourteen from room service (filet mignon, Caesar salad, *crème brûlée* with raspberries, champagne), prompting the credit check that lands Carter temporarily in jail (and thereby disrupts the masquerade he has been using to trick the Tippi Hedren character). Patty also steals an envelope of cash and even the mint off his hotel pillow.

In the course of her maneuverings, Patty comes across a family album filled with photographs and newspaper clippings that detail the wealthy and prominent Danforth parents, the accomplishments of their two sons, and the disinheriting of the oldest son, James, from the family estate. This information hints that Carter / James's pathological invasions of other people's property and identities may be connected to the role played by money and power in the formation and dismantling of his own patriarchal identity. This perspective on her male antagonist echoes Patty's discovery of Drake's unexpected psychic investment in owning and controlling property, an investment at the core of his uncertain masculine identity.

In effect, Patty, like the other female protagonists in this genre, has come to occupy the position, and to engage in the project, that Laura Mulvey aligns with the task of feminist film criticism itself: "As a feminist critic I could become a female spectator, searching for clues or signs with which to piece together the greater mythical narrative, beyond the screen and the story, of the social unconscious under patriarchy."[32] The investigating woman of the woman's psychothriller traces out the "clues" and "signs" that will help her expose and survive the oppressions and violences to which she, and other women, find themselves caught. I myself as a female spectator and feminist critic have been reading these women's psychothrillers for the "clues" and "signs" they offer with regard to one version of such a patriarchal "social unconscious" that casts women into rigidly scripted scenarios of gender, marriage, and family. I also have sought to explore my own investment in the figure of the female protagonist who, by way of her investigations into the performative psycho-male, achieves an empowered selfhood that feels real.

At the same time, I am not anxious to blithely champion all filmic instances of "empowered," "real women" characters, nor to dismiss the performative as necessarily retrograde. It is important to examine, critically and strategically, how particular economies of both the "real" and the "performative" are depicted, gendered, and valued or devalued within a given film, if only because, in an economy, gains are always attended by losses, and someone always pays, and sometimes it may not be the one who can best afford it. Butler writes of the "tacit cruelties that sustain coherent identity, cruelties that include self-cruelty as well, the abasement through which coherence is fictively produced and sustained."[33] The woman's psychothriller is nothing if not a cruel genre. Its battles to construct and impose "coherent identity" are a matter of life or death, its violence far from tacit.

Although I am arguing that there can be a feminist advantage to claiming and valuing a filmic construction of female realness, such constructions are never simply affirmative. Untroubled, unexamined identity claims have a long history of being used to exclude and abject "difference." The woman's psychothriller is underwritten by any number of such exclusions, marked or unmarked.

The unmarked ("natural") whiteness of the investigating woman, for example, functions as an unquestioned authentic identity, implying a certain equivalence between nonwhiteness and performativity.[34] In *The Silence of the Lambs* the emergence of Clarice Starling as a "positive image" is directly contingent on the homophobic coding of Buffalo Bill as a gayish "negative image." Moreover, by having the female protagonist eliminate that figure and be herself warmly welcomed into the FBI, the film deflects its critique away from dominant institutions of patriarchal power. Is this simply a case of the good daughter substituting for the good son?[35]

But the ambivalent nature of these texts ought not to negate what is powerful about how structures of identity and identification work in relation to the female protagonist of the woman's psychothriller. The goal of this essay has been—without losing sight of the larger representational economy of gender—to trace out configurations of coherent female identity and the possibilities for a viewer's self-identification. As Judith Mayne observes, "the competing claims of 'identity,'" have been long "associated with some of the most fervent debates of film studies and related fields in the past two decades."[36] These debates persist as critics speculate about the social, psychical, and political implications of how identity is forged, dismantled, and circulated.

The recent critical emphasis has been on the pleasures and challenges of cinema's performative dismantling of identity. My feeling, however, is that many people walk around feeling "dismantled," feeling fragmented and estranged from a coherent self. I have therefore argued for the feminist importance of also laying claim to coherence, laying claim to the stabilizing avenues of identification (with female realness, for example) that cinema can offer.

Notes

1 In addition to the films discussed in this essay, the book-length version of this project includes discussions of *Eye for an Eye* (1996); *Copycat* (1995); *True Crime* (1995); *Candyman 2* (1994) and *Candyman* (1992); *Deceived* (1991); *Love Crimes* (1991); *Defenseless* (1991); *Blue Steel* (1990); and *Positive I.D.* (1988). The woman's psychothriller is related to, but distinct from, such 1980s and 1990s movies as *Fatal Attraction* (1987), *The Hand that Rocks the Cradle* (1991), and *The Temp* (1993), which critic Julianne Pidduck calls the "fatal femme cycle" in her article "The 1990s Hollywood Fatal Femme: (Dis)Figuring Feminism, Family, Irony, Violence," *Cineaction* 38 (1995): 64–72. In contrast to the female protagonists of the woman's psychothriller, these "fatal femmes" are psychotic, destructive antagonists whom the films violently eliminate. Both film cycles, however, share an ambivalent fascination with powerful women and may perhaps be understood in part as symptoms of North American culture's ambivalence regarding feminism's advances over the past twenty-five years or so.

2 Laura Mulvey, "Visual Pleasure and Narrative Cinema," *Screen* 16, no. 3 (Autumn 1975): 6–18; rpt. in Mulvey's *Visual and Other Pleasures* (Bloomington: Indiana University Press, 1981). This essay has been critiqued in a variety of ways, not least by Mulvey herself in "Afterthoughts on 'Visual Pleasure and Narrative Cinema' inspired by *Duel in the Sun*," *Visual and Other Pleasures*, pp. 29–38.

3 See Butler's *Gender Trouble: Feminism and the Subversion of Identity* (New York: Routledge, 1990) and *Bodies That Matter: On the Discursive Limits of Sex* (New York: Routledge, 1993). Although Butler is a philosopher rather than a film theorist, her conceptualization of gender and sexuality as performative has had a major impact on feminist and queer film analysis, enabling critics to locate the fissures and instabilities in what might otherwise look like seamless, essential identities. For discussions and examples of how poststructuralism and psychoanalysis inform film theory, see *Feminism and Film Theory,* ed. Constance Penley (New York: Routledge, 1988); *Film Theory: An Introduction,* ed. Robert Lapsley and Michael Westlake (Manchester: Manchester University Press, 1988); and *New Vocabularies in Film Semiotics,* ed. Robert Stam, Robert Burgoyne, and Sandy Flitterman-Lewis (New York: Routledge, 1992).

4 One example from my teaching: it was the week before spring break, and my Hitchcock and Gender students were discussing *Marnie* in the context of selected readings in the feminist psychoanalytic theory of masquerade. Sophisticated observations were made about Marnie's theatrical production of an elusive subjectivity that Mark Rutland, for all his prowess, cannot entirely master. (The class had come a long way since the "I like / dislike her" approach that had dominated at the start of the term.) With fifteen minutes of class time left, I decided, as a sort of informal overview, to ask my students which of the Hitchcock films that we had viewed seemed most challenging to the conventions of Hollywood gender representation. That question triggered an instant return to emotional connections with "strong" stars and characters: "I liked the two Ingrid Bergman movies the best." "Yeah—she was really strong and memorable." "It was a good strategy to start the course with *Rebecca*—there was nowhere to go but up from that washout wife." "I agree, I couldn't identify with Joan Fontaine at all." "But Charlie in *Shadow of a Doubt* was the best: she really stood up to her psycho uncle." These responses suggested to me that my students continued to be powerfully drawn to characters whom the films explicitly coded as strong, knowledgeable, and active.

5 With his suspense films, his women's gothics, and *Psycho* (catalyst for the slashers of the 1970s and 1980s), Hitchcock's work has been foundational to women's psychothrillers and is constantly alluded to by them in one way or another. See Tania Modleski's *The Women Who Knew Too Much: Hitchcock and Feminist Theory* (New York: Methuen, 1988) for a provocative discussion of how Hollywood's traditional representation of gender is challenged by certain Hitchcock texts.

6 I am not suggesting that camera techniques have inherent meanings, nor am I implying that a character's predominance within the story depends on literal point-of-view shots, or even necessarily on the optical register. However, both because film is a visual medium and because Hollywood has consolidated a distinct narrational style, when the camera's spatial positioning is correlated with a character's positioning, that character is more likely to serve as what Seymour Chatman calls a "filter"—optical, psychological, emotional—of the story and of the other characters. "Filter, Center, Slant, and Interest-Focus," in *Poetics Today* 7 (2): 189–204. For a useful discussion of film and narrative, see "Film-Narratology," in *New Vocabularies in Film Semiotics,* ed. Stam et al., pp. 69–122. Unsurprisingly, women's psychothrillers feature numerous shots of the female protagonist *looking.* For example, in *Deceived* (1991, directed by Damian Harris), just after female protagonist

Adrienne Saunders (Goldie Hawn) discovers that her husband is not the person she thought he was, a 55-shot sequence of female investigation begins, of which roughly half the shots could be subtitled "Adrienne Looking."

7 "Writers on the Lamb," ed. Lisa Kennedy, *Village Voice*, March 5, 1991, pp. 49, 56.

8 Stephen Harvey, "Critics," in "Writers on the Lamb."

9 Quoted in Michael Bronski's "Reel Politic," *Z Magazine*, May 1991, p. 83.

10 An economy, in the psychoanalytic sense, refers to the organization of psychical processes in terms of "the circulation and distribution of an energy . . . that is capable of increase, decrease and equivalence." J. Laplanche and J.-B. Pontilis, "Economic," *The Language of Psychoanalysis*, trans. Donald Nicholson-Smith (New York: W. W. Norton), p. 127. As historians of sexuality have shown, the displacement of various threatening forces associated with femininity onto scapegoated male homosexuality is a strategy that dates back to nineteenth-century English culture: "The supposed characteristics of homosexuality, 'passion, emotional ill-discipline, and sexual looseness,' were those associated with the Fallen Woman. It was the perceived feminine quality of evil attributed to the homosexual psyche which formed the link to Victorian theories of the prostitute and brought about the conception of homosexuality as a form of female sexual pathology." Morris Meyer, "I Dream of Jeannie: Transsexual Striptease as Scientific Display," *Drama Review* 35, no. 1 (Spring 1991): 33.

11 Although cultural studies is often accused of taking "untheorized" audience response at face value, the field has made important headway in exploring the phenomenon of "self-centered" spectatorship. For example, Lawrence Grossberg's work on rock and roll describes fandom in terms of "mattering maps," the mapping of one's intense affective investments across "potential locations for our self-identifications." "Is There a Fan in the House? The Affective Sensibility of Fandom," in *The Adoring Audience: Fan Culture and Popular Media*, ed. Lisa A. Lewis (New York: Routledge, 1992), p. 57. For a critique of cultural studies' use of audience reception, see Judith Mayne, "Paradoxes of Spectatorship," in her *Cinema and Spectatorship* (New York: Routledge, 1993), pp. 77–102.

12 Janet Staiger, "Taboos and Totems: Cultural Meanings of *The Silence of the Lambs*," in *Film Theory Goes to the Movies*, ed. Jim Collins, Hilary Radner, and Ava Preacher Collins (New York: Routledge, 1993), p. 153.

13 Judith Halberstam, "Skinflick: Posthuman Gender in Jonathan Demme's *The Silence of the Lambs*," *Skin Shows: Gothic Horror and the Technology of Monsters* (Durham, N.C.: Duke University Press, 1995), p. 177. Halberstam's essay eloquently makes the case that Buffalo Bill symbolizes a "literal skin disease" that is shared by "all the other characters in the film" (p. 165). Without denying that insight, I wish to point out that *Silence* nonetheless explicitly and unevenly *codes* and *distributes* identity traits.

Silence has generated a great deal of critical commentary, not surprising given its release at a historical moment of huge cultural and critical interest in gender and sexuality. See esp. Elizabeth Young, "*The Silence of the Lambs* and the Flaying of Feminist Theory," *Camera Obscura* 27 (September 1991): 5–35; Diana Fuss, "Monsters of Perversion: Jeffrey Dahmer and *The Silence of the Lambs*," in *Media Spectacles,* ed. Marjorie Garber, Jann Matlock, and Rebecca L. Walkowitz (New York: Routledge, 1993), pp. 181–205; Julie Tharp, "The Transvestite: Gender Horror in *The Silence of the Lambs*," *Journal of Popular Film and Television* 19, no. 3 (Fall 1991): 106–13. My own essay began as a conference talk entitled "The Girls Against the Boys: Feminist Theory, Queer Theory, and Hollywood Cinema," Modern Language Association, New York, December 1992. A significant number of critical responses to *Silence* cite Carol J. Clover's richly provocative study of gender in the slasher genre, *Men, Women, and Chainsaws: Gender in the Modern Horror Film* (Princeton, N.J.: Princeton University Press, 1992). My thinking about the investigating woman is particularly indebted to Clover's discussion of the "Final Girl," that is, the masculine-coded female victim (modeled on *Psycho*'s Lila Crane) who survives the attacks of the monster.

14 Clarice, like Catherine, is also associated with performative elements that help to anchor rather than to undo her character's realness. Hannibal Lecter notes Clarice's anxious efforts at elevated class performance—the good bag, the career ambitions, the laboriously improved dialect. Motivated by her deeply felt but conflicted efforts toward upward mobility, Clarice's under-stated artifice serves mainly to highlight the "truth" (and transparency) of her *real* inner self. One might argue that Clarice's memory flashbacks and "therapy" sessions with Lecter are themselves a kind of performance: an iteration of stories, images, poses, and gestures, all of which produce simply the *effect* of interiority, of a core self. True enough. But that effect is effective enough, so to speak, that Clarice thereby powerfully inhabits the "realness-slot" in this film's economy of competing identities. Moreover, by bringing her own autobiography to sessions with Hannibal, Clarice comes to signify for him a real person whom later he will not be able to kill. With more space, I would argue that Hannibal Lecter functions as a hinge between Buffalo Bill, the voracious monster, and Crawford, the analytic professional, thus partially undoing another opposition: that between rabid misogyny and institutional patriarchy. Hannibal's own performativity (signaled by a blonde wig, dark glasses, and costumey white suit and hat) is explicitly displayed at the end of the movie when his character gets slotted into the vacancy left by the dead Buffalo Bill.

15 An example of this approach is Sharon Smith's "The Image of Women in Film: Some Suggestions for Future Research," *Women and Film* 1 (1972): 13–21, followed by Marjorie Rosen's *Popcorn Venus: Women, Movies and the American Dream* (New York: Coward, McCann and Geoghegan, 1973) and Molly Haskell's *From Reverence to Rape: The Treatment of Women in the Movies* (New York: Holt, Rinehart and Winston, 1974). For an application and critique of the "images-of-women" approach, see Linda Artel and Susan Wengraf's "Positive Images: Screening Women's Films," followed by Diane Waldman's "There's More to a Positive Image Than Meets the Eye," in *Issues in Feminist Film Criticism*, ed. Patricia Erens (Bloomington: Indiana University Press, 1990), pp. 9–18. Noël Carroll asks for a return to the analysis of images of women within a cognitive theory framework in "The Image of Women in Film: A Defense of a Paradigm," *Journal of Aesthetics and Art Criticism* 48, no. 4 (Fall 1990): 349–60. It is worth noting that cultural studies actually shares an important impulse with 1970s feminist film criticism: a valuing of spectators' potentially self-affirming relations with popular culture.

16 I want to stress that, despite the focus of this essay, potentially stabilizing identifications with the female protagonist of the woman's psychothriller are not restricted to women in the audience. Avenues of identification with realness in all forms of popular culture are variable and unpredictable, crossing lines of gender, race, ethnicity, sexuality, age, and class. A recent example: in his aptly titled *Jackie Under My Skin: Interpreting an Icon* (New York: Farrar, Straus and Giroux, 1995), Wayne Koestenbaum (a gay Jewish man) writes eloquently of his powerful, formative identification with Jacqueline Kennedy Onassis. Koestenbaum asserts, for example, that by emulating ordinary "Jackie activities" ("to walk; to put on sunglasses; to swim; to reflect; to sleep; to read," and so on), one can thereby transform "a media icon into a magical lesson in embodiment, your teacher in the art of training your 'I' to feel like an 'I' " (pp. 283–84).

17 Lacan theorized that we are born into a preexistent order of meaning—what he calls the Symbolic of discourse, laws, codes, prohibitions—in which we have no choice but to assume a sexual identity (through a difficult series of unconscious repressions, identifications, and fantasies). For an explanation of Lacanian psychoanalysis and its significance for film theory, see Lapsley and Westlake's *Film Theory: An Introduction* and *New Vocabularies in Film Semiotics*, ed. Stam et al.

18 Wood, *American Nightmares: Essays on the Horror Film* (Toronto: Festival of Festivals, 1979), pp. 7–28. In her influential essay "When the Woman Looks," Linda Williams agrees that the horror film "permits the expression of women's sexual potency and desire" but clarifies that "it does so . . . only to demonstrate how monstrous female desire can be." *Re-Vision: Essays in Feminist Film Criticism*, ed. Mary Ann Doane, Patricia Mellencamp, and Linda Williams (Frederick, Md.: AFI Monograph Series, University Publications of America, 1984), p. 97. More recently, Carol Clover's study of the slasher admires that genre's various forms of "gender transgression," which Clover interprets as a "brazen tack into the psychosexual wilderness" (*Men, Women, and Chainsaws*, pp. 231, 236).

19 *Attack of the Leading Ladies* (New York: Columbia University Press, 1996), p. 58.

20 Without (as I hope is clear) rejecting critical efforts to undo the narrow ideas of normative identity perpetuated by much popular culture and some traditional academic film criticism.

21 Butler, *Bodies That Matter*, p. 105.

22 It is important to keep in mind that a person's experience of realness might also center on identifications with cultural elements explicitly coded as performative. The documentary *Unzipped* (1995), for example, explores how designer Isaac Mizrahi consolidates his identity in relation to the intensely performative images of high fashion (a message echoed by Sylvester's "You Make Me Feel (Mighty Real)" on the soundtrack).

23 Again, though, I am not suggesting that these camera techniques have inherent meanings. The "Adrienne Looking" sequence from *Deceived* (see n. 6 above) includes numerous close-ups of the female protagonist. However, the meaning is different from the close-ups of Jerry's face, for Adrienne, like Stephanie, is not constructed as image, reflecting back on herself, but is, rather, looking into words and images to decipher their secrets and decode their narratives.

24 Hitchcock's *Rebecca* is famous for inaugurating this cycle of women's gothics. Of the movies under discussion in this essay, *The Stepfather* is most explicit in its debt to Hitchcock, alluding to *Psycho* and *Shadow of a Doubt* in a variety of ways noted by Patricia Erens, "The Stepfather," *Film Quarterly* 41 (Winter): 87–88. *Shadow of a Doubt* stands out for its uncompromising valida- tion of female protagonist Charlie's (Teresa Wright's) suspicions about her deadly Uncle Charlie (Joseph Cotten).

25 Mary Ann Doane, *The Desire to Desire: The Woman's Film of the 1940s* (Bloomington: Indiana University Press, 1987). Diane Waldman, "Horror and Domesticity: The Modern Gothic Romance Film of the 1940s," Ph.D. diss., University of Wisconsin, Madison, 1981.

26 Doane, *The Desire to Desire*, p. 5.

27 Waldman, "Horror and Domesticity," p. 138.

28 Judith Butler's elucidation of the potentially dangerous (even murderous) politics of realness describes very well the stepfather's relation to what I am calling "preconceived Reality with a capital R": "The rules that regulate and legitimate realness (shall we call them the symbolic?) constitute the mecha- nism by which certain sanctioned fantasies, sanctioned imaginaries, are insid- iously elevated as the parameters of realness" (*Bodies That Matter*, pp. 130–31). There is no absolute line that differentiates these "insidiously elevated" "sanc- tioned fantasies" of Reality from those experiences of realness by female protagonists that, I am arguing, feminists should learn to revalue. Indeed, to believe that we could definitively separate (what we might call, loosely speak- ing) "bad" destructive scenarios of realness from "good" empowering feelings of having a stable, real self would be ourselves to participate in the rigid absolutes fantasized by the psychotic antagonists of this genre.

29 Alternatively, one might argue that this sequence represents a woman hopelessly reimmersed in male fantasy. Both readings are available. The film certainly does not critique the status quo; moreover, it presents a nostalgic version of traditional America as its solution to female oppression. But what I am stressing is that this is the route (however unlikely) that Laura takes toward reclaiming her own identity. Indeed, this is a fantasy of a small midwestern town so embracing of people's differing realities that a gay drama teacher, misidentified by Martin as Laura's boyfriend, can cry out: "I live with another man—ask anyone."

30 Another way of getting at how Laura's more flexible relation to realness differs from Martin's reality would be to consider the psychological and emotional effects of the psychothriller genre. That women protagonists in this genre are by definition in jeopardy, their selves threatened, makes their feeling real equivalent to holding onto self at an intensely individual, personal level (rather than legislating reality). In this genre, female realness is literally a form of survival.

31 Is Melanie Griffith's mastery over power tools in this woman's psychothriller a retort to *Body Double* (1984, directed by Brian De Palma), in which Griffith plays a porn star and exotic dancer whose classier female counterpart is horrifically murdered with a power drill?

32 Laura Mulvey's entry in "The Spectatrix," the special issue of *Camera Obscura* 20-21 (May-September 1989): 249. In this short piece, Mulvey also characterizes the critical female spectator as "detective, semiotician and analyst" (p. 250).

33 Butler, *Bodies That Matter*, p. 115.

34 In *Pacific Heights*, two African American male characters (a best friend and a prospective tenant) function mainly as props to—and measures of—the subjectivities of the two white protagonists. In addition, the Asian tenants are there primarily to supply "comic" performativity, engaging in such stereotypical displays as excessive bowing and linguistic confusion. In another chapter of this project, I explore the race and performativity economies of *Candyman* (directed by Bernard Rose, 1992), which opposes (but ultimately connects) a white investigating woman—a graduate student in anthropology—to a monster-ized figure of black male performativity.

35 And in her career has Foster herself assumed the role of Hollywood's good daughter? Nonetheless, as in my discussion of the woman's psychothriller, I think that we ought not be too quick to dismiss the feminist significance of identifications with "role model" women who survive cutthroat situations in patriarchal arenas (like Hollywood). In the course of my research I was struck by the "strong woman" emphasis of articles on Foster in woman-targeted magazines: "Jodie Rules: Jodie Foster Lets Loose and Takes Control" on the cover of *Vanity Fair*, May 1994; "Wunderkind" (in a section titled "Breaking New Ground"), *Harper's Bazaar*, November 1991, pp. 124–25; "The Power of Women: Ten Women to Watch," *Working Woman*, November 1991, pp. 87, 92; and "Jodie Foster's 'Best Performance': An Actress Calls Her Own Shots," *Maclean's*, September 1991, pp. 48–49.

36 Mayne, *Cinema and Spectatorship*, p. 101.

Conspiracy Theory and Political Murder in America:

Oliver Stone's *JFK* and the Facts of the Matter

Christopher Sharrett

The numerous, unremittingly hostile attacks by the mainstream media that greeted the release of Oliver Stone's *JFK* (1991) are not unfamiliar to those who have followed the coverage of such "conspiracy theories" of the domestic political murders of the 1960s and other crimes associated with state power. Indeed, it is useful to understand the outrage toward Stone's film within the context of the 1990s, an especially reactionary epoch that readily accepts state doctrine as authorized history.

The Kennedy assassination has, after all, been represented in popular culture hundreds of times in the past quarter century, often in outlandish or degraded ways.[1] Even as Warren Report supporters continue to have their say (often in highly showcased media venues) in order to provide "balance" to the debate over the Kennedy assassination, rarely has an assassination narrative met the kind of vituperative onslaught that attempts to make *JFK* such a marginal discourse about the event, this despite its overwhelming acceptance by the general public. Unlike the 1970s, when a post-Watergate consensus of dissent emerged concerning the cover-ups of the Kennedy and King assassinations, the complacent official culture following the Reagan-Bush era causes a film such as *JFK* to be regarded as a fascinating product of a paranoid malcontent. It appears that Stone's principal sin is his rejection of the official public version of the assassination in favor of one developed from the late New Orleans District Attorney Jim Garrison's "thoroughly discredited" 1967–69 investigation.

In his afterword to Garrison's *On the Trail of the Assassins* (a principal source for Stone), Carl Oglesby remarks that the radical conclusion of

Garrison's argument (the assassination as coup d'etat) is almost impossible for the citizenry to contemplate, even in an age rife with bizarre conspiracy narratives. In its refusal of bogeymen / scapegoats (KGB, Mafia, et al.) and its insistence on the assassination's continuity within the development of the intelligence apparatus and the postwar clandestine state, Garrison's thesis undermines the very notion of constituency-based, representative democracy.

Garrison's investigation was roundly condemned not for legal impropriety, but for its assertions about the legitimacy of the state. Perhaps more important, this investigation (and those of many independent researchers) ultimately forces us into a reassessment of some commonly and blithely held assumptions about the political-economic order. Students of this matter cannot help but intuit John Dewey's assertion that government is but the shadow cast by business, thus assassinations, coups, and other forms of political violence flow from economic assumptions. Garrison's later writing placed the JFK assassination within the context of the CIA support of coups in Guatemala, Iran, Chile, the Congo, and elsewhere; this work, largely unknown to Stone's audience, stands with the most important progressive indictments of the real dynamics of contemporary state power as it serves specific class interests.[2] Stone's adaptation of Garrison's work prompted media commentators to suggest that further conspiracy talk might push a nation already suffering a profound legitimation crisis into catastrophe.[3]

The radical aspect of the Stone / Garrison approach to the assassination is its insistence on the murder's central political moment, something most contemporary JFK historians (Herbert Parmet comes to mind) deny, to the point of suggesting that the assassination has no relationship whatever to Kennedy's life or administration. Not only does JFK refute this position, but its explanation makes us contemplate political assassination in clandestine America (that is, the state apparatus from the dawn of the Cold War through Iran-Contra to the present) rather than various, cabalistic notions of "conspiracy" frequently traded on by the commercial media in a grab-bag, inchoate form.

Representative are the numerous, frequently snide surveys of supposed candidates for conspiracy scenarios imputed to assassination hobbyists. In the shopping list format of these presentations, the KGB, the Mob, renegade CIA agents, Castro, Texas oilmen, Cuban exiles, Freemasons, the military, and the far right enjoy equal time in competing narratives that reduce the case to absurdity and refuse any coherent methodology that places the assassination in a political-economic context. The assassina-

21 Jim Garrison. Photo courtesy of the Office of the District Attorney of Orleans Parish.

tions of the Kennedys and King now inhabit the same landscape as *The X-Files;* people learn to "trust no one" as a climate of paranoia creates a cult of antipolitics furthering an atomized, asocial nation, preventing any reasonable discussion of the relationship of state authority to the economic authority on which its power depends. Political paralysis, the fragmentation of the left, and a growing sense of the ineffectuality of alternative politics occurred in the aftermath of the Warren Commission, as many progressives felt obliged to support Earl Warren and the liberal sectors of state power, since Warren appeared to forestall a new witch hunt by dismissing claims that Oswald was an authentic leftist (although commissioners like Gerald Ford continued to insist that Oswald's Marxism was a central reason for his decision to murder Kennedy).[4]

When the media discuss the CIA, the military, the Mafia, and anticommunist exiles as conspiracy candidates, the assumption under this list is that no commonalities of interest exist among these groups. Critics of Garrison and *JFK* assert that conspiratorialists create an enormous submarine sandwich without rhyme or reason and that their arguments represent the general chaos, illogic, and disagreement within their circles. The central dishonesty of this notion is its failure to recognize that we live in an Orwellian predicament concerning governmental verdicts on the assassination. The executive branch (the Warren Commission) asserted, without adversarial process, Oswald's guilt (and no evidence of conspiracy). The legislative branch (the 1976–79 House Select Committee on Assassinations) concluded otherwise, that conspiracies were "highly probable" in both the John Kennedy and Martin Luther King assassinations. The schizoid nature of official response to these matters over the past thirty years informs Stone's disjunct visual and moral landscape.

In fact, *JFK* rearticulates what most legitimate researchers have long believed about the real politics of the assassination. The murder was part of, in Nixon's famous expression, "The whole Bay of Pigs thing," the crucial episode in a series of clandestine and not-so-clandestine joint CIA / military / Mafia activities aimed at destroying the Cuban Revolution and assassinating its leaders. In his Watergate memoir *The Ends of Power*, H. R. Haldeman makes a rather forthright suggestion that the real worries of Nixon and the CIA centered on possible new revelations about the Kennedy assassination.[5] More significantly, the assassination was the culmination of an intense period of internecine conflict over the most cost-efficient way for the United States to maintain its hegemony in the colonial arena while also recuperating after capitalism's delegitimation during the Depression, a crisis alleviated only by the Keynesian economic formula that made the state a major client of the private sector through the manufacture of weapons.[6]

While Stone's sketch of this terrain is vague at times, enough in his film is drawn from the actual historical moment to raise various hackles in and out of state power. The film's much-maligned reconstructions take liberties with history only in the sense that they compress certain episodes that actually took place over a longer span (for example, the male hustler portrayed by Kevin Bacon is a composite of several informants who helped Garrison understand the relationship of Oswald to Clay Shaw and David Ferrie). Most disconcerting in the media assault on Stone is the implicit notion that *JFK* is a perverse occurrence within film history, and that American popular art is usually reverential toward the truth of historical events. The notion is accurate, with the proviso that the truth usually supported by Hollywood is one comforting to specific concepts of race, gender, and class interest. A variety of films, from *The Birth of a Nation* (1915) to *They Died with Their Boots On* (1942) to *The Green Berets* (1968) to *Mississippi Burning* (1988), have taken far greater liberties with historical evidence than does *JFK*, and they have not only suffered less calumny but even enjoyed approbation from contemporary reviewers who felt that such renditions of American history did no disservice to our collective sense of the Real.

It is unnecessary to make a point-by-point comparison of the film's various assertions about the assassination and their correlation with evidence (the single-bullet theory, the framing of Oswald, etc.), since Stone and coauthor Zachary Sklar in 1992 published a fully annotated screenplay with 340 research notes accomplishing this task.[7] Instead, I will discuss those moments of the film demonstrating why its basic premises ran afoul of dominant ideology, and also why some of those premises fall short of being a comprehensive criticism of the true nature of the political-economic structure.

Perhaps the greatest criticism of *JFK* centered on the meeting between the film's Garrison (Kevin Costner) and a Deep Throat-style informant called X (Donald Sutherland). Critics of the film harped on the idea that Garrison's investigation showed no record of a secret meeting. While it is true that Garrison never made a trip to Washington, D.C. (where the scene takes place), he did in fact have a secret meeting in New York with a CIA defector named Richard Case Nagell, an individual who became useful to Garrison and to twenty years of subsequent research on the operations of the intelligence organizations.[8] This meeting was not, however, the text for the scenes with Garrison and X.

The Washington scene, which provides the film's thesis, is drawn from a variety of sources, including government documents and informed accounts of the Cold War years from 1945 to 1963. The narration by X also draws on the work of L. Fletcher Prouty, a retired Pentagon liaison officer to the CIA whose analysis of the postwar intelligence and foreign policy structure coincides very well with the work of former CIA agents Philip Agee, Victor Marchetti, and John Stockwell, all of whom produced information extremely important to a radical reading of the Cold War.[9] Prouty was involved in 1970s' efforts to reopen the Kennedy investigation— achieved with the House Select Committee on Assassinations—and in educating the remaining fragments of the New Left on the operations of clandestine state power (which ran rampant after the 1947 National Security Act). One of Prouty's thinly veiled assertions, replicated in *JFK*, is that his former boss, Major General Edward Lansdale, was a key player in the Kennedy assassination.[10] The media made no mention whatever that this allegation was given some substantiation by information gathered by two congressional studies, the 1975 Schweiker-Hart Subcommittee of the Senate Intelligence Committee, and the 1976–79 House Select Committee on Assassinations (HSCA). Lansdale, a sometime legendary figure in intelligence circles and the inspiration for the books *The Ugly American* and *The Quiet American,* was the architect of the low-intensity counterinsurgency approach to the policing of the U.S. colonial domain.

22 Major General Edward Lansdale (*second from left*) in Saigon in 1962.

Lansdale achieved an early and spectacular success in the postwar Philippines by co-opting and destroying the leftist Huk guerrillas and installing Ramón Magsaysay, a figure sympathetic to U.S. interests, in an act that made Lansdale a featured player in intelligence actions for more than twenty years.[11] What is especially germane here, however, is Lansdale's abandonment of counterinsurgency strategy (meaning the use of indigenous proxy forces backed by American "special operations") during the Kennedy era in favor of massive troop commitment in Southeast Asia, a change of heart that represented the internal state warfare overtaking Kennedy.[12] None of Lansdale's malevolence is a figment of Oliver Stone's imagination, and in fact Stone could have educated the viewer further by adumbrating Lansdale's crimes in order to widen the narrative's context.[13]

Lansdale, with colleagues Theodore Shackley, William Harvey, E. Howard Hunt, and David Atlee Phillips, was an overseer of Operation Mongoose, the CIA / Pentagon attempt to assassinate Fidel Castro and overthrow the Cuban Revolution; this effort continued even after the Cuban missile crisis and Kennedy's orders to curtail anti-Cuban activities (culminating in Kennedy's attempt to begin back-channel negotiations with Castro through emissaries Jean Daniel and William Attwood).[14] The CIA-funded training centers and safe houses in Florida and New Orleans, where people such as Lee Harvey Oswald and David Ferrie cavorted, as depicted in *JFK,* were part of the Mongoose operation. The existence of

23 David Atlee Phillips. Photo courtesy of the U.S. Congress.

these centers and training camps was hardly a secret; it was documented by photographs published at the time in *Life* magazine. The Mongoose operation included a variety of "buffer" groups protecting the CIA and the military; Mongoose has been well-documented for its employment of Mafia types, Cuban exiles, and American mercenaries of an extreme rightist stripe. Looking at this single operation alone, one can appreciate the disingenuousness of the media for their insistence that "conspiracy buffs" create an improbable scenario by their scrutiny of so many groups, conveniently overlooking the long history of congenial and mutually supportive relations that these groups have afforded each other out of their common ideology.[15] The Iran-Contra episode is only one of the more recent in a long history of operations providing a model of explanation for understanding the commonalities of reactionary interest among various elements of state and private power.

As noted, in covering *JFK* the media were relatively oblivious to the 1979 conclusion of the House Select Committee on Assassinations that a probable conspiracy murdered Kennedy. This compromised body was almost as dishonest as the Warren Commission through its statement of the obvious (as state power was in a mea culpa mode in order to regain legitimacy post-Watergate) without any attempt to focus on the real nature of the coup. The former chief counsel for the committee, G. Robert Blakey, is a proponent of the "Mafia Did It" theory. His is one among many fallback positions that reduces the case to a grotesque aberration carried out

Conspiracy Theory: Oliver Stone's *JFK* **223**

24 David Ferrie (*in helmet at extreme left*) and a teenaged Lee Oswald
(*far right*) at a Civil Air Patrol picnic in 1955. Photo by John Ciravolo.

by archetypal bad guys unrepresentative of authentic American interests
(Blakey, for the record, denounced *JFK* as "leftist fantasy"). For all of its
failings, however, the HSCA provided information that not only gives X
more authority and resonance, but also points to, in the words of British
researcher Anthony Summers, "the heart of the matter."[16]

Staffers for HSCA strongly believed that an Edward Lansdale colleague,
David Atlee Phillips, using the pseudonym "Maurice Bishop," was one of
a number of mentors instructing Lee Harvey Oswald in his provocative
activities. An HSCA witness to this effect was Antonio Veciana, organizer
of the anti-Castro paramilitary group Alpha 66. Veciana claimed Phillips / Bishop (Veciana's CIA case officer) tried to interest him in assist-
ing with the framing of Oswald through the bolstering of Oswald's false
pro-Castro credentials. Veciana declined, largely because the operation
was an inconvenience to him, although he turned informant to the HSCA
because Phillips later betrayed him on a drug-running operation. Although Veciana nervously declined to identify Phillips as Bishop publicly,
chief investigator Gaeton Fonzi believed Phillips and Bishop to be the
same man and thus believed Veciana's account to be accurate.[17] The
wording of the HSCA final report supports, with the committee's typical
caution, its staff's research of the issue.[18] (Phillips in the late 1960s became
chief of Western Hemisphere Operations and assisted the orchestration of
the 1973 coup against Chilean President Salvador Allende.)

Among the film's more nitty-gritty issues that provoked protest-too-
much media outrage are the connections alleged between Lee Harvey
Oswald, Clay Shaw, Guy Banister, and David Ferrie. To suggest that these

connections actually existed, and were of a political rather than merely a personal nature, is to vindicate Garrison entirely and allow a view of Shaw, Banister, and Ferrie that the media and the U.S. Justice Department (of both Johnson and Nixon) refused mightily from the first moments of the New Orleans investigation. To this day, Clay Shaw is in many circles an unjustly maligned, Kafkaesque figure (Shaw's own favorite expression), a benign philanthropist, a victim of a McCarthy-style persecution that had a nasty antigay aspect. (While Shaw and Ferrie were indeed gay, Garrison never allowed mention of their sexuality in the preliminary hearings, at the trial, or in subsequent interviews, often against the advice of staff lawyers and researchers who felt that he should exploit this information.)[19]

A few points need to be made country-simple. First, under Louisiana law at the time, a prosecutor could indict a suspect directly, without preliminary hearing. Because of the controversy surrounding his investigation and indictment of Shaw, Garrison took the unprecedented step of placing his case before both a grand jury and a three-judge review panel, thus far exceeding the requirements of state law. After both the grand jury and the review panel returned a true bill, Garrison was *forced* to proceed with his case against Shaw.[20] Contrary to popular anti-Garrison narratives that were revived with the release of *JFK*, the trial of Shaw was postponed for two years not by Garrison (who was forced to wait and watch his witnesses and evidence disappear), but by federal officials who refused to serve or answer subpoenas (including former CIA Directors Allen Dulles and Richard Helms, and a deputy director, General Charles Cabell), and conservative governors (including Ronald Reagan) who refused to extradite witnesses. As the film suggests, by the time the trial finally took place, Garrison's case had been so picked apart and compromised that he used the trial chiefly to demonstrate the nature of the Kennedy execution based on the evidence developed by his office, in effect throwing in the towel in the prosecution of Shaw. (Only the longer videocassette version of the film known as "the director's cut" refers to some of the more compelling evidence that Garrison had against Shaw specifically. Stone has remarked that some material was deleted from the movie's theatrical version because of time problems with an already three-hour commercial film.)[21]

Among Garrison's grand jury and trial evidence were the testimonies of many witnesses who placed Oswald in the company of Clay Shaw, David Ferrie, and Guy Banister in the summer of 1963, including a group of Congress of Racial Equality (CORE) volunteers who witnessed Shaw, Ferrie, and Oswald's disruption of a voter registration drive in the small town of Clinton, Louisiana, in one of the most bizarre incidents of the JFK assassination story.[22] These same witnesses were deposed by HSCA staff lawyers in 1978 and found credible.[23] More significant, abundant research, including that produced by congressional investigators, proves beyond question that 544 Camp Street, the New Orleans address from which Oswald operated his one-man Fair Play for Cuba Committee (in his role as provocateur), had long been an intelligence safe house supervised by Ferrie and Banister under high-level CIA authority. Research also proves unequivocally that Guy Banister, a picaresque figure in the intelligence community, had a close association with Oswald during the accused assassin's last summer in New Orleans.

To this day, certain proponents of the Warren version of the assassination, particularly Gerald Posner in his much-ballyhooed post-Stone book *Case Closed*, insist that Oswald never even *met* the rightist fanatic David Ferrie, notwithstanding the photographic evidence that shows them together at a Civil Air Patrol picnic in the mid-1950s, when Oswald was a high school student becoming interested simultaneously in Marxism and James Bond.[24] A long documentary and eyewitness record persuasively demonstrates that Oswald was cultivated as one of many throwaway intelligence functionaries in one of the most heinous counterintelligence operations of the Cold War, something *JFK* touches on only briefly in its discussion of Oswald's background.[25]

Although Shaw's jury acquitted him, few commentators have paid close attention to the trial's full outcome and aftermath. At the conclusion of *JFK*, we hear the jury foreman saying that the jurors felt there was a conspiracy to kill Kennedy but were not convinced beyond a reasonable doubt of Shaw's involvement. We learn in the epilogue that late-60s' CIA Director Richard Helms, later admitted under oath to Shaw's affiliation with the Agency.[26] The epilogue fails to mention that two alternate jurors voted for conviction and that the sitting jury felt Shaw lied on a number of issues. Judge Edward Haggerty, who heard the case, felt Shaw lied on *all* substantive issues.[27] In an unprecedented move, the federal courts blocked Garrison's prosecution of Shaw for the obviously perjured testimony that he never knew David Ferrie, a matter about which Garrison had extremely compelling eyewitness testimony.

The jury was unable to convict Shaw for his role in the assassination principally because Garrison failed to adequately demonstrate motive, which for the jury (based on polling done by Mark Lane and others) meant proving Shaw's association with the intelligence community. Since the early 1970s a wealth of information has become available proving

Shaw's long-term CIA affiliation, dating nearly to the Agency's inception. In 1973, former agent Victor Marchetti revealed that at the time of the Garrison investigation, then-Director Helms regularly queried senior officers at morning staff briefings about whether "we are giving them [Clay Shaw and David Ferrie] all the help we can."[28] Present at these meetings were Deputy Director Admiral Rufus Taylor and Deputy Director of Plans Thomas Karamessines. When he asked why the Agency should be interested in a case in the domestic courts, Marchetti was informed that both Shaw and Ferrie had been contract agents, and that further revelations about them could be "very embarrassing to the Agency." Both Helms and William Colby (who succeeded Helms as Agency head) admitted to Congress and in depositions at a lawsuit brought in the late 1970s by E. Howard Hunt, that Shaw indeed was an employee of the CIA's Domestic Contacts Division, but that he was merely an occasional informant.[29]

In fact, research shows that Shaw was far more than an international businessman giving the odd tip to the CIA, nor was he merely the shadowy proctor, a la Monks in *Oliver Twist*, observing the Ferrie / Banister gang of young anticommunist, anti-civil rights provocateurs, which is the main role that the film ascribes to him. Cumulative study, including work done by the Italian and Canadian media, suggests that Shaw worked for U.S. intelligence since his service on the staff of General Charles Thrasher, deputy commander of the Western theater of operations during World War II.[30] There is compelling evidence that Thrasher and Shaw were among the U.S. Army officers and other officials responsible for constructing Operation Paperclip, which created the "rat lines" central to the migration of Nazi military brass, intelligence officials, and scientists, including Reinhard Gehlen, who orchestrated the "Gehlen Org," a powerful arm of Western intelligence within the Eastern bloc during the postwar years; Klaus Barbie, the notorious Butcher of Lyon; and Walter Dornberger and Wernher von Braun, the scientists who pioneered the V-2 "buzz bomb" ballistic missile at Peenemünde (murdering many slave laborers at the Nordhausen concentration camp in the process) and became central to the construction of the National Aeronautics and Space Administration (NASA).[31] The "rat lines" project is chronicled in documentary filmmaker Marcel Ophüls's *Hotel Terminus* (1988), among other sources. During these operations, General Thrasher was simultaneously responsible for the gratuitous murder of ordinary German POWs (mainly old men and boys) while their officers actually became part of the U.S. state apparatus.[32]

In the 1950s, with colleague Theodore Brent, Clay Shaw oversaw con-

25 Clay Shaw

struction of the International Trade Mart in New Orleans, which had myriad intelligence functions, including a role in an obscure CIA operation called QKENCHANT, a strategy for monitoring Latin American traffic in and out of the Port of New Orleans. Documents declassified in 1992 go far beyond Marchetti's initial claims about Shaw's intelligence role as an informant to the Agency's Domestic Contact Division. Shaw filed more than thirty separate reports to a case officer between 1948 (the year after the CIA's inception with the passage of the National Security Act) and 1956; Shaw was regularly debriefed following his various European and Latin American business junkets, during which he cultivated the friendship of the heads of U.S. client regimes, including the Somoza family in Nicaragua.[33] An important Shaw role was the guidance of various overseas dummy corporations, the most notorious of which was Permindex and its spin-off, Centro Mondiale Commerciale (CMC). These organizations, whose boards of directors were composed almost exclusively of French, Italian, Hungarian, and German neo-Nazis and old-line fascists, were implicated by the government of France for support of the attempted military coup against Charles de Gaulle following the French-Algerian war (1954–62).[34] Indeed, Guy Banister's personal attorney, Maurice Gatlin (an official, with Banister, in something called the Anti-Communist League of the Caribbean, one of many intelligence-funded rightist organizations), hand-carried $500,000 to the Paris offices of Permindex in support of Jacques Soustelle, the key figure in the Secret Army Organization (OAS) that plotted the attempts on de Gaulle.[35] (We must be mindful of the specific context. At the time of these events, such figures as McGeorge Bundy and Dean Acheson denounced de Gaulle in Congress for his equivocation on a valuable part of the Western colonial domain.)[36]

Shaw was eventually considered persona non grata; he, Permindex, and CMC were expelled from France, Italy, and Switzerland. They found hospitable lodgings in South Africa before their operations became obsolete and the companies dissolved.[37] Shaw's antics now seem relatively quaint in light of the intelligence community's more recent use of fake corporations and even reasonably legitimate businesses to facilitate its operations. Iran-Contra, the savings and loan catastrophe, and the financial machinations of the BCCI affair all contain instructive examples. The important point finally is that during the trial Shaw adamantly denied *any* CIA involvement. The major media never challenged Shaw, but instead merely posed Garrison's assertions as risible.

There are indeed elements of Tommy Lee Jones's portrayal of Clay Shaw (a performance that significantly helped make Jones a major star) that are troublesome, including an overly effete and sinister demeanor that one would not readily associate with the usually gregarious Shaw. Jones's commanding screen presence, cultivated in a number of tough-guy / psycho roles throughout the 1970s and 1980s (*Jackson County Jail* (1976), *Rolling Thunder* (1978), *The Eyes of Laura Mars* (1978), seemed well-applied to the villain that Stone constructed, and, to be fair, Stone did not misperceive Shaw's villainy. The problem is that the villainy of Stone's Shaw flows too much from his manner rather than from his acts—Garrison's information about Shaw and Centro Mondiale Commerciale is mentioned only fleetingly. The film's representation of Shaw's sexual liaisons, although accurately based on a solid evidentiary record, contradict Garrison's policy of keeping the sexuality of Shaw and other suspects off the table. While Stone is not off-base in suggesting the sexual / emotional associations among Shaw, Ferrie, and Oswald, Stone's critics are also apt in noting the excessive malevolence given to the gay orgy and the homosexual subculture in general.

The Shaw case was significant to an understanding of the Kennedy murder and useful as a narrative focal point of the film, but Stone's handling of it comes close to deflecting attention from a basic Garrison contention, one that Stone certainly embraces as well. Garrison asserted both in court and in his interviews and memoirs that the New Orleans phase of the case was marginal to the politics of the assassination, and that Shaw, Ferrie, Banister, and the others involved in the grooming of Oswald

26 Jim Garrison (Kevin Costner) confronts Clay Shaw (Tommie Lee Jones) in Oliver Stone's *JFK* (Warner Bros., 1991).

were very small fish in the large ocean of Cold War clandestinity. While the Washington interlude and the investigations / narrations by Costner / Garrison and his staff address this concern and flesh out the narrative, there is a weakness to Stone's linking of the Shaw matter to the larger dynamics described within Stone's source materials.

Suffice it to say that *JFK* is well-grounded in fact, even if events are compressed or changed in sequence; one might paraphrase the old saw that the truth, in this case particularly, is far stranger than Stone's dramatic representations. As X / Fletcher Prouty puts it, wondering about whether or not Oswald, Ruby, and Ferrie met together is far less germane than allowing an understanding of the case as a political assassination. The media emphasis on "The CIA Did It," or "The Mafia Did It," or "The Cubans (which ones?) Did It" gives the assassination a murder mystery patina, with the audience invited to immerse itself in endless minutiae, solving it within the dissolved, subjectivist realm of the entertainment industry. ("You be the judge" was a blurb for an edition of the CBS news-magazine *48 Hours* that attempted to respond to the new dissent following the release of Stone's film.)

The media compartmentalization of assorted bad guys in this case constantly obscures the essential fact of the cooperation and mutual support among all sectors of the ruling class and those who reflect and represent its interests. When right-wing billionaires such as Howard Hughes or H. L. Hunt give their own money to anticommunist efforts—both supported the Operation Mongoose attempts on Fidel Castro—we see merely the private sector exerting its influence within the halls of state. The "privatizing" of the Contra war, a key issue of the Iran-Contra criminality, is only a more recent model.[38]

By imputing to Stone's film a hodgepodge, loosely constructed, and baroque conspiracy scenario, the media proceed largely out of their own deep complicity in this issue, either because of a refusal to investigate, or through very manifest, conscious attempts to obscure the truth. JFK touches on media responsibility only tangentially, as when it reconstructs the NBC "White Paper" on Garrison's case that made the investigator, not the crime, the focus of investigation. A central icon in the film referring the viewer to media complicity in covering up the truth is the famous photograph of Lee Harvey Oswald standing in his backyard, wearing a sidearm and holding a rifle and the newspapers of the Communist Party and the Socialist Workers Party. (As researcher Sylvia Meagher once remarked, Oswald's leftism seems to have been extremely catholic.)

Let's put issues of that photo's provenance aside for a moment (the evidence seems clear that it is a montage) and consider the media's use of this image against the accused—who was murdered while in police custody and therefore unable to offer any adversary response. On the cover of its February 24, 1964, issue, Life published the notorious "backyard photo" with the caption "Lee Oswald with the weapons he used to kill President Kennedy and Officer Tippit." Jim Garrison opined that Life might have saved the taxpayer considerable money, since the Warren Commission's conclusion was not issued until that September. Life was not alone in its rush to judgment. A month earlier, the New York Daily News published a piece on the assassination that included the remark, "The only good murderer is a dead murderer, and the only good communist is a dead communist."[39]

Perhaps more telling, the December 7, 1963, memorial issue of Life, released two weeks after the assassination and the autopsy, contained a piece, authored by Paul Mandel, entitled "First Answers to the Nagging Rumors: What Lay Behind Six Crucial Seconds." In this article the author deals with the forensic data of the assassination, attempting to account for the number and trajectory of the shots fired at Kennedy, with the presumption of Oswald's guilt. Life had by this time bought exclusive rights to the Zapruder film, a detailed chronicle of the shooting itself, and Mandel availed himself of this valuable evidence in the process of researching his article. We know this to be the case since Mandel mentions an 8mm motion picture (without naming the filmmaker) of the assassination. Mandel and *Life* note that President Kennedy had a wound of entrance to his throat, a peculiar phenomenon since the Zapruder film shows that at the time of the shooting Kennedy's back is always to Oswald's "sniper's lair," high in the Texas School Book Depository. How does an assassin produce a wound of entry in the front of his victim's thorax when the assassin is at a stationary location behind him?

Life's answer comes from examination of the Zapruder film. Mandel writes that at the time of the shooting, Kennedy is shown "turning his body far around" to wave to someone, thereby exposing his throat to the sixth-floor window "just before he clutches it." Since the Zapruder film was not available to the public for more than a decade after the assassination, it was difficult to criticize *Life*'s perception of this photographic representation.[40] With the Zapruder film now in fairly common circulation, thanks to Jim Garrison and researchers such as Richard E. Sprague and Robert Groden, cursory examination of its frames shows that no such turn took place and that Kennedy was always facing to the front throughout the shooting sequence. Similarly, we know now that Kennedy was rocketed backward at terrific velocity when his head was struck by the fatal shot, not forward as suggested by *Life* and by CBS newsman Dan Rather, the first journalist to see the Zapruder film and comment on it in the electronic media.[41] With the publication of the Warren Report, entrance wounds were transformed into exit wounds; the new official version was carried with gushing praise by all the mass media without their bothering to reconcile the extraordinary contradictions of this new narrative with preceding versions of the murder. Knowledge of the perversity of the official channels of discourse concerning this matter makes the viewer of Stone's film feel that its critique of the media is reserved rather than excessive.

It is probably not especially relevant to an understanding of the dynamics of this case (and to why its rendering by Stone was so severely traduced) to know each instance of media chicanery or incompetence relative to their coverage of the assassination data. That the media have no adversarial relationship to the dominant order has been amply and convincingly argued by such scholars as Noam Chomsky, Michael Parenti, and Ben Bagdikian.[42] The media's complicity in protecting state crime and their function as an ancillary arm of state power, especially with regard to the JFK assassination, has been well-demonstrated by the research of Jerry Policoff, Richard E. Sprague, and others.[43] It is instructive to know that media moguls Henry Luce and Clare Boothe Luce walked out on a Kennedy luncheon, irked at Kennedy's vacillation on Cuba policy (to which the Luce family donated its own money).[44] It is also instructive that Clare Boothe maintained regular association with prominent CIA officials and was an executive on the board of the Organization for Former Intelligence Officers. At the time of the congressional study of the assassination, Clare Boothe Luce provided investigators with bogus leads almost certainly designed to discredit the probe (which nearly happened shortly after the committee's inception).[45]

In a 1977 issue of *Rolling Stone,* Carl Bernstein published a long article on the CIA's long-standing relationship with virtually all the major media, whose arms were hardly twisted in becoming servants of the state and the private sector interests it represents.[46] The CBS chief executive, William Paley, himself an OSS veteran (the World War II precursor to the CIA), turned his network into, in Bernstein's words, "the CIA's most valuable asset." Similar accolades were afforded the Sulzbergers at the *New York Times,* bellwether syndicated political columnists Stewart and Joseph Alsop, Hedley Donovan at *Time* (who with McGeorge Bundy and Zbigniew Brzezinski participated in a propaganda project funded by the CIA), Philip Graham at the *Washington Post,* David Sarnoff at NBC, and of course the Luce family; in short, virtually *all* the major media have complex relationships with the state that severely compromise their ability to report fairly and comprehensively.

A two-part 1977 *New York Times* article not only verified Bernstein's piece but did it one better by including a sidebar on the CIA's use of the media to discredit critics of the Warren Report.[47] Such audacity must be associated not just with the news industry's desire to relegitimate both itself and state authority in the wake of profound challenges, but with a need to engage in a strategy of "inoculation" (in Roland Barthes's term), wherein the critical faculties of the public are acknowledged and a great deal of evil is admitted in order to conceal the fundamental, systemic evil that is the substance of contemporary democratic society under standing economic relations.

No student of the Kennedy murder, including Oliver Stone, has argued that all media magnates, CIA agents, exile leaders, oil millionaires, or Mafia dons fired rifles in Dealey Plaza or knew who loaded them. We understand, however, particularly in light of the half-baked Watergate and Iran-Contra investigations, why the cover-up of such a crime is as important to an understanding of the current order of things as the crime itself. A genuine knowledge of the assassination, of the gangsterism and contempt for a social contract that has ruled state power throughout this century, upsets the apple cart of the standing political economy and all ideological assumptions supporting it. Stone's radicalism is suggested in the montage that opens the film, with the deluge of images that place the assassination in a Cold War context. The speech by X has a similar function, describing as it does the economic relationship between state power and the private sector in the years leading up to the full-fledged Vietnam invasion. There are difficulties here, however, related to the limits of montage.

While Stone is more akin to MTV than to Eisenstein, his use of the montage developed by television advertising turns his strategy on its ear by undermining any stable notion of truth rather than offering merely a disjointed visual universe whose appeal is emotional. Stone has gone so far as to state that the film is "more philosophical than political," a work intent on shooting "splinters into the brain" as it reminds the spectator of the role of representation and narrative in history.[48] Stone's use of montage seems modernist as it assaults linear reasoning, postmodernist as it reminds us that in the media age we must reflect on a "montage consciousness" that has overtaken us. It seduces us with effect, even as we are reminded constantly of the role of representation—in particular the easy juxtaposition of images and ideas on film and TV—in constructing consciousness and determining concepts of history.

In JFK, Stone's handling of montage is often remarkably effective in reconstructing certain moments of the assassination scenario. The montage sequences—several of which intercut found footage (mostly from TV) with original material shot by Stone in a variety of formats (8mm, 16mm, video)—underscore his sense of the assassination as a turning point for history in that an understanding of the event depends so heavily on media representation. The downside of Stone's dependence on montage is that it comes perilously close to the postmodernist conceit of presenting all experience as Nietzschean "perspectival illusion." The style also has limits as a way of educating the viewer about the political and economic particulars of the given historical moment.

Although the opening montage makes reference to corporate displeasure with Fidel Castro and the rising tide of Third World revolt, Martin Sheen's voice-over and the quick inserts of banana shipments and soft drink logos are not adequate ways to describe the roles of the United Fruit Company and Pepsi-Cola in the colonization of large parts of Latin America. The hurricane of images, a storm at whose center stands Kennedy, tends, in Ralph Schoenman's words, to replace the lone assassin narrative with that of the lone conspiracy, as much as Stone contextualizes the events of his film.[49] It is extremely easy to create a narrative about official murder that treats this type of crime as aberrant and scandalous, outside the normal, healthy workings of our democracy (a basic assumption of most conspiracy narratives), and *JFK* sidesteps these traps at times by the skin of its teeth. Stone's responses to some of his more vitriolic critics on the left suggests that he indeed subscribes to an essentially liberal, libertarian vision, albeit one that is still being refined (tending, haltingly, toward radical directions).[50] Although Stone has stated that "JFK" serves as a kind of free-floating signifier suggesting an absent presence (indeed, we can ask where the "real" Kennedy is, since his role in the narrative is marginal, the film focusing on his "erasure" not just as a living human being but as a significant force in the narrative), the film uses Kennedy to express a political vision that ultimately casts the assassination in a rather simplistic framework.

In *Born on the Fourth of July* (1989) and *Nixon* (1995), Stone places the male subject very much at the center of the narrative, but he shows him to be the consequence of a number of social, political, and cultural assumptions (and, as a result, the locus of a critique of those assumptions). In *JFK*, the hero stands outside such assumptions. Kennedy, although absent, is portrayed as a courageous visionary challenging the status quo of postwar ruling-class demands. At no point is he associated with the concerns of the class that produced him.

Jim Garrison, linked in various ways to Kennedy, is similarly drawn, with Stone depending a great deal on the conventions of Frank Capra's films for the characterization of a single, inspired man against the system. Although this is certainly Kevin Costner's most energized performance, his flat, monotone adaptation of James Stewart does little service to Garrison, an almost folkloric figure at six feet six inches, with a booming, mellifluous voice and a frontier marshal's tendency to take on all comers single-handedly. At certain points (the staff firings, for example), Costner captures some of the real Garrison's tilt toward arrogance and bravura gestures. But Costner's nice-guy demeanor is just not a good correlate for the real Garrison's remarkable self-deprecating humor and erudition. More important, the white-knight approach to the character misses the radicalization process that transformed Garrison's view of this case. If Stone wanted his Garrison to function as a symbol of all those who bravely undertook research into covert operations at home and abroad, as he so states in interviews,[51] the film might have done more with the role of the assassination in Garrison's political awakening.

27 Jim Garrison, Capra-esque hero, fighting the system in *JFK* (Warner Bros., 1991).

The limitations of Stone's character construction are indicative of the limits of his moralistic political vision. For Stone, the assassination of JFK is finally an epic struggle of good vs. evil, a clash of hawks and doves. Despite the eloquent courtroom summation by Costner / Garrison detailing the political system's servitude to the enforcement apparatus and "their hardware manufacturers," the film offers inadequate analysis of its compelling preamble, Eisenhower's extraordinary farewell address that cautions against the growing "military-industrial complex." By proceeding with the assumption that the state's warmakers are at odds with the political system, Stone separates the state's warmaking ability from the precise context of the assumptions of postwar U.S. capitalism. He creates a Manichaean struggle that avoids a view of the assassination as internecine warfare within state power, a standard occurrence within history, and substitutes a battle of a superior moral vision against an arcane eruption within a previously benign government.

This approach is suggested in the opening montage, as Stone shows images of rural and suburban America being overtaken by the arms race and "duck and cover" nuclear anxieties, images that in another framework would look like parody. (Some of the shots were used in the satiric documentary *The Atomic Café,* 1982.) But there is little indication that Stone means any irony, any criticism of the suburbia spawned by the postwar boom. This society was, after all, accepted uncritically as its luxuries provided compensation for the sacrifices of the Depression and World War II. Stone's establishing sequence, while hurling us into the maelstrom of recent history, offers the problematical liberalism off-putting to the progressive community.

Eisenhower's concerns about a military-industrial complex were certainly not born out of a conversion to pacifism, nor of any particular moralism related to a desire to end the Cold War or to educate the American public on the realities of the postwar condition. The Eisenhower address is indeed a landmark document. But in *JFK* the context becomes lost. Stone makes compelling use of Eisenhower as a portentous chorus setting the stage for the drama to follow. In his farewell to public life, Eisenhower showed a deep concern shared by many within sectors of U.S. finance capital about the consequences of linking the state to the private sector with the classic Keynesian pump-priming method of turning the state into a ready-made consumer of munitions manufacture. The deficit spending built into this policy, which significantly weakened the dollar in relation to other Western currencies by the late 1950s, began to set off major internal debate against the backdrop of the apparently quiescent period of recovery and U.S. military and economic hegemony.[52]

The Marshall Plan, conceived as a bulwark against socialism, was also designed to rebuild the devastated economies of the Western European nations, the friendly rivals of the United States, allowing them to absorb U.S. manufacture. The downside of this strategy was manifest in Europe's inclination to rebuild its industrial base and a diverse GNP unburdened by such a heavy commitment to military production. Europe and Japan emerged as serious competitors (for territories, markets) as inter-imperial rivalry was reinstituted. This battle became intense in the post-1960s period with the collapse of the Vietnam invasion and with the "Nixon shocks" to the economy ending the U.S. role as international banker established by the Bretton Woods plan of 1944. The struggle between protectionism and open markets first brought Trilateralism and the attempt to build a coalition among the capitalist friendly rivals, then the "free trade" ideology of global corporatization that has dominated the 1980s and 1990s,[53] although not without dispute within the Reagan, Bush, and Clinton regimes. The point is that throughout the postwar epoch internecine conflict within the capitalist states has been a constant. The Kennedy assassination is merely its most dramatic and instructive representation.

At the start of the Kennedy period, the concern for the economy's dependence on weapons production was profound within certain Keynesian circles, even as Kennedy, entrenched in Eastern banking capital, ran for office on a false "missile gap" platform designed to fan more Cold War flames and justify a military budget many times more than that of the Eisenhower years. The space race, a centerpiece of the Kennedy-Johnson years, was always a military boondoggle. Kennedy's wariness of the "military-industrial complex" eventually showed in his approach to the policing of the vast colonial domain that the United States picked up from Europe and Japan in the postwar years. The Central Intelligence Agency was conceived to effectuate what has been termed "counterrevolution on the cheap,"[54] with the United States depending on indigenous forces of rightist regimes throughout the Third World. Backed by U.S. intelligence support, the policy was commonly called counterinsurgency.

David Atlee Phillips, who, as mentioned, figures prominently in the assassination, noted that Eisenhower was pleased that the CIA could carry out coups in Iran and Guatemala "almost without effort."[55] Kennedy, who counted among his favorite authors the pop culture spy novelist Ian Fleming, and who inaugurated the Green Berets, recognized the cost-efficiency of counterinsurgency, the real basis of his reluctance to make direct, overwhelming military incursions into Cuba and Vietnam.[56] Many sectors of U.S. capital (oil, aerospace, munitions) that boomed with the implementation of the Keynesian formula argued vehemently against such programs as the Alliance for Progress (which called for strong, indigenous bourgeois regimes in Latin America that could maintain a democratic patina and do their own policing) and demanded "frank" rather than "surreptitious" involvement in patrolling the colonial domain.[57] The Kennedy approach, like that of Eisenhower before him, was insightful in its recognition before the fact of Che Guevara's call for "two, three, many Vietnams" that the United States could be bankrupted by overtaxing its policing resources, thus delegitimating the state before its people, provoking a rise in U.S. domestic resistance.

The evidence is clear that by the time of the Kennedy presidency, virtually all sectors of U.S. capital were moving toward a policy of direct military incursion, something demonstrated by the Bay of Pigs. In *JFK*, X correctly notes that "Dulles and the CIA wanted Kennedy to invade, but Kennedy wouldn't invade." The Bay of Pigs, planned during the Eisenhower tenure, was ostensibly rather unorthodox counterinsurgency, but counterinsurgency nonetheless, with a small army of Cuban exiles backed by the U.S. landing on the beaches of Cuba with the intention of acting as a flashpoint for an internal revolt against Castro. This plan, as several scholars have noted, was designed to fail without direct and massive U.S. military support, which Kennedy, feeling misled by the CIA, was unwilling to commit.[58] The call by people such as Edward Lansdale for troop commitments in Southeast Asia is another indicator of the serious disharmony within state power.[59] That counterinsurgency should be abandoned by the very people who created it as a low-profile means of patrolling U.S. economic interests indicates something of the panic provoking the internecine conflicts of the times.

It indeed seems evident that in the wake of the Cuban missile crisis of 1962, a shockingly gratuitous and dangerous operation on the part of the Kennedy brothers, JFK had a change of heart about the prosecution of the Cold War. His speech at American University, which actually asked of us that "we reexamine our own attitudes" toward the Evil Other and consider that both the United States and the Soviet Union "breathe the same air," and that both "cherish our children's future," are words that would be unthinkable to almost all politicians even after the Cold War. Our leaders, perfect representatives of ourselves, are not noted for introspection and the ability to reexamine strategies, much less ideological and economic assumptions. As Fidel Castro noted, Kennedy seemed to be a man capable of changing his mind, of reexamining policy. Such a tendency toward even a small degree of self-doubt and reflection cannot recommend a state functionary to dominant economic interests.[60]

While *JFK*'s radical vision could be further developed, it at least makes such a vision available to the viewer, and it provokes a reassessment of the assassinations of the 1960s and their context. Particularly in his interviews, where he draws parallels between the assassination and subsequent events such as Iran-Contra, Stone manifests the ambitions of sectors of the New Left that saw the Kennedy assassination as a means of mobilizing an issue-oriented population into a radical critique of the American political economy. When the Vietnam War and the Watergate crisis ended, so did much public concern for issues of social and economic justice.[61] Many hoped that the assassinations, involved as they are in basic assumptions of foreign policy and the development of the clandestine state, would provoke serious, continued investigation and resistance resulting in political change. By suggesting the apocalyptic aspect of the JFK, RFK, and King assassinations, *JFK* does not adequately demystify conspiracy by showing that such crimes are perfectly contiguous with, rather than aberrant to, standing assumptions. These crimes need to be associated with the mundane reality of the worthlessness of the two big, propertied political parties. The complacency about the assassinations of the 1960s on the part of both parties is of a piece with both parties participating in the neoconservative economic game plan of the Reagan-Bush-Clinton epoch, containing the

28 Jim Garrison (Kevin Costner) presents the case for conspiracy (Warner Bros., 1991).

assault on the public sector and the "downsizing" and "re-engineering" that creates enormous corporate profits while working people and the poor face a future without prospects. This is an outlook embraced by the center, the right, and the Democratic "neoliberals."

JFK appears to resurrect a populist plea associated with Depression-era cinema, as well as a belief in the adversarial role of art itself toward dominant ideology, a fairly extraordinary identity for a mainstream Hollywood film of the 1990s. It is disturbing, although hardly ironic in the climate of postmodernity, that the Kennedy assassination is retained in the realm of the spectacle. As if to admit to the collapse of (in Fredric Jameson's words)[62] civic idealism and the death of politics itself after the public execution of so many liberal or progressive leaders, the film also seems to acknowledge that, in the current moment, public outrage is discovered and retained in the channels of private fantasy.

The debate about *JFK* continues as its video and laser disc versions circulate. But it is a debate within a thoroughly atomized, dissolved, politically ignorant society. We get myths and countermyths, not political activity. But *JFK*, in igniting a remarkable furor, enjoys a position shared by few works of art in recent times, and as such it poses a serious challenge. To radical academics who examine "sites of resistance" within the consciousness industry, *JFK* points to the error of replacing the political with the cultural, chiefly through the urgency and passion of its message even as it trades in a romanticized liberal vision. This film's implicit notion of the collapse of a social contract tends to obviate discussion of "contradiction" (basic in any event to all experience and relevant to our society only as people are prepared to address contradiction politically) in mainstream, mass-produced art. Close analysis of a film raising the kind of fundamental questions about power relations that one finds in *JFK* cannot substitute for the further education and mobilization of our people and the construction of truly alternative progressive politics.

Notes

1 A survey of JFK assassination imagery is in Art Simon, *Dangerous Knowledge: The JFK Assassination in Art and Film* (Philadelphia: Temple University Press, 1996).

2 Jim Garrison, "The Murder Talents of the CIA," rpt. *The JFK Reader* (Santa Barbara, Calif.: Prevailing Winds Research, 1992).

3 See Tom Wicker, "Does *JFK* Conspire Against Reason?" *New York Times*, December 15, 1991, sec. 2, p. 1.

4 Sylvia Meagher discusses the shabby response to the Warren Commission from some prominent sectors of the left in her magisterial *Accessories After the Fact: The Warren Commission, the Authorities, and the Report* (New York: Bobbs-Merrill, 1967), pp. 462–63. A fuller discussion of the relationship of the American left to the Kennedy assassination is E. Martin Schotz, *History Will Not Absolve Us: Orwellian Control, Public Denial, and the Murder of President Kennedy* (Brookline, Mass.: Kurtz, Ulmer and DeLucia, 1996). Noam Chomsky discusses the antipolitical thinking fostered by current conspiracy theory in *Class Warfare* (Monroe, Maine: Common Courage Press, 1996), pp. 83–84, 111, 118–20. Chomsky is a leading figure of the contemporary left dismissive of the political utility of the ongoing debate concerning the domestic assassinations of the 1960s. He conflates this debate, very wrongheadedly, with the disaffected paranoid style analyzed by Richard Hofstadter and others.

5 H. R. Haldeman, *The Ends of Power* (New York: Dell Books, 1978), pp. 68–69.

6 John Maynard Keynes is, of course, this century's central theorist of capitalism. His idea of linking the state to the private sector was one among many formulations crucial to capitalism's postwar recovery within the United States and much of the West. The books on the subject are plentiful, including Kees van der Pijl, *The Making of an Atlantic Ruling Class* (London: Verso, 1984), pp. 16–18, 25, 81, and passim; Marty Jezer, *The Dark Ages: Life in the United States, 1945–1960* (Boston: South End Press, 1982), p. 122 and passim. A detailed discussion of the Keynesian public subsidy of the private sector, particularly as related to military production, is Noam Chomsky, *World Orders Old and New* (New York: Columbia University Press, 1994).

7 Oliver Stone and Zachary Sklar, *JFK: The Book of the Film* (New York: Applause Books, 1992).

8 Garrison discusses Nagell in *On the Trail of the Assassins* (New York: Sheridan Square Press, 1988), pp. 212–16 and passim. A fuller account of Nagell is Dick Russell, *The Man Who Knew Too Much* (New York: Carroll and Graf, 1992).

9 L. Fletcher Prouty, *The Secret Team: The CIA and Its Allies in Control of the World* (New York: Prentice-Hall, 1973). Although written from a politically conservative perspective, this book is an important insider account of the workings of the postwar clandestine apparatus. It is complemented by the extremely radical Philip Agee, *Inside the Company: CIA Diary* (New York: Stonehill, 1977). Victor Marchetti's and John D. Marks's, *The CIA and the Cult of Intelligence* (New York: Alfred Knopf, 1974), a book censored by the CIA under court order, established Marchetti as an important "Deep Throat" on clandestinity, his work finding its way into *JFK*. See also John Stockwell, *The Praetorian Guard: The U.S. Role in the New World Order* (Boston: South End Press, 1991). Stockwell, a former CIA agent in Africa, has been an important source on the politics of official murder. An important book informing the X / Garrison dialogue is William Blum, *Killing Hope: U.S. Military and CIA Interventions Since World War II* (Monroe, Maine: Common Courage Press, 1995), originally titled *The CIA: A Forgotten History* (London: Zed Books, 1986).

10 Prouty discusses Lansdale at some length in *JFK: The CIA, Vietnam, and the Plot to Assassinate John F. Kennedy* (New York: Citadel Press, 1992), pp. 33, 39, and passim.

11 Ibid., pp. 33, 35–36. The definitive book on Lansdale's exploits is Cecil Currey, *Edward Lansdale* (Boston: Houghton Mifflin, 1988).

12 John Newman, *JFK and Vietnam: Deception, Intrigue, and the Struggle for Power* (New York: Warner Books, 1992), pp. 40–41.

13 Stone has stated that he did not want to name people like Edward Lansdale outright since he wanted to focus on the essential forces involved. Gary Crowdus, "Clarifying the Conspiracy: An Interview with Oliver Stone," *Cineaste* 19, no. 1 (1992): 26–27. A careful reading of the film suggests that he accepts Lansdale's culpability, perhaps because the notion is shared by L. Fletcher Prouty.

14 A discussion of the Mongoose program and Kennedy's changing Cuba strategy is in Warren Hinckle and William Turner, *Deadly Secrets: The CIA-Mafia War Against Castro and the Assassination of JFK* (New York: Thunder's Mouth Press, 1992). This book was originally published in 1981 by Harper and Row under the title *The Fish Is Red*. In the early 1990s the Cuban government released its own study of the attempts on Fidel Castro and the assassination of Kennedy. See Claudia Furiati, *ZR Rifle: Cuba Opens Secret Files* (Melbourne, Australia: Ocean Books, 1994).

15 A fascinating sketch of these relations based on research by the Christic Institute is the graphic novel by Alan Moore and Bill Sienkiewicz, *Brought to Light* (Forestville, Calif.: Eclipse Books, 1989).

16 Anthony Summers, *Conspiracy* (New York: McGraw-Hill, 1980), p. 262. The conclusions of the HSCA are in *Report of the Select Committee on Assassinations* (Washington, D.C.: U.S. Government Printing Office, 1979). This document was issued simultaneously by Bantam Books under the title *The Final Assassinations Report*.

17 The Veciana-Phillips / Bishop affair is masterfully documented in Gaeton Fonzi, *The Last Investigation* (New York: Thunder's Mouth Press, 1993).

18 See *Final Assassinations Report*, p. 162.

19 Garrison's adamant refusal to exploit the sexuality of the individuals he investigated is discussed by Zachary Sklar in an interview with Gary Crowdus, "Getting the Facts Straight," *Cineaste* 19, no. 1 (1992): 31.

20 See Carl Oglesby, "Who Killed JFK?: The Media Whitewash," *Lies of Our Times,* September 1991, pp. 3–5.

21 Crowdus, "Clarifying the Conspiracy," pp. 25–28.

22 The Clinton, Louisiana, episode is recounted in Garrison, *On the Trail of the Assassins,* pp. 122–25.

23 *Final Assassinations Report,* p. 170. While the HSCA clearly recognized Oswald's associations, the committee refused to pursue the implications of this material.

24 A photograph of Oswald with David Ferrie at a Civil Air Patrol picnic in 1955 is reproduced in several formats in Robert J. Groden, *The Search for Lee Harvey Oswald: A Comprehensive Photographic Record* (New York: Penguin Studio Books, 1995), pp. 19–20, 228.

25 An excellent overview of Oswald's intelligence affiliations is Philip H. Melanson, *Spy Saga: Lee Harvey Oswald and U.S. Intelligence* (New York: Praeger, 1990). The fact that U.S. intelligence ran a false defector program at the time of the Oswald affair is established by the existence of *four* ex-military men who gave up their citizenship to live in the Soviet Union, only to have a sudden change of heart and return without incident. One of these men, Robert Webster, actually bore a close resemblance to Oswald. See *Warren Commission Hearings,* vol. 18, p. 115. Peter Dale Scott discusses Webster and the false defector program in "The Dallas Conspiracy" (unpublished manuscript, 1970), sec. 2, p. 12.

26 Helms's admission is discussed in Mark Lane, *Plausible Denial* (New York: Thunder's Mouth Press, 1991), pp. 223–24.

27 Haggerty commented to documentary filmmakers Barbara Kopple and

Danny Schecter that Shaw "put a good con job on the jury." The remark is contained in the film *Beyond "JFK": The Question of Conspiracy* (Warner Home Video, 1992).

28 Marchetti's statement was printed in the *Boston Phoenix, True,* and other publications and has since been reprinted many times. It first appeared in book form in the introduction to the paperback reissue of Mark Lane's *Rush to Judgment* (New York: Dell Books, 1975), pp. xxvi–xxvii. See also Lane, *Plausible Denial*. Marchetti discusses Helms's remark in the video documentary *The JFK Conspiracy* (BMG Video, 1992).

29 Shaw denied any CIA affiliation in an interview with former *Saturday Evening Post* writer James Phelan, published in the British edition of *Penthouse,* September 1970, pp. 26–30.

30 Shaw's record is discussed in James DiEugenio, *Destiny Betrayed: JFK, Cuba, and the Garrison Case* (New York: Sheridan Square Press, 1992), pp. 215–25. See also Henry Hurt, *Reasonable Doubt: An Investigation into the Assassination of President Kennedy* (New York: Holt, Rinehart and Winston, 1985), pp. 265–83. A prescient early discussion of Shaw is in Harold Weisberg, *Oswald in New Orleans* (New York: Canyon Books, 1967), pp. 207–51.

31 The definitive study of U.S. protection of Nazis is Christopher Simpson, *Blowback* (New York: Weidenfeld and Nicolson, 1988).

32 James Bacque, *Other Losses* (New York: Prima, 1991), pp. 92, 96.

33 The best and most recent research into Shaw's history is William Davy, *Through the Looking Glass: The Mysterious World of Clay Shaw* (Sherman Oaks, Calif.: privately published, 1995). See also Davy, "Clay Shaw's DCS Career: An Analysis of a Recent File Release," *Probe,* May-June 1996, pp. 6–7.

34 Shaw's connections with these organizations were first documented in the Italian Communist Party newspaper *Paesa Sera* in a series that ran from March 4 until March 18, 1967. Similar stories appeared in the conservative Italian *De la Sera* and in Montreal's *Le Devoir,* on March 16, 1967. A full discussion of these stories and Shaw's association with Permindex and Centro Mondiale Commerciale is in Paris Flammonde, *The Kennedy Conspiracy: An Uncommissioned Report on the Jim Garrison Investigation* (New York: Meredith Press, 1967), p. 215 and passim.

35 CIA support for the OAS attempt on de Gaulle was reported in *Le Devoir,* March 4, 1967. In 1975 the Senate Intelligence Committee received information on CIA support of the French rightists, carried on the front page of the *Chicago Tribune,* June 15, 1975. On Jacques Soustelle and the OAS, see Andrew Tully, *CIA: The Inside Story* (New York: William Morrow, 1962).

36 The Acheson and Bundy remarks about de Gaulle are in the PBS documentary *De Gaulle* (1993).

37 Flammonde, *The Kennedy Conspiracy,* pp. 221–23.

38 Indeed, many personnel supporting the Contra war began their career at the Bay of Pigs. See Hinckle and Turner, *Deadly Secrets,* pp. 410–14.

39 The *Daily News* remark was reprinted in a special edition of the *Los Angeles Free Press*, March 1978. This issue of the *Free Press* was reprinted in its entirety in *The JFK Reader* (Santa Barbara, Calif.: Prevailing Winds Research, 1991).

40 This issue of *Life* was reprinted as a souvenir on the twenty-fifth anniversary of the assassination, November 22, 1988.

41 Rather's comment on the film is presented and discussed in *JFK Assassination: The Jim Garrison Tapes* (Vestron Video, 1992).

42 Edward S. Herman and Noam Chomsky, *Manufacturing Consent: The Political Economy of the Mass Media* (New York: Pantheon, 1986); Michael Parenti, *Inventing Reality: The Politics of the Mass Media* (New York: St. Martin's Press, 1985); Ben Bagdikian, *The Media Monopoly* (Boston: Beacon Press, 1983).

43 Jerry Policoff, "The Media and the Murder of John F. Kennedy," *New Times*, August 8, 1975, pp. 28–37. An expanded version of this historic piece, coauthored with Robert Hennelly, appeared in the March 31, 1992, *Village Voice* as "JFK: How the Media Assassinated the Real Story." The piece is reprinted in Stone and Sklar, *JFK: The Book of the Film*, pp. 484–99.

44 See Hinckle and Turner, *Deadly Secrets*, p. 186.

45 Fonzi, *The Last Investigation*, pp. 53–59.

46 Carl Bernstein, "The CIA and the Media," *Rolling Stone*, October 20, 1977, pp. 55–67.

47 John M. Crewsdon, "CIA: Secret Shaper of Public Opinion," *New York Times*, December 27, 1977, p. 1.

48 Oliver Stone, "Splinters to the Brain," *New Perspectives Quarterly* 9, no. 2 (Spring 1992): 51–54.

49 See Schoenman's taped lecture, "The Real Politics of the Kennedy Execution," available from Prevailing Winds Research, Santa Barbara, Calif. Schoenman, former secretary to Bertrand Russell and organizer of both the Bertrand Russell Peace Foundation and the Who Killed Kennedy? Committee in Great Britain, has been a leading critic of the assassination cover-up from the left. The present article is much indebted to Schoenman's pioneering research and writing and to the similarly groundbreaking and politically insightful critique of the Warren Commission by Vincent J. Salandria.

50 Stone's liberal sentiments are evident in his nevertheless apt response to a very wrongheaded Alexander Cockburn, "A Stone's Throw," *Nation*, May 19, 1992, rpt. in Stone and Sklar, *JFK: The Book of the Film*, p. 519.

51 Stone discusses his conception of Garrison as a kind of Everyman representative of the large cadre of assassination researchers in "Oliver Stone Talks Back," *Premiere*, January 1992, rpt. in Stone and Sklar, *JFK: The Book of the Film*.

52 A good overview of the difficulties within postwar U.S. capital resulting from the linkage of the economy to weapons production is Jezer, *The Dark Ages*, pp. 24–76. The best contextualization of the JFK assassination within the postwar economic circumstances is Ralph Schoenman, "Who Killed Kennedy and Why?" *The Organizer*, January 1992, pp. 18–19. See also Schoenman's

"Official Murder and Capitalist Rule in America: The Continuing Dynamic," *Prevailing Winds*, no. 2 (1996): 30–41. Also illuminating in its discussion of internal state and private sector conflicts and their relation to the assassinations of the 1960s is Carl Oglesby, *The Yankee and Cowboy War: Conspiracies from Dallas to Watergate* (Kansas City: Sheed Andrews and McMeel, 1976). The difficulty with Oglesby's much-acclaimed book is basic to its conception; Oglesby tends to see conflicts within U.S. capital as associated with specific geographic and ideological groupings. By seeing the Sunbelt nouveau riche "Cowboys" as having a special stake in weapons manufacture, he overlooks the enormous role of the Eastern "Yankees" in crafting this economy and upholding it over the last half-century. By suggesting that the JFK assassination was largely an expression of the "Cowboys," Oglesby also sets aside the simple fact that the Warren Commission was conceived, constructed, staffed, and administered not by Lyndon Johnson but by men deeply entrenched in Eastern finance capital, including Eugene Rostow, Dean Acheson, Nicholas Katzenbach, John McCloy, and Allen Dulles. See Donald Gibson, "The Warren Commission A.K.A. The Eastern Establishment," *Probe* 3, no. 5 (July-August 1996): 12–15. The real issue here is the change of heart on the part of the Eastern gray eminences of U.S. capital toward counterinsurgency, then toward the Vietnam incursion itself (turning it into "LBJ's war") as banking capital saw the real disaster of enormous deficit spending that was flowing from the arms race and direct military action within the colonial arena.

53 These economic ruptures and transitions are described in Chomsky, *World Orders Old and New.*

54 Schoenman's very apt description of counterinsurgency of the Truman-Eisenhower-Kennedy period appears in "The Real Politics of the Kennedy Execution."

55 Phillips made the remark in a 1975 interview, contained in the documentary *Inside the CIA: On Company Business* (MPI video).

56 A fine sketch of the rise of the counterinsurgency and John Kennedy's and Robert Kennedy's fascination with it is in Richard Slotkin, *Gunfighter Nation: The Myth of the Frontier in Twentieth-Century America* (New York: Atheneum, 1992), pp. 441–511.

57 The remark was made by Socony Mobil executive William Henderson in a paper published as "Some Reflections on United States Policy in Southeast Asia," in *Southeast Asia: Problems of United States Policy,* ed. William Henderson (Cambridge, Mass.: MIT Press, 1963), p. 263.

58 The Bay of Pigs is discussed in Prouty, *JFK*, pp. 121–29 and passim. Insightful, if largely hagiographic studies of Kennedy's conflicts with the CIA over the Bay of Pigs invasion are Evan Thomas, *The Very Best Men: Four Who Dared—The Early Years of the CIA* (New York: Simon and Schuster, 1995), pp. 237–60. Richard Bissell, director of plans at the CIA until his dismissal over the invasion, is quoted (p. 266) as saying "I thought Kennedy was tough, that he wouldn't cancel the air strikes and lose his first major effort." See also Peter Grose, *Gentleman Spy: The Life of Allen Dulles* (Amherst: University of Massachusetts Press, 1995), pp. 519–56. Grose is surprisingly forthright about Dulles's manipulative role on the Warren Commission.

59 Newman, *JFK and Vietnam*, pp. 36, 40–41, 59. An important book that illuminates the internecine struggles of the Kennedy era and the context of the assassination (without discussing the assassination itself) is Francis X. Winters, *The Year of the Hare: America in Vietnam, January 25, 1963–February 15, 1964* (Athens, Ga.: University of Georgia Press, 1997). Winters argues that Kennedy supported the coup against Diem in part out of a conviction that reformism within the Saigon government and increased indigenous popular support were essential to the success of U.S. counterinsurgency. Winters accepts the view that Kennedy was not inclined toward long-term U.S. commitment to Vietnam and was in fact planning on troop withdrawals after the 1964 elections. The belief in democratic reformism was counter to what occurred in the Johnson administration, which dismissed the "do-gooder" impulse of such notions (p. 116).

60 Perhaps the most cogent analysis of the Kennedy assassination, one that nearly obviates thirty years of assassination research, is a speech by Fidel Castro to the Cuban people on November 23, 1963, the day after the murder. Fidel Castro, "Concerning the Facts and Consequences of the Tragic Murder of President John F. Kennedy, November 22, 1963," rpt. in Schotz, *History Will Not Absolve Us*, pp. 51–87.

61 Such a view was advanced by the Boston-based Assassination Information Bureau in the 1970s. The organization's governing board was very much informed by New Left ideology of the era. See Jeff Cohen and David Williams, "A Radical Analysis of Political Assassinations and Conspiracies Is Needed," *People and the Pursuit of Truth* 2, no. 6 (October 1976): 6.

62 Fredric Jameson, "Periodizing the Sixties," in *The 60s Without Apology,* ed. Sohnya Sayres et al. (Minneapolis: University of Minnesota Press, 1984), pp. 182–83.

Zooming Out:

The End of Offscreen Space

Scott Bukatman

You remember the shot . . . I'm sure you do. The introductory titles are disappearing into the astronomical distance, and the camera tilts down to reveal a planetary surface from a low orbit. The bombastic promise of the music briefly calms to a quiet, modulated scene-setting, but almost at once the strings begin again their insistent call, and a spaceship enters the frame. An interesting ship, slightly blocky but otherwise normal—whatever "normal" means here—and it is being fired upon. As it disappears into the distance, its pursuer enters from above and behind the camera. A triangular prow glides into view. And glides. And glides. And glides. Its angular insistence penetrates and obliterates the frame's rectangular regularity, even as its uncompromising bulk—which is still appearing, its expanse is still sliding over us—throws our learned sense of scale onto the scrap heap.

This is not an essay about *Star Wars* (1977), but some of its aspects interest me. With that first shot, new cinematic technologies redefined space, displaced narrative, and moved cinema into a revived realm of spectacular excess.

As a cinematic object, *Star Wars* is less a movie than an extended multimedia universe. Like its predecessor, *Star Trek,* one does not so much watch *Star Wars* as inhabit it. In both of these senses at least, although in no others, *Star Wars* exploded the frame of narrative cinema, referring back to early cinematic and precinematic spectacles and pointing forward to later forms of hypercinematic entertainments such as simulator theaters and IMAX films.[1]

There is no coincidence in this conjunction of science fiction and new media. I suspect that the return of science fiction as a film genre was partly a function of the move to summer blockbusters for younger audiences and partly a showcase for new cinematic technologies (primarily computer-controlled cameras). *Star Wars* opened in May 1977 and quickly became one of the most popular films in Hollywood history. While its initial success was predicted by no one, the history of this saga exemplified the strategies of the new, postclassical Hollywood film industry.

In 1975, *Jaws* had remade the marketing wisdom of Hollywood. The summer always had been a slack period for major film releases, which were generally held for autumn and, of course, Christmas. *Jaws*, however, found and exploited a summer audience with uncanny dexterity. First of all, the tale of a beach resort and its annoyingly hungry shark was specifically tailored to the summer vacation crowd. But the film was also a blockbuster, heavy on thrills and marketing savvy, designed to appeal to a family audience (but especially its adolescent members, who became heavy repeat viewers). *Star Wars* also reaped the benefits of this new cinematic season. Its combination of old-fashioned romantic swashbuckling and new computer-driven camera effects proved irresistible to older and younger audiences, while its innate gentleness was acceptable to mainstream audiences of both genders.[2] Technically innovative but ultimately (very) reassuring and familiar, it was a canny blend. This combination of narrative conservatism and technical wizardry had predecessors at other points in Hollywood history, most evidently in the films produced in the late 1930s and early 1940s at the Disney studio, including *Snow White and the Seven Dwarfs* (1937) and *Pinocchio* (1940).

Moreover, like so many examples of recent Hollywood cinema the film has had an extended afterlife. *Star Wars* expanded beyond the screen through such successive narrative and multimedia incarnations as sequels, computer games, internet discussion groups, novels, comic books and comic strips, screen savers, simulator theaters, and as a radio serial.[3] So, while the effects in the film extended what cinema could do, and while that first shot promised an expansion beyond the parameters of human scale, the film itself extended outside the cinema into a multimedia, global consciousness.

Unlike the western or, say, the film noir, science fiction is not really a director's genre, but some authorial consistency can be found among the visual designers and special-effects supervisors. The film work of Douglas Trumbull has been particularly distinctive and influential over the last three decades, so he will serve here as a recurrent point of reference. From

29 Technically innovative, yet reassuring and familiar: George Lucas's *Star Wars* (Twentieth Century–Fox, 1977).

the psychedelic effects of *2001*'s Stargate sequence to the direction of the prototypical VR film, *Brainstorm* (1983), to the development of simulator theaters (or, as his company called them, *ridefilms*) and his involvement with IMAX 3-D, Trumbull has emphasized a highly experiential cinema. The trajectory of his career demonstrates that he has been one of the primary forces "exploding the frame" beyond the boundaries of a primarily narrative cinema.[4]

The meaning of science fiction films is often to be found in their visual organization and in their inevitable attention to the act of seeing, and this is where special effects begin to take on a particular importance. The special effects of contemporary cinema are a recent version of centuries-old spectacular technologies that moved toward immersive and apparently immediate sensory experiences, such as monocular and elevated perspectives, panoramas, large-scale landscape paintings, kaleidoscopes, dioramas, and the early "cinema of attractions."

Urban disorientation at the turn of the twentieth century led to entertainments that permitted a cognitive and corporeal mapping of the individual into an inhuman and overwhelming space. What Wolfgang

Schivelbusch called "panoramic perception" was a fundament of the Machine Age. With the rise of railway travel, vision was put in motion. The replacement of horse-drawn coach by speeding train transformed travelers into spectators, separated from the world by velocity, closed compartments, and a sheet of glass. Attention shifted from proximate objects to distant panoramas. Commodification was as important as speed; goods and citizens both circulated through the city, displayed and objectified.[5] Telescopes and microscopes continued the extension of vision, as the cosmos and microcosmos became available to panoramic contemplation. Meanwhile, travel became the metaphor for a continuing dedication to "progress." Journeys—to new heights, new perspectives, and new worlds—defined packaged tours, lectures, panoramas, and world's fairs. Susan Buck-Morss notes that new modes of conveyance at the end of the nineteenth century became linked to new fields of knowledge and the extensive possibilities for human advancement.[6]

Panoramas of one's home city allowed a controlling vision to encompass an alienating, yet familiar, immensity. The visual was now separated from the confirming experience of the haptic (bodily orientation, the physical), and the visual became hyperbolically self-sufficient. Moreover, a related set of entertainments recalled the multisensory body into an ersatz existence. Panorama and diorama came to incorporate lighting, sound, and temperature effects, and perhaps motion platforms to actually rock or move the audience.

Cinema extended panoramic perception through its own emphasis on objects and movement. Of course, the panoramic has played an important part in the cinema since its inception. As John Belton has noted in his discussion of the persistent overlap between extended widescreen projection and travelogue (even in such narrative films as *Around the World in 80 Days*, 1956; and *2001: A Space Odyssey*, 1968).[7] All of these spectacular displays incorporated the audience through a direct address emphasized by the use of new technologies.

Belton further points to American postwar spending habits that emphasized participation and recreation. He cites a 1953 article from the *Hollywood Reporter* in which Darryl Zanuck, head of Twentieth Century-Fox (the studio that pioneered CinemaScope), borrowed from "the recreational language of the day" to describe widescreen and 3-D technologies as "participatory events."[8] Belton further notes the obvious: that with Cinerama and the more familiar CinemaScope, "Spectacular excess carefully regulated the sorts of narratives selected" rather than the reverse.[9] Large-screen formats "recaptured the experience of . . . early films and restored affective power to the motion picture," always at the expense of narrative centrality.[10] The publicity for *The Search for Paradise* (1957), a Cinerama release, explained: "Plot is replaced by audience envelopment." In such large-format events, the causal chains of narrative are displaced by a more participatory, bodily engagement.[11]

Spectatorial experience has been similarly central to Trumbull's projects. In all their manifestations, his effects have emphasized a massive technological object or environment: the Stargate (*2001*), the mothership (*Close Encounters of the Third Kind*, 1977), V'ger (*Star Trek*, 1979), the city (*Blade Runner*, 1982), or the space under a pyramid (*Secrets of the Luxor*, 1993). The camera is emphatically subjective, moving around and through space in an extended process of kinetic exploration and revelation. Language yields to mute fascination. Delirious, kaleidoscopic images bombard the audience, yet Trumbull's epic luminism also offers an unexpected, contemplative aspect.[12]

A reliance on vision within a detailed simulation of the real world has been supplemented by bodily experience in order to produce a more deeply rooted understanding of the world. So the address to the body that marks the panorama, the amusement park ride, and hypercinematic attraction were—and are—means of inscribing new, potentially traumatic phenomena and perspectives onto the familiar field of the film spectator's body. It is a holdover from a time when, as Barbara Stafford writes, "spatial and kinesthetic intelligence were not yet radically divorced from rational-linguistic competence and logical-mathematical aptitude."[13] Immersive media serve as both a physical and (therefore) a conceptual interface with new technologies and the life-world they produce.

If cinema was a reflexive product of industrial culture, well, so was science fiction. The genre was simply predicated upon continuous, perceptible change; it narrated a world that would become noticeably different over the course of a single lifetime. Those changes were part of the profound philosophical and political shifts of the nineteenth century, but they were most clearly connected to the rapid (and increasing) pace of technological development. Science fiction narrative might easily be seen as another "immersive" medium, interfacing with the lived complexities of the technosphere (so it is hardly surprising that both fairground attractions and early cinema dramatized Jules Verne's *A Trip to the Moon* and similar panoramic science fictions).

Science fiction has been an essential part of technological culture for more than a century. It also has served as a vehicle for satire, social criticism, and aesthetic estrangement. In its most radical aspect, science fiction narrates the dissolution of the most fundamental structures of human being. By positing a world that behaves differently—whether physically or socially—from this world, ours is denaturalized. Science fiction even denaturalizes language by emphasizing the normally implicit processes of making meaning. The variable distance between the language of the text and the reader's lived experience can be seen as the genre's ultimate subject. What science fiction offers, in Fredric Jameson's words, is "the estrangement and renewal of our own reading present."[14]

The shift from the expansionist and highly visible machineries of the Industrial Age to the more imploded, invisible technologies of the Information Age has created a crisis for science fiction. As John Clute has elegantly argued:

No longer has information any tangible, kinetic analogue in the world of the senses, or in the imaginations of writers of fiction. Gone are the great arrays of vacuum tubes, the thousands of toggles that heroes of space fiction would flick *almost* faster than the eye could see as they dodged space "torpedoes," outflanked alien "battle lines," steered through asteroid "storms"; gone, more importantly, is any sustained sense of the autonomy, in space and time, of gross visible individual human actions. And if "actions" are now invisible, then our fates are likewise beyond our grasp. We no longer feel that we penetrate the future; futures penetrate us.[15]

The purpose of much science fiction, and especially cyberpunk in the 1980s, was to construct a new human (or posthuman) subject that could interface with the global, yet invisible, realm of data circulation—a new being to occupy the emerging cyberscapes. I call this new position *terminal identity*, which refers both to the end of the traditional self and the emergence of a new self-definition constructed at the computer station or television screen. Through the language, inconography, and narration of science fiction, the shock of the new is aestheticized and examined. Science fiction constructs a *space of accommodation* to an intensely technological existence, and this space has continued through the present electronic era.

Science fiction film also uses a complex "language," but it is a special case because of its mainstream positioning and big-budget commodity status. Science fiction novels or comics need sell only a few thousand copies to recoup their costs, so experimentalism is not discouraged, but the Hollywood blockbuster must find (or forge) a mass audience. Science fiction cinema's mode of production has committed it to proven, profitable structures, and so it is necessarily a more conservative form. But even such massive successes as *Star Wars* and *Terminator 2* (1991) are riven with hidden complexities and profound contradictions regarding the intertwined status of technology and the definition of the human. And although the narratives are conservative, and they almost always are, the delirious technological excesses of these films and their spectacular effects may "speak" some other meaning entirely. Often, the most significant "meanings" of science fiction films are found in their visual organization and their emphasis on human perception—this is the importance of special effects.

Tom Gunning has described early, prenarrative film as a "cinema of attractions," a form that he characterizes as exhibitionistic, a cinema in which "energy moves outward towards an acknowledged spectator rather than inward towards the character-based situations essential to classical narrative." In the silent era, cinema quickly passed from a presentational style of direct address to a representational system of illusionistic storytelling—a shift accompanied by an increased separation of the theatrical space from the space on-screen.[16] Gunning acknowledges that the "attraction" still survives, in a tamed and contained form, "as a component of narrative films, more evident in some genres (e.g., the musical) than in others."[17] But he does not specifically consider how later cinematic technologies, such as color and widescreen, provided the ground for the attraction's return. Belton, for example, argues that "Cinerama and other multiple-screen systems constitute the most extreme instances of cinema as pure spectacle, pure sensation, pure experience"—in other words, as attraction.[18]

Through its special effects, presented in color and widescreen, science fiction film participates in a presentational, participatory mode of address. Douglas Trumbull's most extended sequences, in *2001, Close Encounters of the Third Kind, Star Trek,* and *Blade Runner,* all present human figures as provisional guides through an alien space, but they do not defuse the experience of the film viewer. The passage into the kinetic lights and amorphous shapes of the Stargate in *2001* is directed *right at* the viewer: the inserted (still-frame) close-ups of the astronaut do not reintegrate the viewer into a fictional representational space, but rather emphasize sensation, immersion, and spectacle.

Brooks Landon has argued that science fiction film's "sense of wonder" actually "derives from 'a new way of seeing.' "[19] By emphasizing cinema's visuality, special effects foreground principles of perception, while their reflexive, presentational quality emphasizes the technologies of their production. In a sense, science fiction cinema mediates between human vision and the wide and growing range of technological enhancements.

Effects present the once-inconceivable in detailed, experientially convincing forms. This has important ramifications for the Information Age. The invisible workings of electronic technology are made perceptible and physical, and are figured in metaphorical but embodied terms. The anxieties and desires of the era take on concrete, literal form. The non-visibility of the data strata of society has led to a series of attempts to refigure the workings of the computer as a space that could be perceived by the human sensorium.

Despite a spate of cyberthrillers (including the less than thrilling *Lawn-mower Man* (1992), *The Net* (1995), *Johnny Mnemonic* (1995), and others), after more than a decade *TRON* (1982) still represents the most sustained cinematic attempt at mapping cyberspace: first by envisioning it *as* a space, then by thrusting the human inside it. In *TRON,* effects construct new objects and spaces that concretize abstract cultural concerns and produce a tentative reembodiment of the human, refigured in response to those concerns.

It is possible to exaggerate the educational aspects of immersive media and special effects, which are, when all is said and done, really pretty cool. After all, the experience produced by many effects sequences in science fiction cinema is one of hallucinatory excess as the narrative yields to an abstract, kinetic spectatorial experience that exists apart from its representational function. The descriptions surrounding such immersive environments as fairgrounds, amusement parks, and virtual reality are rife with a sense of delirious liberation. Fairs have always emphasized stimulus and overload—a too-muchness was the whole idea. In the 1980s the magazine *Mondo 2000* served as a similarly delirious guidebook to the dizzying world beyond the computer screen. World's fairs and virtual reality are all about losing your bearings.

The discourse of modernity continually evoked (and continues to evoke) the interwoven phenomena of delirium, immersion, and kinesis— phenomena associated with the lived experience of urban concentration and industrial expansion. While the panorama may have provided one of the dominant metaphors for spectacular visual culture in the eighteenth and nineteenth centuries, it is worth remembering that the obscurity of the phantasmagoria was nearly as frequent a point of reference. As Terry Castle notes: "This association with delirium, loss of control, the terrifying yet sublime overthrow of ordinary experience, made the phantasmagoria a perfect emblem, obviously, of the nineteenth-century poetic imagination."[20] Modernist painter and filmmaker Fernand Léger held that lived experience had, in the twentieth century, become more rapid and fragmented for everyone: "A modern man registers a hundred times more sensory impressions than an eighteenth-century artist."[21]

The dark distortions of the phantasmagoria are related to the brightly colored fragments seen tumbling in a kaleidoscope. The kaleidoscope itself was another important model of modernist perception, and its ephemeral collages possessed an immediate metaphorical value. Baudelaire, for example, was fascinated by them because, as Jonathan Crary reminds us, the kaleidoscope "coincided with modernity itself." To "become a 'kaleidoscope gifted with consciousness' was the goal of 'the lover of universal life,'" the flâneur. The kaleidoscope became a machine that disintegrated any fixed perspective; "shifting and labile arrangements" became the new point of view.[22]

Kaleidoscopic perception—composed of equal parts delirium, kinesis, and immersion—characterized many popular entertainments in the later years of the nineteenth century, from expositions to magic lantern shows and phantasmagoria. Cities were becoming kaleidoscopic environments, for which not everyone was equally suited (Harriet Beecher Stowe wrote that New York "always kills me—dazzles, dizzies—astonishes, confounds and overpowers poor me").[23] While panoramas and museums organized experience and emphasized the latent order underlying urban chaos, other media produced what should be called an "ordered chaos." Amusement parks and fairs were immersive environments that distilled the essence of the modern city at the turn of the century. For the architect Rem Koolhaas, Coney Island was really an experimental laboratory for the city at large, providing the key to decoding New York's deep irrationalism: "Coney Island is a fetal Manhattan."[24]

The emphasis on urban chaos in spectacular visual culture allowed—still allows—mass entertainment to calm some fears by serving as an entertaining and embodied interface with the complexities of the emerging technoculture. But it also demonstrates a developing taste for delirium, kinesis, and immersion. In the later nineteenth century, at a time of increasing concentration of bureaucratic power and control, popular recreations offered a licensed escape into the irrational.[25] Technology was almost always the vehicle for these transporting effects. As noted, panoramas and dioramas used sensory effects, moving platforms, and even photography and cinema. Of all the arts, none combined spatiotemporal solidity with metamorphic fluidity as sustainedly as the cinema, and cinema documented and, in some ways, liberated the ephemerality that lay latent in the urban field.

The immersive attractions represented by panoramas, dioramas, early cinema, and Hale's Tours[26] have made a strong comeback in the simulator-theater systems that combine high-resolution film imagery with platform motion to produce a striking sense of immersion in a technological space. Technology becomes an enveloping, inescapable phenomenon, incomprehensible and overwhelming.

But maybe it is not so bleak—perhaps this was (and is) a technology redirected against itself and against the rationalist control that usually adhered to it.[27] The special effects of science fiction cinema, nested as they are within the rational discourse of narrative, might possess a similar antirationalism (and are therefore not as "tamed" as Gunning has argued). The rationalist, mechanical visions of photography and cinema have always been undermined by a pervasive interest in, and attraction to, the uncanny.[28] As Geoffrey O'Brien says so perfectly, "Upon the motion picture—the most alluring mechanism of the age of mechanical reproduction—would devolve the task of reconstructing the imaginary worlds it had helped to dismantle."[29]

Science fiction is traditionally regarded as the rationalist literary yield from the Age of Reason and the Industrial Revolution, but the linguistic play of the literature and the kinetic delirium of cinematic special effects push past these constraining bounds. The Stargate sequence from *2001* is exemplary (but also unique—its impact has never been surpassed within narrative cinema), and begins to indicate how central Trumbull's work has been to this push. Before *2001*, Trumbull had been working for a company that produced educational, promotional, and training films for the National Aeronautics and Space Administration (NASA) and the military. These films' attention to scientific detail captured Stanley Kubrick's attention at the 1964 World's Fair. They had provided designs for Kubrick's science fiction project before some technicians, including Trumbull, "defected" to work directly with the filmmaker. To produce the visual onslaught of the now-legendary, psychedelic Stargate sequence, Trumbull helped develop "slit scan" photography: a variation of single-frame animation that involved sliding artwork behind a moving screen in which there was a single slit. The results are streaky, variable patterns of light, and if the camera simultaneously tracks in, a vertiginous sense of penetrating motion results. Slitscan might be regarded as "a precursor of the electronically controlled camera systems that began to proliferate in the 70s."[30]

Building on *2001*'s experiential emphasis and his background with film-

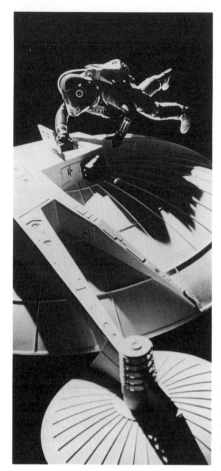

30 Lost in space: Before *2001*, special effects designer Douglas Trumbull produced films for NASA (MGM, 1968).

making for exhibition spaces, Trumbull began to develop ShowScan, a hyperreal film process that exposed 65mm film at 45–60 frames per second. To fund it, he handled the effects for Steven Spielberg's *Close Encounters of the Third Kind,* taking advantage of new computer-controlled camera movement (developed by John Dykstra for *Star Wars*) to produce a set of luminous UFOs composed more of light than of matter. The film climaxed (there ain't no other word for it) with an astonishing scalar redefinition at the coming of the mothership.[31]

Trumbull's work on *2001* and *Close Encounters* reminds us that science fiction film incorporates at once the magisterial gaze of the panorama, the sublime obscurity of the phantasamagoria, and the "shifting and labile arrangements" of the kaleidoscope. Like kaleidoscopes, science fiction films are "toys" that distort and remake human perception. The science fiction cinema's kaleidoscopic worlds comprise inventive set design and pervasive motion. A dizzying kinesis fills the screen, whether the motion belongs to an alien artifact or to the camera that investigates it. Effects sequences perform a process of revelation through a sustained kinetic exploration. Of course, *2001* sustains the most incredible sense of panoramic revelation. Especially in the outer space sequences, the drifting camera keeps moving in a single direction, as if engaged in elegant and casual description, but at the same time each shot leads to *something*— objects, locations. Each shot is the film in miniature. Again and again, the camera drifts, seemingly without purpose, only to arrive somewhere. This is the film's most insistent visual motif; movement seems at once under- and overdetermined. Panoramic perception that surrenders to kaleidoscopic perception as "the ultimate trip" of the Stargate sequence begins. Without a doubt, *2001* features the most vertiginous effects since *Le Ballet Mécanique*, the 1924 film by Fernand Léger and Dudley Murphy, of which Jean Epstein wrote, "Anyone enchanted by this abstract cinema should buy a kaleidoscope."[32]

In spectacular visual media, movement has become more than a tool of bodily knowledge; it has become an end in itself, and it is often accompanied by a utopian sense of possibility. Movement has become a passage across borders, and it holds the promise of a resistance to external control.

In *The Right Stuff* (1983), Philip Kaufman's film based on—but very different from—Tom Wolfe's history of the Mercury space program, movement is linked explicitly to transcendence and implicitly to the spectacularity of cinema. The film begins in black and white, with an Academy ratio screen image (1.33:1) that soon expands into color, multiple-channel sound and Panavision. *Superman* (1978) begins with a similar shift, and both films recall Belton's description of a 1926 screening of *Old Ironsides:* "During two sequences . . . the Rivoli's curtains and screen maskings, opened, on signal, beyond their usual position for 1.33:1 projection. Scenes of the launching of the USS *Constitution,* moving under full sail directly toward the camera, and the climactic sea battle were projected on this expanded screen with the wide-angle Magnascope lens." Belton's evocation is relevant to this entire essay: "The cumulative effect was that of an illusion of the image's movement into the space of the theater auditorium, breaking down the spectator's sense of the barriers of the proscenium."[33]

The central problem confronted by *The Right Stuff* is how to resist corporatism within a fully technologized world. What becomes of the individual, and what "space" becomes proper to its survival? The film depends upon an alternately ironic and beatific evocation of American frontier mythology to produce a sense of embattled masculinity in a world that has displaced power from people to systems (technocracy, mass media, politics). The rugged, individualist test pilot Chuck Yeager (played with astonishing cowboy conviction by Sam Shepard) nearly "pushes the envelope" to earth orbit, but his is no longer the way. The film really culminates with the orbital flight of John Glenn, whose squeaky-clean, "Dudley Do-Right," team-playing character is constantly mocked, yet who is nevertheless permitted to literally reach the heights that Yeager was denied. What the film describes is a massive technological, techno-cratic system that solely exists in order to launch one man beyond its reach—beyond itself. Thus, Glenn's transcendence is not a transcendence of self, but a movement beyond the authority and rationality of systems; unlike a western however, this frontier can be crossed or occupied only through the resources and control provided *by* that very system. The film, which has been criticized for its inconsistent tone, is actually more in-coherent than inconsistent—the result not of artistic failure, but of the incoherence of its foundational mythos.[34]

Even *The Right Stuff*'s spectacular, kinetic effects are diverse, not uni-fied, in style. Hyper-realistic earthscapes and miniatures combine with abstract imagery by California-based experimental filmmaker Jordan Bel-son. Belson's trancelike films, including *Re-Entry* (1964), *Samadhi* (1967), and *Cosmos* (1969), encourage a meditative posture; his shifting lumines-cences quietly distort perception while becoming a compelling point of focus.[35] He was a fascinating choice to contribute special effects to a mainstream Hollywood production, both because of his relatively low-tech relation to cinema and his notably nonrepresentational approach. The sequences conveyed the transcendent world that lay outside the envelope—that domain lying on just the other side of the possible—the world beyond the sound barrier, the upper atmosphere, outer space. The gorgeous, streaky, but intense blue that pilot Yeager sees after surpassing the speed of sound fully conveys the transcendental mythologism of the test-pilot sequences. At first, Belson's footage (it is hard to call it imag-ery) is framed by the controls and cockpit windshield, but after a cut to Yeager's beatific face the blue is unbounded—unmediated—extending to the edge of the Panavision frame.

The sequence not only represents (*and* presents) a distorted perception, it also transforms the film into a more heterogeneous and self-aware text. Belson's work, which can (to my eyes) become monotonous on its own, is reinvigorated by its narrative "setting" (as in the setting of a precious stone). If, in the narrative, it represents a utopian transcendence and the penetration of a new frontier, as filmmaking it presents a utopian transgression of the realist, representational codes that dominate narrative cinema. The film viewer—this one, at least—is swept away in an aural and visual crescendo. The technological, indeed technocratic, system of the postwar, postclassical Hollywood cinema has produced an emotionally moving evocation of a temporary transcendence, a sublime but ephemeral movement beyond its own boundaries.

At the end of the 1970s science fiction reemerged as a showcase for new cinematic technologies, which were still cushioned by a set of fairly conservative narrative structures. By the 1980s, however, Hollywood saw the proliferation of effects in nearly every film produced. This is not an exaggeration: almost every major production now contains some digital effects work that is used to add, alter, or even remove visual elements. (*Forrest Gump,* 1994, is the most familiar example.) Once-exotic computer technologies had become less novel; indeed, they were almost embarrassingly familiar. At the same time, new forms of hyper-sensational cinema were becoming more popular. IMAX films moved from science museums to Broadway while simulator theaters, once restricted to theme parks and world's fairs, became staple attractions at hotels and lowly state fairs.

Movie theaters and screens may have become smaller through the 1970s and 1980s, and most movies were encountered in the form of videocassettes. But cinematic spectacle has not entirely vanished. In a way—in many ways—a branch of cinema, no longer tied to the necessities of narratives and explanations, retook the status of untamed attractions, once more combining immersion, technological novelty and direct address.[36] *Brainstorm,* directed by Trumbull, narrated this very shift through its tale of a new technology that directly records sensory experience (and even memory). Trumbull intended to use his high-speed ShowScan system to present this hyper-experience, but MGM was, reasonably, unwilling to burden its exhibitors with the associated costs, and instead, during the "experiences" the film shifted from 35mm to widescreen, wide-angle 70mm. Anticipating the hyper-real technologies of virtual reality while quoting the participatory address of *This Is Cinerama* (1952), *Brainstorm* also invokes the imaginary immersion in the screen that André Bazin called "the myth of total cinema."[37]

The phenomenology of science fiction cinema and special effects, ex-emplified by *Brainstorm*'s emphasis on sensory experience as the basis of self-definition (not to mention cinematic pleasure), continues in new hyper-cinematic, exhibitionistic forms. Special-effects sequences in sci-ence fiction films are often about revelation, and so they evoke a strong sense of presence. Film noir or horror might play with absence and the mystery hiding in the shadows, but science fiction shows us stuff. More is added to the visual field; nothing is taken away.[38]

Belton makes a similar point with reference to the psychoanalytic con-cept of "suture," which has been adopted by some film theorists. He argues that with large-screen film formats: "suture . . . remains less pronounced than that in traditional films: takes tend to be longer, the viewer's ability to exhaust the details contained within them tends to be reduced, and shot / reverse shot editing patterns tend to give way to 'the-atrical,' single-perspective (that is, unedited) modes of dramatic presenta-tion." The process of spectatorship is not fixed or static. It has historical, changing, components.

Blade Runner, for example, presents itself as an inexhaustible world of constant fluidity and passage; it offers infinite perceptibility—a "utopian vision," so to speak. As the camera traces a detailed path across scalar levels, the city becomes a complex, self-similar space—a fractal environ-ment. The film begins with an extreme long shot of Los Angeles. Vision reflects and penetrates the space, moving forward to locate the Tyrrell corporate headquarters (note its microchip design—similarity across scale). A little later, the camera locates Deckard by submerging to street level; and here there are neon signs, futuristic attire and lighted umbrella handles. A streetside vendor uses an electron microscope and the drama becomes molecular. Deckard's electronic inspection of a photograph transforms its visual field, as the screen, that frontier separating terminal and physical realities, becomes permeable. Infinite complexity structures perceived urban reality as *Blade Runner* introduces us to fractal *geography.*

The film also changes the terms of "panoramic perception" itself: a mode that was partly predicated on anxiety: about the train traveler's sense of hurtling out of control in a speeding projectile and the inexorable grasp of merchant capitalism. Withdrawal into numbed alienation provided some psychological protection: the complexities of panoramic perception were founded on separation, abundance, anxiety and alienation.

No better cinematic example exists than the trolley ride in *Sunrise* (1928).[39] Having nearly strangled his wife, the man begs for forgiveness, but she runs, in terror and grief, from lakeshore to woods, boarding an improbable city-bound trolley. He follows, and as they stand at the front of the moving car the world beyond the windows slides laterally past them (and us). They seem fixed and centered in space, but the world itself has become unmoored.[40] The sequence speaks, with eloquence, of separation and disconnection, even as this most literal of tracking shots links rural and urban spaces in a fluid continuum. Despite its evident solidity, the world takes on an hallucinatory mutability. Our attention is divided between their unspeaking trauma and the complexly shifting space beyond the windows; we are divided between alienation and exploration. Some of the poignance comes from the inescapable sense of loss built into the very aesthetic of the shot. Whatever comes into view will just as surely be removed, to be replaced, and replaced again. Absence and distance structure the shot.

Leo Charney has located variations on this absence as fundamental to the fascination that cinema exerted on such philosophers and theorists as Henri Bergson, Walter Benjamin, and Jean Epstein. Characteristically, "cinema transformed the hollow present into a new form of experience, as the vacated present opened space for the viewer's activity. Experience arose in—was defined by and in—the space vacated by the present's movement away from itself."[41] In *Sunrise,* the movement of the world past the fixed frame of the camera becomes a metaphor for cinematic experience in which plenitude and loss coexist in an unresolved dance of marked but unfulfilled desires.

Compare this to the expansive hovercraft flight above Los Angeles in *Blade Runner.* Again the world slides past the window, but now we can unreservedly revel in the revelation of urban immensity and complexity (it is mysterious to me that anyone could consider this film dystopian). At the same time, though, a moment's reflection demonstrates that *Blade Runner,* with its retrofitted future built on the debris of the past—our present—*does* insist on a sense of absence and separation similar to *Sunrise.* This is also in keeping with the narrative's emphasis on the passing of human centrality. Does memory remain the residue of something now absent, or has it become a simulation, and thus a false presence? Charney notes (without reference to *Blade Runner*), that "if the present disappears, and thereby hollows out presence, this shift also hollows out the subject who constructs that presence." The modernist human is hollow, while replicants, interestingly enough, with their totemistic family photos and implanted memories, are whole.

The hypercinematic experiences of IMAX films and simulator theaters (and even interactive museum displays) continue to make quite a fetish of physical presence. An environment, such as the rain forest or outer space, is presented, seemingly without boundaries: the frame either wraps around the viewer or is simply too big to encompass. In the simulator theaters, movement is synchronized with what is seen, further reinforcing a sense of fullness and plenitude.[42]

Special-effects sequences are splendidly exhibitionistic and can even contribute to a reintegration of fictive and theatrical spaces. This is one obvious impetus behind the simulator theaters that Trumbull and others have designed. Following *Brainstorm,* Trumbull abandoned a Hollywood that was, as he said, "multiplexing itself to death," and continued to develop technologies to expand and exaggerate cinematic experience. He achieved great success with his *Back to the Future* ride for the Universal Studios tour, the first of the simulator-theater attractions, and he has since become a vice president of the IMAX Corporation, which has moved into large-format 3-D production and exhibition. IMAX films and simulator theaters both emphasize a sense of first-person experience not unlike the fantasied hypercinema of *Brainstorm.*[43] The theater becomes a vehicle traveling through some kind of space, while the individual psychology of an on-screen character is superseded by the collective presence and participation of the film's audience.[44]

These collective, immersive experiences stand in strong contrast to the other dominant technological interface of the present: the hypertext-based language of the World Wide Web. While its promise remains extraordinary, and its value undeniable, the existing experience of the Web is somehow less than inspiring. In some ways this is as far as one can get from the sensuous delights of a fantasied virtual reality. Despite its graphical interface and multimedia extensions, the Web is profoundly monotonous. It is a nonimmersive system in which screen after flat screen displays endless amounts of information, mostly in the same font, mostly on the same gray background. The flatness of the screen is emphasized by that wonderful act of scrolling, which recalls nothing so much as the pre-electronic clunkiness of microfilm projectors (except perhaps unrolling paper towels or bathroom tissue). While IMAX films present an "unframed" image, the Web (like GUI's—graphic user interfaces—in general) is *all* windows, frames, borders, and boundaries. And if simulator theaters provide a kinetic, sensory address, the Web is profoundly nonsensory, restricted as it is to static visuals, occasional sounds, and the tactile limitations of the mouse. Pointing and clicking our way to a better tomorrow is hardly as seductive as the experiences of Norman Bel Geddes's Futurama at the 1939–40 New York World's Fair or William Gibson's zippy cyberpunk novel, *Neuromancer.*

Immersive entertainments also differ from the virtual-reality interface of Eyephones and Bodysuits, which was discussed to death in the last decade. (Is VR the first technology to become obsolete before it was even invented?) In theaters the interface becomes invisible; the illusion of an immediate experience is unhampered by a bulky and unwieldy personal prosthetics. Virtual reality could hardly hope to succeed at projecting the body into another reality while the flesh remained so firmly tethered to this one. The recession of VR as the ideal interface raises the question of whether anybody learned anything at all from the failed experiment with 3-D movies in the early 1950s. Here, again, enhanced experience was acceptable only when the apparatus remained transparent; the illusion of a natural body negotiating unnatural surroundings could be preserved. Of course, IMAX films are now being produced in 3-D, using impressive new electronic headsets with alternating polarized states and built-in sound systems, but I still see no indication that this technology will prove any more successful now than it did then. The dream remains—3-D without the glasses.[45]

The proliferation of special-effects films in the 1990s was accompanied by a related reduction in narrative sophistication. Steven Spielberg's films, for example, including *Schindler's List* (1993), seemed frightened by the possibility that someone, somewhere, might miss the point (*any* point). Tropes of the classical Hollywood cinema, such as cross-cutting, subjective camera movement (now provided by the astonishingly ubiquitous Steadicam), or intensely emotive close-ups dominate today's film aesthetics with a vengeance. To me, this reduction, combined with the emphasis on effects-centered films, speaks to an anxiety about the very status of narrative as an explanatory system. Narrative implies history, depth, purpose. So, while Hollywood cinema continued to revel in the sensational, sensual realm of visual, auditory, and kinesthetic effects, the devaluation of narrative was hidden within a desperate overvaluation of overly explicit storytelling; a denial of its own undeniable supersession.

At the same time, Hollywood developed what might be called the "theme park movie"—a set of overdesigned, hermetically sealed, totalizing environments masquerading as movies. *Dick Tracy* (1990), *Batman* (1989), and *Jurassic Park* (1993) all could be considered examples of this phenomenon. It is hardly accidental that these films feature extensive licensing agreements or that these agreements often predated their filmed incarnations either as comics or novels. Today, these theme park movies penetrate the so-called real world in the form of toys, games, clothing, "making-of" documentaries, and rides. The rides are important—in the 1980s and 1990s, films became rides, which is to say that they became less narrative than they used to be and more spectacular, with their spectacles more compressed one atop another but also more extended, hammering across an entire two-hours-plus film with scarcely any let-up. Meanwhile, theme park rides and attractions became more narrative than, say, roller coasters had been. These also were extended. Waiting on line for Star Tours was part of the ride, as elaborate sets and amusing droids entertained but also grounded the spectacle. The expansion of cinema into the theme park attraction resulted in some wonderful paradoxes, such as full-size miniature sets.

Film has become something to inhabit rather than watch, and new means of extending the cinematic experience have become available. To some degree, as with role-playing games for example, the extra-cinematic experience is a means of extending the narrative, but most often—in theme park attractions, downloadable illustrations, fan activity and Internet discussion groups (around *Star Trek*, *Blade Runner*, *Pulp Fiction* [1994], or *The X-Files*, for example), and (if you are in town stop in at) Tokyo's *Alien*-inspired Giger Bar—it becomes a means of inhabiting the world of the film.[46] Note also the almost obsessive interest in special effects on the part of adolescent male audiences. While this might first indicate a healthy interest in technological modes of production, it actually fetishizes the prosthetically built-up body and reifies other aspects of production. (When adolescent males start reading magazines dedicated to the production of ideological meaning, *then* I will sit up and take notice.)

The film and its narrative project a world. But while cinema remains the initial ground for this world projection, there is now a global multimedia culture in place to absorb and extend that experience. Some potential good might be found in the shift from a profoundly individual cinematic experience to one based on ecstatic group experience and participation,[47] but the Reagan-era return of vertical integration and centralized corporate control largely predetermined the paradigms of participation. Nevertheless, the expansion of cinema into a more environmental, even ambient form might reflect some nostalgia for the combination of control and freedom represented by the act of *play*. These textual worlds are hyper-complete and exhaustive (every doorknob, every shoelace, every musical motif seems overdetermined, oh-so-carefully designed and placed) yet also incomplete and even inexhaustible. They are at once closed and open, both predetermined and playful.

The rise of effects-centered films, the decline of narrative, the return of the cinema of attractions, the sequels and simulations and spectaculars— all of this suggests that what we seem to be witnessing is the *end of offscreen space*. The space that can be viewed on-screen has expanded through widescreen effects, IMAX films, simulator theaters, and all of their forms of direct address, while real space is increasingly penetrated by filmic realities in the forms of merchandising, Web sites, fan conferences, and theme parks. Physical and spectacular spaces commingle. Of course, exhibitions have always emphasized presence in the face of a dialectic of disappearance or absence, and world's fairs have always been merchandised to the max.

But an emphatic sense of presence is now the rule, not the exception, and I think this marks a qualitative difference. There are other differences as well. As cyberspace has replaced urban space as the primary locus for confusion, dislocation and alienation, so it has become the new referent for the immersive entertainments founded on some form of kaleidoscopic perception. And if the anxiety provoked by electronic culture is a function of its nonvisibility, then an overemphasis on visibility and immersion is only to be expected.

What interests me about these effects, then, is that they are addressed *to me:* they reinstate and acknowledge my presence in the theater and emphasize the experience of viewing. They license an adventure of perception that is not voyeuristic but exploratory. They reintroduce cinema as a public space (especially the simulator theaters). They also provide a release from such causal structures as narrative. Gaston Bachelard held that the poetic image spoke more to the soul than to the mind and functioned as an antidote to rationalist, causal biases.[48] If most of these filmic experiences have yet to achieve the condition of poetry, their potential to do so is evident.

But at the moment, most of what is offered is less poetry than play. Immersive environments offer something less than that transcendence of the flesh so often envisioned by New Agers and cyberpunks. In actual practice, this is much more of a return to the body, which is justified by a rhetoric of spiritual transcendence. The fantasied "escape velocities" of various transcendentalist and millenialist schools translates, in mass culture, to a hunger for vertiginous physical sensation—a cheap and semi-licensed high.[49] Cosmic connectedness yields to a playful group participation that hints at the survival of the ecstatic and uncanny—the nonrational—within dominant, rationalist culture.

Special effects and the hypercinematic experiences of Las Vegas and other theme parks are not a substitute for either rationalism or that liberation which situationists called "the revolution of everyday life."[50] These are not environments where such needs are fulfilled. They alone will not "free" us from rationalist structures of control. But it is not so important for mass culture to show us how to attain liberation, release, and the rest of it. It is enough—actually, it is more than we might expect and more than many scholars will admit—for them to recall for us the possibilities of irrationality and its continued presence in a supposedly rational world. When Koolhaas refers to Coney Island as "a fetal Manhattan," he is not pointing to Coney as a sanctioned release or safety valve for culture-at-large, he is arguing that these environments reveal the irrationalism that is always an imperfectly suppressed part of ostensibly rational cultural systems. Science fiction films, simulators, and unframed cinemas are not revolutionary; they are *playful*.

Ridefilms are a fetal cyberspace.

Notes

My thanks to Ben Shedd and Thomas Elsaesser for some frighteningly helpful discussions.
1 The very useful phrase, "exploding the frame," belongs to filmmaker Ben Shedd.
2 My mother watched the end credits for *Star Wars*—highly unusual behavior.
3 It is worth remembering that in the 1980s the Disney organization came to dominate this extended and increasingly globalized entertainment market.
4 For more on the aesthetics of Trumbull's effects work, see my "The Artificial Infinite: On Special Effects and the Sublime," in *Visual Displays: Culture Beyond Appearances,* ed. Lynne Cook and Peter Wollen (Port Townsend, Wash.: Bay Press, 1995). The present essay condenses a number of my other works in its early sections; see esp. "The Artificial Infinite" and *Terminal Identity: The Virtual Subject in Postmodern Science Fiction* (Durham, N.C.: Duke University Press, 1993) for more extended discussions.

5 Wolfgang Schivelbusch, *The Railway Journey: The Industrialization of Time and Space in the Nineteenth Century* (Berkeley: University of California Press, 1986). See esp. chap. 4.

6 Susan Buck-Morss, *The Dialectics of Seeing: Walter Benjamin and the Arcades Project* (Cambridge, Mass.: MIT Press, 1989), p. 91.

7 John Belton, *Widescreen Cinema* (Cambridge, Mass.: Harvard University Press, 1992), pp. 92–93.

8 Darryl Zanuck, "Entertainment vs. Recreation," *Hollywood Reporter*, October 26, 1953; cited in Belton, *Widescreen Cinema*, p. 77.

9 Belton, *Widescreen Cinema*, p. 195.

10 Ibid., p. 93.

11 Ibid., p. 95.

12 As with the panoramas and cineramas and other visual displays of the last two centuries, these sequences reveal an ambivalent relation to the new technologies upon which they depend. See "The Artificial Infinite."

13 Barbara Stafford, *Artful Science: Enlightenment Entertainment and the Eclipse of Visual Education* (Cambridge, Mass.: MIT Press, 1994), p. 3.

14 Fredric Jameson, *Postmodernism, or, The Cultural Logic of Late Capitalism* (Durham, N.C.: Duke University Press, 1991), p. 285.

15 John Clute, "Introduction," in *Interzone: The Second Anthology*, ed. John Clute (New York: St. Martin's Press, 1987), p. viii.

16 Tom Gunning, "The Cinema of Attractions: Early Film, Its Spectator and the Avant-Garde," in *Early Cinema: Space, Frame, Narrative*, ed. Thomas Elsaesser (London: British Film Institute, 1990), p. 59; Miriam Hansen, *Babel and Babylon: Spectatorship in American Silent Film* (Cambridge, Mass.: Harvard University Press, 1991), p. 34.

17 Gunning, "Cinema of Attractions," p. 57. While Gunning refers to those surviving forms as *tamed* attractions, I am not sure that is always the case.

18 Belton, *Widescreen Cinema*, p. 97.

19 Brooks Landon, *The Aesthetics of Ambivalence: Rethinking Science Fiction Film in the Age of Electronic Reproduction* (Westport, Conn.: Greenwood Press, 1992), 94. This is sometimes echoed by the narrative, as when computer-generated special effects dominate films about the experience of computers (as in *TRON* or *The Lawnmower Man*), but it need not be (*Toy Story*, for example, is not about computer games).

20 Terry Castle, *The Female Thermometer: Eighteenth-Century Culture and the Invention of the Uncanny* (New York: Oxford University Press, 1995), 159.

21 Cited in Stephen Kern, *The Culture of Time and Space: 1880–1918* (Cambridge, Mass.: Harvard University Press, 1983), p. 118.

22 Jonathan Crary, *Techniques of the Observer: On Vision and Modernity in the Nineteenth Century* (Cambridge, Mass.: MIT Press, 1990), pp. 113–14.

23 Cited in David McCullough, *The Great Bridge* (New York: Touchstone, 1972), p. 123.

24 Rem Koolhaas, *Delirious New York* (New York: Monacelli Press, 1994), p. 30.

25 See Alan Trachtenberg's *The Incorporation of America: Culture and Society in the Gilded Age* (New York: Hill and Wang, 1982).

26 Hale's Tours were a fairground attraction, most popular from 1904 until 1906, that simulated a railway journey. Audiences boarded a mock train carriage, and while they watched a film of exotic locales (often filmed from the front of a train), the coach would be rocked back and forth and a breeze might even blow over them.

27 And my thanks to the nameless student in Cologne who brought up the phenomenon of raves, where some variation of techno provides an unrelenting stimulus and permits a trancelike (dis)engagement.

28 Castle, *The Female Thermometer*, pp. 137–38. Also see Tom Gunning, "Phantom Images and Modern Manifestations: Spirit Photography, Magic Theater, Trick Films, and Photography's Uncanny," in *Fugitive Images: From Photography to Video*, ed. Patrice Petro (Bloomington: Indiana University Press, 1995), pp. 42–71.

29 Geoffrey O'Brien, *The Phantom Empire* (New York: W. W. Norton, 1993), p. 109.

30 Christopher Finch, *Special Effects: Creating Movie Magic* (New York: Abbeville Press, 1984), p. 124. The success of Kubrick's film, at least from a technical standpoint, marked the beginning of a vogue for special effects "stars" and led to Trumbull's directorial debut on *Silent Running* (1971). The success of *Easy Rider* (1969) had led Universal to produce low-budget films by unknown directors. *Silent Running* was significant for its overall quality, despite a very small budget, and for its opening sequence, in which a natural forest is gradually situated aboard the hyper-technologized space of a vast spacecraft (the ship, a mass of jutting details, influenced all subsequent science fiction design work).

31 Trumbull had nothing to do with the reshot ending within the ship, but his unit's work anticipates the design of *Star Trek* and the movements in depth that dominate *Blade Runner*. His next major effects projects, *Star Trek: The Motion Picture* and *Blade Runner*, both involved "visual futurist" Syd Mead as a designer.

32 Jean Epstein, *"Bonjour Cinéma* and other Writings," trans. Tom Milne, *Afterimage* 10 (1981): 18.

33 Belton, *Widescreen Cinema*, pp. 36–38.

34 And aren't many American epics somehow incoherent? I think of *Leaves of Grass*, *Moby-Dick*, and *The Wild Bunch* as exemplary of a vast and inclusive incoherence.

35 Malcolm Le Grice uncharitably compared Belson's work to the car chase in *The French Connection* (1970), directed by William Friedkin, a comparison with some ironic relevance to this essay. See Le Grice's *Abstract Cinema and Beyond* (Cambridge: MIT Press, 1977), p. 83.

36 In the 1990s, movie theaters began to expand again, and people again seemed to take some pleasure in "going to the movies."

37 See Bukatman, *Terminal Identity,* for more on virtual reality and the myth of total cinema.

38 Belton, *Widescreen Cinema,* p. 197.

39 *Sunrise* was written by Carl Mayer, photographed by Karl Struss, and directed by F. W. Mumau, all of whom deserve credit for this extraordinary piece of cinema.

40 "Tracking shots are a question of morality," Godard once pronounced.

41 Leo Charney, "In a Moment: Film and the Philosophy of Modernity," in *Cinema and the Invention of Modern Life,* ed. Leo Charney and Vanessa R. Schwartz (Berkeley: University of California Press, 1996), p. 292.

42 IMAX filmmaker Ben Shedd has referred to the IMAX as an "unframed" cinema. By the way, isn't it odd that astronauts are America's foremost IMAX cinematographers? Clearly, in films like *The Dream Is Alive,* hypercinematic spectacle is intended as public relations, as a means of reinstating the spectacular effects of the space program itself.

43 It also is interesting that both *Silent Running* and *Brainstorm* feature anthropomorphized movie cameras.

44 Annette Michelson has noted the homology between the shape of the Super Panavision 70 screen on which *2001* played, and the helmet visors of the astronauts in the film. See her "Bodies in Space: Film as 'Carnal Knowledge,' " *Artforum International* 7, no. 6 (1969): 54–63.

45 No doubt VR will overcome its early technological limitations; that goes without saying. My comments pertain to existing VR systems and their contrast with immersive cinematic systems.

46 I always get in trouble when I talk about fandom, so I instead defer to these far more knowledgeable and interesting authors: Henry Jenkins III, "*Star Trek* Rerun, Reread, Rewritten: Fan Writing as Textual Poaching," in *Close Encounters: Film, Feminism, and Science Fiction,* ed. Constance Penley et al. (Minneapolis: University of Minnesota Press, 1991), pp. 171–202; Michael Jindra, "*Star Trek* Fandom as a Religious Phenomenon," *Sociology of Religion* 55, no. 1 (1994): 27–51; Constance Penley, "Brownian Motion: Women, Tactics, and Technology," in *Technoculture,* ed. Constance Penley and Andrew Ross (Minneapolis: University of Minnesota Press, 1991), pp. 135–61.

47 As Siegfried Kracauer once optimistically noted of the "mass ornament" of chorus lines, sporting events, and parades in Weimar Germany. See *The Mass Ornament: Weimar Essays,* trans. Thomas Y. Levin (Cambridge, Mass.: Harvard University Press, 1995).

48 Gaston Bachelard, "Introduction," *The Poetics of Space,* trans. Maria Jolas (Boston: Beacon Press, 1964).

49 See Mark Dery's *Escape Velocity: Cyberculture at the End of the Century* (New York: Grove Press, 1996).

50 The Situationist International was a loosely constituted gathering of political radicals who, between the late 1950s and mid-1960s, shared a distrust of the labor orientation of both capitalist and communist modes of production. Through a series of manifestos, slogans, and collages of advertisements and comic strips, they called for a thorough rethinking of the relation between the individual and society. One of their slogans, made popular again during the May 1968 uprisings in France, was "Under the pavement, the beach!"

Independents and Independence

John Cassavetes:

Amateur Director

Ivone Margulies

I want my films to reflect a truly democratic spirit and I find myself siding with the lone minority. John Cassavetes[1]

John Cassavetes' cinema summons a mythical image of America, one that mixes inexact quantities of jazz, scotch, and emotion. Cassavetes added to this romantic cocktail an insistence on complete creative freedom. And he looked at failure as an amateur would: it allowed him to avoid sophistication and to preserve his original freshness.[2] Losers constantly start anew— a romantic myth of self-construction that resides at the center of Cassavetes' filmmaking. Part of a classic sixties' dropout sensibility, this anticareerist creed was expressed in Cassavetes' depiction of tenacious losers as well as in his wary positioning vis-à-vis Hollywood. The director's romantic liberalism, his fondness for marginal, eccentric characters, tempered with a complex and rhythmic energy generated, on-screen, a harsh image of the American middle class. His characters' idiosyncrasies are simultaneously celebrated and challenged, and it is the particular Cassavetian shape of this ordeal that interests me here.

This essay investigates the elements that define Cassavetes' insistence on individual expressivity—both his own as a filmmaker, and that of his characters and actors. Cassavetes professed that art and authorship are effective antidotes to the mediocrity of the Hollywood product, an auteristic view that made him the object of both the "Movie Journal" column of Jonas Mekas, the organizer and defender of alternative film,[3] and the cinephile pantheon of first-generation film school filmmakers like Martin

Scorsese.[4] *Shadows,* Cassavetes' first film, made in 1958–59, was championed by New York experimental and avant-garde artists, and his later failure to adapt to Hollywood, where he was invited to work with promises of creative freedom, assured his status as an auteur. Cassavetes' contribution to contemporary filmmaking is widely acknowledged. Jonathan Demme, Scorsese, Robert Altman, and Francis Coppola all owe a debt to him.[5] He paired individual vision with communal work. Like Altman, and to a lesser extent Coppola, Cassavetes worked with a repertory company of actors, crew members, and producers. But while Altman's and Coppola's success in Hollywood and their desire for autonomy have been channeled into their own high-profile production companies, Cassavetes' mode of production retained a home-movie quality, which, particularly in such close contrast to Hollywood, constantly aligned him with the underground family-and-friends filmmaking economy.[6] This search for an alternative communal affiliation operated not only on the level of production but emerged as a constant theme in his work (*Husbands,* 1970; *The Killing of a Chinese Bookie,* 1976; *Love Streams,* 1984). While Cassavetes shared a history of alternative practices with other filmmakers in the sixties and early seventies, the unique and idealized aspect of his work lies elsewhere.

Cassavetes' belief in artistic expression is cathetically fixated not on his autonomy, but on that of his actors. He posited a moment-to-moment self-creation, based mostly on a collapse of distinctions between acting and being. This notion complicated issues of film construction (as well as of film authorship), inviting into the director's work a continuous tension between spontaneous and scripted actions; natural, authentic movements; and a dramatic structure that is as strong as in any conventional stage play.

This celebration of creative self-expression is transposed from actor to character. Cassavetes' narratives of individual autonomy emerge through a thematic constant: the friction between social scripts—notions of proper behavior—and the alternate corporeal inventiveness of characters / actors. The intensity of this clash is, in turn, negotiated in the encounter between Cassavetes' clear dramatic armature and his seemingly random visual and aural style. While his handheld camera registered an emotional expressiveness, Cassavetes, the scriptwriter, created dramatic backbone—a screenplay with precise, though unexpected, dramatic undulations. He added momentary pockets of theatrical excess—scenes in which the characters can exercise their creativity—to an otherwise clear and compact dramatic design.

This essay will examine Cassavetes' individual auteristic expression,

31 A housewife (Gena Rowlands) loses her footing in John Cassavetes' suburban nightmare, *A Woman Under the Influence.* Photo courtesy of the Museum of Modern Art.

which emerged out of a highly complex communal interaction. I will describe his vision of the artist as an amateur; his respect for pro-filmic reality—mainly that of acting (in *Shadows*); his home-movie style of production (in *Faces,* 1968); his thematics of fall and regeneration (in *Husbands,* 1970, and in *Opening Night,* 1978); and, finally, I will turn to a more detailed analysis of Cassavetes' *A Woman Under the Influence* (1974), an avowedly more ambitious project about a woman who, "like so many others," is driven "crazy trying to play a role she can't fulfill."

The Artist in Breakdown Mode

Cassavetes defined himself as a professional actor and amateur director.[7] In many ways this assertion is undeniable. The director of *Shadows, Too*

Late Blues (1961), *Faces, Husbands, Minnie and Moskowitz* (1971), *A Woman Under the Influence, The Killing of a Chinese Bookie, Opening Night, Gloria* (1980), and *Love Streams* is known to the general public primarily as an actor in Roman Polanski's *Rosemary's Baby* (1968), Robert Aldrich's *The Dirty Dozen* (1967), Martin Ritt's *Edge of the City* (1957), Elaine May's *Mikey and Nicky* (1975), and in numerous television roles, including the lead in the detective series "Johnny Staccato."[8] The irony that a maverick director such as Cassavetes might be known first as an actor was not lost on him. The director / actor gave to the seeming paradox—that his all-consuming passion for directing looked marginal to his acting career—a positive, romantic valence.

It is curious that through a genre film—the film noir,[9] a traditional Hollywood domain—Cassavetes affirms the amateur as the bearer of artistic independence. In *The Killing of a Chinese Bookie,* Cassavetes creates a character whose poise in pursuit of his artistic and showbiz beliefs stands as a mirror image for an independent filmmaker living in Los Angeles. Cosmo Vitelli, played by Ben Gazzara, is the owner and manager of the Crazy Horse West Club, a cheap strip joint. According to Cassavetes, the club is "his world, a world he recreates every night: he writes, he directs, he choreographs, and announces." He constructed "his life as many American men do. He has defined himself in terms of his work; it is his entire existence—paid for in monthly installments [to the mob]."[10] Confronting the mob creditors to whom he might lose his club, Cosmo shows grace under pressure—a hallmark of film-noir heroes. His show must go on. In one of Cassavetes' typical touches, when Cosmo leaves on his killing assignment, his car breaks down, and as he waits for a cab, he calls his club from a pay phone near a gas station out of an Edward Hopper painting. Cosmo indicates by singing to his manager which song should be performed. His concern with his nightclub's routine, in the midst of such tension, is as misplaced as it is dramatically justified. For Cassavetes, one's own work comes first. Detours, such as the mob's murder assignment, exist mainly to assert the amateur's ultimate professionalism.

Cassavetes underlines this link by crosscutting between a moment of crisis at the Crazy Horse West and Cosmo's dilettante murder in Chinatown. Mr. Sophistication, the sadly clownish nightclub MC (played brilliantly by Meade Roberts), refuses to go on stage. Meanwhile, Cosmo murders the Chinese gangster in the barest of shooting scenes. The fragile gangster wades in the pool and whispers to Cosmo, just before being shot, "Sorry, I don't feel so well." This scene seems to unravel in internal slow motion.[11] Cosmo's escape is similarly understated; he flees the Chinese gang first by foot, then by bus.

Intercutting a caricature of MGM musicals with a botched version of Bogart's casualness,[12] Cassavetes suggests the appropriate register for his art. Clean, organized, and polished strip shows do not interest him. As a director and screenwriter, Cassavetes multiplies the occasions for small, pathetic spectacles—on and off stage. He encourages circumstances in which one can make a fool of oneself (and survive), and this awkward chutzpah is the real gauge of one's creative energies. The courage to confront an audience that is visibly bored (as Hugh does in *Shadows*), annoyed (in *The Killing*), or utterly embarrassed (in the various performances in *Husbands* and in *Faces*) is analogous to Cassavetes' notion of the filmmaker's task vis-à-vis his audience. He feeds off the breakdown of polished showmanship and polite behavior, and his characters' ultimate dignity is measured precisely in challenging received ideas of what is proper.

Cosmo offers an allegory of the romantic artist facing the pressures of capital. Blind to practicalities, oblivious to the rules of money, Cosmo, like Cassavetes, displays what seems like an anachronistic tenacity. He bets on individual style to beat the system—a scenario common in mainstream movies in the 1970s. The stakes of individual self-expression are represented by loners who refuse being pinned down—by marriage or job— curiously allegorizing the filmmakers' own projects in romantic visions of a threatened, creative autonomy. The glories and failures of the hippie movement (Arthur Penn's *Alice's Restaurant*, 1969), the struggle of small communities (Penn's *Little Big Man*, 1970), or of small business entrepreneurs (Altman's *McCabe and Mrs. Miller*, 1971), in short the grandeur of individuals or close-knit groups, is sung in a balladlike rhythm that counters a nostalgia for a pre-Vietnam, pre-Watergate innocence with a purposeful ironic deflation.

Various other seventies' films—Bob Rafelson's *Five Easy Pieces* (1970), Coppola's *The Rain People* (1969), for example—portray drifters who do not go too far in their escape from family strictures. Despite gestures toward independence, these characters are shown continuously calling and visiting their relatives. And even in a woman's film like Scorsese's *Alice Doesn't Live Here Anymore* (1975), we see Alice, the protagonist, split, but not for long, between a desire for self-expression, family responsibility, and the urge for romance. Coppola's last scene in *The Conversation* (1974) figures, sublimely, this romantic bankruptcy. After an orgy of corrupt corporate politics, Harry Caul (Gene Hackman) paranoically retreats from his position as a surveillance expert to that of an artist; from a deft reader of voices, he becomes passionately deaf to sounds other than his own tenor saxophone.

Raymond Carney appropriately titles one of his books on Cassavetes *American Dreaming*, and, indeed, one of the issues posed by the filmmaker's romantic investment in individual self-expression regards the extent to which it feeds on or questions a dream whose shaky status during the late sixties and early seventies prompted a series of narratives of rebellion—from mild leftist critiques of the golden promise of capitalist society (*The Graduate*, 1967; *Easy Rider*, 1969) to right-wing allegories of violence bringing order to a chaotic pluralistic society (*The French Connection*, 1971; *Dirty Harry*, 1972). Cassavetes' work is often cited in passing as somehow marginal to a discussion of the complex negotiations of cinema and political disaffection in the early seventies.[13] With the exception of *Shadows* and *A Woman Under the Influence*, Cassavetes does not seem to directly engage social issues. His films partake of a general mistrust of social conventions. As Cassavetes sustained, with a great ambivalence, his faith in the little man, his retreat to what some critics deemed a micro social scale—the cinéma-vérité renditions of family, small business, and friendship—allowed him to consider individualism—as well as a correlate democratic acceptance—in a different register.

The filmmaker operates through hyperbole. He takes individual particularities to an extreme. He films idiosyncratic portraits of unique people in what could be seen as a representational pluralism. The main characters in his work behave in unexpected ways.[14] They change their minds and moods because, as Ben in *Shadows*, they might not know where they are going. Defined by their immediate responses to events and by extreme temper swings, Cassavetes' characters are given room to explore the full range of their emotions, and the camera finds its mobile rhythm by following the varying performances. From *Faces* to *Love Streams*, the characters frame their own action. They either perform in or "direct" discrete shows that parade their irreducible and unique existential conditions. They display amazing aplomb in the face of threats, like Gloria in *Gloria*, or Cosmo Vitelli in *The Killing of a Chinese Bookie*, or, more frequently, they show, by way of contrast, how stuck-up and repressed are those who look on and refuse to join in song, dance, or play.

In *Husbands*, Harry, also played by Ben Gazzara, insists that each person at his bar's roundtable sing a song. When one of the women starts singing, Gus (Cassavetes), Archie (Peter Falk), and Harry join forces as an intransigent directing unit. In turn, they claim her singing to be "unreal," "horrible," or they urge "please don't sing," "please, once more, from the heart!," in commands bordering on psychological torture. Her failure to express herself in the song, as well as the trio's demands for authenticity, are of course another image for Cassavetes' directorial creed. While directing *Faces* he recalled, "I kept telling them—we've got to go further, and we've got to go underneath. I was giving these amateurish directions. I would stand there like some tyrant—to the point where everyone would want to quit—waiting with great faith and apprehension for this miracle to take place."[15] Cassavetes' films exhibit scenes in which the spectacle, on the verge of collapse, finally takes place. In *The Killing of a Chinese Bookie,* and in *Opening Night,* both audience and backstage personnel anxiously wait for such a miracle to happen. Delays and breakdowns are as important as the play or show that is finally staged. The preference for the unfinished, for the process over the result, situates Cassavetes far from Hollywood and its penchant for special effects—and happy endings.

Beginnings:
The First New American Cinema

John Cassavetes' respect for actors' working methods and existential conundrums can be explained by his experience as an actor. He trained at the American Academy of Dramatic Arts in New York City, where many faculty members were heavily influenced by Stanislavskian and Method acting techniques. Cassavetes' approach to acting, although different from the approaches of famous disciples of Lee Strasberg or Stella Adler—for example, Rod Steiger, Marlon Brando, James Dean—retained the same impulse toward precision and personal expression.[16] After graduating from drama school, Cassavetes had occasional parts in movies and on television before acting in leading roles in the major TV drama series of the time (Kraft Theatre and Lux Playhouse). In 1956 he starred in director Don Siegel's *Crime in the Streets,* and the following year in Martin Ritt's *Edge of the City.* Interviews were conducted with him for such large-circulation magazines as *Look* and *TV Guide,* and he was soon recognized in Hollywood as a rising star.[17] In 1957 he founded, with Burt Lane, the Variety Arts Studio, a drama workshop on West 48th Street.[18] The workshop was the training ground for scenes developed in *Shadows,* and it was in part because he felt passionate about certain improvisatory exercises that Cassavetes decided to make a film.

Shadows was produced independently. Its initial backing came in a scene, seemingly out of a movie, one charged with the populist belief in individual entrepreneurship that characterized both Cassavetes and Frank Capra, his favorite director.[19] On the occasion of a casual discussion about cinema on the radio show "Jean Shepherd's Night People's Story," airing at 1 A.M., Cassavetes was asked how he intended to raise money to make the kinds of films he would like to see on-screen. The soon-to-be film-maker answered that if people wanted to see a film "about people," they should contribute money. When, in a replay of the penny-donation scene in *Mr. Smith Goes to Washington* (1939), two thousand listeners responded with dollar bills, the film's production was launched.[20]

Shadows depicts a few days in the lives of two brothers and a sister played by Ben Carruthers, Hugh Heard, and Leila Goldoni. In the spirit of authenticity suggested by the film's cinéma-vérité style, all three use their real names in the film. The plot moves in fits and starts, matching its improvisatory dialogues and situations to background music composed by Jazz legend Charles Mingus.

Shot in real time and cut in a loosely structured, episodic way, *Shadows* exemplified a new approach for fiction filmmaking and a true alternative to the feature product of Hollywood. The film seemed to document as well as provoke the events it showed. *Shadows* proudly announces in its closing credits: "The Film You Have Just Seen Was an Improvisation."

Because of its improvisatory nature, *Shadows* was routinely associated with *Pull My Daisy* (1960), the Beat film by Alfred Leslie and Robert Frank that featured Gregory Corso, Allen Ginsberg, and the seemingly ad-lib voice-over narration of Jack Kerouac. Part of an emerging cinema called the American New Wave, or the New American Cinema,[21] both films adopted the free style of Beat poetry and the rhythms of black music to ground their pulse in authentic American forms.[22] They recorded artists and intellectuals who, like the filmmakers, were in search of authentic expression[23] and who reshaped their identities in constant acts of artistic self-creation. Kerouac's exhortation on personal writing injected the project with immediacy: "Never afterthink to 'improve' or defray impressions, as the best writing is always the most personal wrung-out tossed from cradle warm protective mind—tap from yourself the song of yourself, blow, now! your way is the only way."[24]

Added to the urge for individual expression from the Beat front was the influence of the relatively late production and release of American neorealist films.[25] For Cassavetes, these films' use of location shooting, non-professional actors, and episodic structures promised a fresh cinematic narrative, a direct confrontation with reality.[26] In this new form of cinematic realism, which marked Cassavetes' contribution to the American New Wave, editing and cinematography are supposed to react with utmost sensitivity to the rhythms and necessities of life—that is, people or actors, in front of the camera. This approach to film was common to an entire generation of independent, low-budget filmmakers ranging from Morris Engel (*Lovers and Lollipops*, 1955, and *The Little Fugitive*, co-directed with Ruth Orkin and Ray Ashley, 1953), and Shirley Clarke (*The Connection*, 1961, and *The Cool World*, 1963), to documentarists such as Lionel Rogosin (*On the Bowery*, 1956, and *Come Back Africa*, 1960).

Clarke's and Rogosin's work particularly interested Cassavetes because they were "really interested in their subjects, and in finding out about what they think and feel."[27] The beauty of their shot composition did not overwhelm subject matter; the unexpected movement and impromptu actions of actors as characters were the sole motivation for camera movement and cuts. Clarke's *The Cool World*, for instance, displays jagged editing and a handheld camera that is relentless and inconclusive in registering the musings of its protagonists. Although it is debatable whether Clarke's rhythm was determined by her object or by her camera, this sensitive interdependence between form and content was what Cassavetes aspired to in his films.

In 1960 the New American Cinema Group was formed. A manifesto launched by twenty-three producers, filmmakers, actors, and theater managers, including Clarke, Rogosin, Peter Bogdanovich, Robert Frank, Alfred Leslie, Emile De Antonio, Ben Carruthers, and Gregory Markopoulos, and organized by Jonas Mekas and Lewis Allen (a stage and film producer), was the group's institutional marker. The manifesto was followed by concrete attempts to organize cooperative production and film distribution for an independent cinema.[28] What set this eclectic group—comprised of avant-garde documentary and fiction filmmakers, producers, and actors—apart was the members' ardent desire for expression unencumbered by prevailing norms. If Hollywood was a clear counter-model, neither the American avant-garde nor neorealism were the exclusive ideals. The insistence on artistic and personal authenticity and the respect for reality, the impatience with the medium's stagnation, and the desire for an expressive niche created this new filmic development.

"The First Statement of the New American Cinema Group," published in *Film Culture* in 1961, echoed the contempt displayed by other new wave cinemas around the world for their established national cinemas. The New American Cinema coincided with the French New Wave and Brazilian Cinema Novo, among others, in its break with the "slickness of the product film." These groups concomitantly defended the new auteurs.[29] "The official cinema . . . is running out of breath, is morally corrupt, aesthetically obsolete, thematically superficial, temperamentally boring. . . . The very slickness of their execution has become a perversion covering the falsity of their themes, their lack of sensitivity, their lack of style."[30]

Cassavetes' own aversion to "slickness of execution" is crucial to all of his work. Although he was not officially part of the New American Cinema Group, he knew most of the people in it and defended similar ideas in relation to creative freedom. In a short article, "What's Wrong with Hollywood," published in *Film Culture* in 1959, he stated one of his oft-repeated creeds: "the producer intimidates the artist's new thought with great sums of money and . . . clings to past references of box-office triumphs and valueless experience. The average artist, therefore is forced to compromise. And the cost of the compromise is the betrayal of basic beliefs . . . the artist is thrown out of motion pictures, and the businessman makes his entrance."[31]

Shadows was championed by Jonas Mekas as an example of how small budgets—Cassavetes' film was shot for $20,000—could generate great work. Screened at three free midnight screenings at the Paris Theatre on West 58th Street to packed houses at the end of 1958, *Shadows* received *Film Culture's* First Independent Film Award, with this citation by Mekas: "*Shadows* was able to break out of conventional molds and traps and retain original freshness . . . [in it] an atmosphere of New York night life [is] vividly, cinematically, and truly caught. . . . It breathes an immediacy that the cinema of today vitally needs if it is to be a living and contemporary art."[32]

Even though *Shadows* had been heralded by the American experimental film community, Cassavetes thought the film indulgent.[33] The thrill of fluid images was for him a consequence of falling in love with his hand-held Arriflex camera. "I got a pleasing rhythm, but which had nothing to do with my characters."[34] As a result, the "emotional expressiveness of the first version . . . was dissipated in its generality—the emotions were not precise and particularized."[35] In an indicative revision, Cassavetes decided to shoot additional scenes and reedit the film, making his priorities clear. In the second version, Cassavetes' subordinates shot composition, plot, and shooting technique to his characters' development in an effort to find the right way to communicate.[36] The second and existing version of the film was then attacked by Mekas, who felt betrayed by what he saw as a "bad commercial film."

Because and despite of his experimental drive, Cassavetes' work does not fit commercial or avant-garde film agendas. In this respect, his association with the New York school of filmmaking cannot be underestimated. In "What's Wrong with Hollywood," Cassavetes selected two films that he believed would "affect strongly the future of American motion pictures"[37]—*The Goddess* (1958; written by Paddy Chayefsky of *Marty* and later of *Network* fame) and director Sidney Lumet's *12 Angry Men* (1957). Not only because he had acted in their films, but because of his own preference for screenplays that depend on and enhance gritty, strong acting, Cassavetes regularly associated himself with filmmakers like Martin Ritt, who directed television drama, was also an actor (at the Group Theatre), and had a reputation as an actors' director (in the Actors Studio where he worked with Rod Steiger, Paul Newman, and Joanne Woodward); Don Siegel, who directed movies for television in the sixties and was known for his penchant for on-the-spot improvisation; and Lumet, whose speed and efficiency in completing a film—he did *12 Angry Men* in twenty days—was a consequence of his TV experience.

After Cassavetes finished *Shadows,* he accepted work as the lead in a detective series for NBC, *Johnny Staccato,* and the money he earned in the series helped him pay off loans that financed his first film. He also got to direct some of the series' episodes.[38] When one watches Cassavetes' later work, simply lighted and shot in extremely mobile long takes so as not to hinder the flow of performance, this TV apprenticeship appears to inflect a highly cinematic style. The plasticity of Cassavetes' spaces is one of his trademarks. But the characters are still boxed in a single set and in one extended shot. Even his love of close-ups seems a hyperbolic comment on television's constraints. His emphatic denial of studio pyrotechnics comes close to the economy of the TV studio or the stage, and he uses the single set's centripetal energy to have characters collide and performances bounce off each other.

His only truly improvised film is a rightful marker of a new cinema. *Shadows* gave room to young people's contradictory desire for the absolute and the transitory, and the film displays the central existentialist ethic of the fifties: to improvise, to choose one's action and behavior on a moment-to-moment basis, which allows one to hold bad faith at bay, to fully pursue one's authentic self.

These principles informed the two realist schools close to Cassavetes. On the one hand, the experimental docudramas of Clarke or Rogosin provide a realist model based on the analogy of camera and editing with the texture of locales and actual, real-time acting. Cassavetes' handheld camera works as an emotional seismograph registering and creating screen correlates for his characters' sensations, translating an empathetic, action-filming record.[39] On the other hand, films and TV dramas by actor-directors are strong reference points. Cassavetes' work registers a modern sense of drama, with a character's mood swings minutely recorded. The focus on the authenticity of performances and the contradictions of complex characters, a strong element in Cassavetes' dramaturgy, constitute the filmmaker's share in the realist aesthetics issuing from theater and TV in the early sixties.

Home Movies in Los Angeles: Style and Mode of Production

Cassavetes wrote *Faces* originally as a play, at a time when he thought that the only free form of expression left for the actor was the stage.[40] He started writing it in 1962 after being consigned to Hollywood limbo, and he began shooting in 1965. For six months he briefly interrupted production and took a job running a TV package company, in partnership with Screen Gems (the television production company associated with Columbia), in order to finance the picture.[41] In Cassavetes' own words, *Faces* evolved into a "movie about the middle aged, high middle income bracket people . . . the white American society that certain social groups talk about all the time." He knew there was "something to be said about these people's . . . insular existence."[42] *Faces* mixed professional and nonprofessional actors, and with it Cassavetes was able to prove his "poor man's philosophy."[43] In a mode that became the norm for his future productions, all actors and crew worked with deferred salaries. As an artist, Cassavetes seemed spurred on by practical difficulties. He filmed all of his very long first screenplay draft, and during the three-year editing period he acted in five films to finance the project. As Cassavetes was proud to state, the film was made with only one professional technician (Al Ruban, who had worked as a cameraman on some films in New York), and its many technical mistakes—they had to painstakingly synchronize each line of a dialogue in postproduction after a sound recording disaster—were (almost all) solved with artisanal dedication.[44]

Faces became, for Cassavetes, a sort of private manifesto. More than a film, "it became a way of life."[45] Shooting nearly every day for six months, taking shooting shifts with the camera operator and the producer, and moving fast in order to follow the actors, the filmmaker gradually instituted his working signature.

Cassavetes' films are characterized by the thematic complexity of his changeable characters and, on the level of style, by an action-painting energy that imprints on the screen an absorption in the scene. His fluid camera seems to follow the actors' high levels of energy and as a result energizes the space around them. The combination in his films of a great number of close-ups, handheld camerawork, deep space mise-en-scène, and distanced telephoto shots produces an elastic shot space, molded, so it seems, by a tactile impulse. While characters touch each other, bump into or move away from each other, the camera closely registers their minute expressions, but from a distance it also depicts the changing space around bodies. Cassavetes avoids formalized shots so that his actors stay in character when not on screen. "Even though we sometimes shoot very tight," he says, "they never know when the camera is going to swing onto them, so everyone has to play every moment." The "fluidity of the camera really keeps it alive."[46]

Because his camera is so supple, and his characters and actors express themselves so imaginatively, Cassavetes' films give the impression of being improvised. But, in fact, except for *Shadows*, all of his work is scripted. He filmed only after extensive rehearsals in which actors were encouraged to bring up suggestions.[47] When he was filmed in the process of directing *Husbands* in a documentary for BBC, Cassavetes is seen arguing with Peter Falk for reducing one-and-a-half pages of script to a few words.[48] When the question about improvisation is asked of Cassavetes, he answers in a mystifying way: "The emotion was improvisation. The lines were written."[49] With this paradox, he suggests that the relationship between script and improvisation is not reducible to an easy polarity. What one perceives as improvisation corresponds to a process where unexpected reactions are scripted, and at the same time the acting is unspoiled by clichés. The uncontrolled event, the excessive gesture, the spilling out of emotions is always written in, and these free forms are the very stuff of Cassavetes' scripts. If the actors give depth to their characters, the filmmaker helps them on two fronts—with his script and with his ferocious demand for an extreme degree of inventiveness in making the characters' actions seem original.

In *Faces*, for instance, Cassavetes discussed with John Marley, the lead actor, what his character would do if he found his wife in the middle of an affair. Until the actual scene was filmed, nothing was decided. When the scene was finally shot, Marley decided that the character would find it difficult to merely leave his wife. In one of the greatest endings in cinema, the husband goes up the stairs, sits down, and as his wife sits some steps higher, they connect by passing cigarettes and matches over a silent and exhausted relationship. One coughs, the other coughs, one of them leaves, then comes back and sits down. One, then the other, leaves, and the set, at the end, is empty, establishing a clear, almost theatrical, closure.

Cassavetes' method of filming corresponds to his intended open narratives. He is a "slave to his actors," to their timing and their "comfortableness" in embodying a character. He detests the idea that a film is made by camera composition or cinematographic effects. He does not like "perfect" or "beautiful" shots.[50] Technical errors are irrelevant if a scene has strong inner logic. For example, when in *Faces* some of the scenes were overexposed, he interpreted this photographic lapse as an expressive addition to a strong scene depicting weariness, sadness, and sterility.[51] To protect his actors from technical interruptions such as lighting changes or makeup corrections, Cassavetes moved fast. "We light the entire room and have the camera available for the actors using the entire depth of the set. . . ."[52] Once the actors get started, the handheld camera, usually operated by the filmmaker himself, rushes after them in extreme long takes.[53] A second agile unit, using a telephoto lens, keeps different planes of the action alternately in focus, linking them in the fast motility of his deep-space mise-en-scène. Long takes in wide-angle and long lenses are used so as not to cramp the actors' time and space.

Cassavetes' preference for long takes and his use of two simultaneously running cameras send his filming ratio to a level well above average Hollywood productions of the period. His desire to catch the right moment, the very instant when a character says the scripted words as if from his or her own heart, motivates endless shooting and a vast expenditure on film footage. For *A Woman Under the Influence* he shot 600,000 feet of film, then finally used only 14,000.[54] This surplus of footage eventually forces an elliptical narrative style. When editing, Cassavetes never cuts bits and pieces. He prefers instead to eliminate entire shot sequences, leaving at times a mere twenty seconds at its tail, or instead he leaves sequences whole. In *Faces*, for example we see the end of a scene when Richard and Freddie are with Jeannie without any prior exposition. The result is a narrative so elliptical that it can only be read retroactively in the very way that people interact. Instead of being introduced to characters, we are thrown into the heart of their awkward dealings or into their tired familiarity.[55]

In accordance with his economic needs as well as his antiprofessional

32 An actress on the verge: Gena Rowlands in John Cassavetes' *Gloria* (Columbia Pictures, 1980).

ethos, Cassavetes encouraged his cast and crew to participate in *Faces* as full partners. Relying mainly on family members and friends and on actors who became friends, Cassavetes created a work situation in which people joined in a creative and communal atmosphere (sharply distinguishing his working methods from Hollywood's).⁵⁶ In any given Cassavetes film, the credits consist of permutations of a regular cast of friends. Gena Rowlands, his wife; Katherine Cassavetes, his mother; and Lady Rowlands, Gena's mother, all appear in the films. Peter Falk and Ben Gazzara, who, after their work in *Husbands,* alternate in Cassavetes productions; Fred Draper, who appeared in *Faces* and *A Woman Under the Influence;* Seymour Cassel, who joined Cassavetes at the time of his acting workshop, acted in *Faces, Minnie and Moskowitz, The Killing of a Chinese Bookie,* and *Love Streams,* and was associate producer of *Shadows;* Al Ruban, who worked as director of cinematography, editor, and producer in *Faces,* was the director of photography for *Opening Night* and *Love Streams,* and produced *Minnie and Moskowitz, Faces,* and *Husbands,* also acting as one of the mob creditors in *Chinese Bookie;* Sam Shaw, who produced *Husbands, A Woman, Opening Night,* and *Gloria;* Maurice McEndree, who produced *Shadows* and *Faces;* and Bo Harwood, who wrote the score for *Minnie, Chinese Bookie, A Woman,* and *Opening Night.* Without any of the comforts of a subsidized repertory company, Cassavetes gathered around him a group of people who, like him, placed their artistic commitment above their ongoing careers. They interrupted their work with Cassavetes to take acting jobs, and they returned to act in *Faces* or *A Woman Under the Influence* as one comes back to one's own life routine.

Communitas: Cassavetes' Ground Zero

Cassavetes' 1972 film *Husbands* starts with pictures of the filmmaker, Falk, Gazzara, and David Rowlands clowning for the camera, showing their muscles. Gathered around the pool, family poses show Gena Rowlands with Cassavetes and their child and with other women, men, and children huddled together. This credit sequence, made up of still images, introduces us to Cassavetes' filming universe. Are these photos of the characters' families (almost completely absent from the film itself), or are they snapshots of the actors' own relatives taken during a break? This imaginary convergence of actor and character, of course, is an idealized vision of the Cassavetes artistic community. But given this perfect family picture (he despised an atmosphere of awe toward the director),[57] one asks why his characters, usually shown in family situations, are instead in constant search of another community, another living configuration?

Cassavetes shows the sterility of long-standing relationships; then, in contrast, he suggests fresher modes of being. In *Husbands*, the only marital relationship shown reveals a repeated scene of abuse. The film's title names the social role that Gus, Archie, and Harry evade. After their fourth buddy, seen in the credit photos, unexpectedly dies, they need to bring themselves to their most basic core of being. Suspending for two days their jobs and societal links, the three buddies undergo a rite of passage. They mark their friend's death by inventing their own mourning ritual—drinking themselves sick, gambling, not showering, not sleeping, going to London.

Anthropologist Victor Turner (following Van Gennep's influential work, *The Rites of Passage*) calls this process of losing oneself in meaningless behavior, in anonymous, rankless play, *liminality*.[58] Liminality corresponds to the transitional period in a rite of passage during which a group of individuals undergoes a series of humbling measures that equate them in terms of status and property. Turner suggests that in liminal rites a new, unstructured society, a "communitas" emerges.[59] Rites of passage, demarcating the passage between two major moments in life—for instance, before marriage, childbirth, and puberty—involve the stripping away of one's public identity, staging a sort of social invisibility, a form of death that is followed by rebirth.

Given that for Cassavetes an individual's creativity can flourish only within an ideal state of communitas, it is no wonder that the most intense, dramatic moments in his films depict liminal stages of existence. Drinking and falling are Cassavetes' preferred representations of liminality. In *Love Streams,* whenever Sarah, played by Gena Rowlands, suffers a disappointment, she reacts by lying on the floor in a sort of broken-down faint. Myrtle Gordon in *Opening Night* drinks continuously, and in rehearsals she is so out of joint with the role she has to play that she anticipates a slap, constantly falling before she is even touched. When she falls, Cassavetes frames an abstracted patch of unfocused red, and this blurred image of the red carpet before Myrtle rolls into the frame stands for the coming into being of the character. Later in the film, Myrtle, whose erratic behavior has everyone guessing if she will be able to perform, appears for the opening night. She is so drunk that she can barely stand up. Her eyes roll, she falls on all fours, she crawls. Manny Victor, the play's director, discloses the affinities of Cassavetes and Method acting; he repeats director Elia Kazan's gesture in *On the Waterfront* by forcing Myrtle, like Brando's Terry Malloy in that film's final scene, to walk on her own.

Cassavetes' work thrives on such bracketed instances of crisis in which characters try, with middling success, to venture into a freer mode of being, to literally change their life scripts. This change is always qualified. In *Faces,* Chet (played by Seymour Cassel), the gigolo who entertains a group of high-class women on their night out, rescues Maria Forst from her attempted suicide. With Maria sitting by his side, her mascara streaking her face, Chet imitates an automaton, compassionately demystifying the image of a supposedly liberated man. All one can do, Cassavetes seems to be saying, is to try, furiously and tenderly, to grasp the few real moments of love—love being the single most-repeated word in his films.

Such poignant moments of bonding often come after an extended break in which the characters exile themselves from their usual social roles, drink themselves to the ground, or, like Maria, will themselves into nonexistence. For the "husbands," this moment comes after their buddy's funeral, a formal and insufficient way of marking death. As Cassavetes explained, "the characters weren't vomiting just because they happened to be drunk: they got drunk so they could vomit—vomit for their dead friend."[60] But their attempt to lose themselves in male routines is marred by the absent scene of their waiting families. At the end of the film, Gus, played by Cassavetes, hesitantly arrives home to a not very welcoming family. This image of a reluctant return home is as important to Cassavetes' oeuvre as the characters' prolonged and haphazard escape. Attempts at independence are often shown to be very short respites from a world structured around gelled responses. In Cassavetes' films, falls mark the dramatic curve announcing the passage from the spectacle of identity crisis to some form of resolution in the shape of compromise or smuggled resistance.

Cassavetes' Humanist Theater: *A Woman Under the Influence*

Cassavetes' films are often described as theatrical, and he directly deals with theater in *Opening Night* and a cabaret show in *The Killing of a Chinese Bookie*. In other films, his ideal of fresh, original action is realized in characters who step out of their everyday lives in momentary spotlighted performances.[61] This theatricality, however, is not restricted to the liberating exhilaration of showbiz or everyday performance. His free-form cinema is traversed by a clear dramatic design,[62] as well as a theatrical space, a rift within his films between those who are on display and those who look on judging.

It is no small matter that in the mid-seventies, at the height of the feminist movement, Cassavetes consciously confronts the "Woman" question precisely through this judging frame. In *A Woman Under the Influence* he depicts Mabel as a woman who is crazy but who in many ways is also perfectly functional as a wife and mother. By providing this dual characterization, Cassavetes felt he was addressing the actual cause of women's fragility—their creative and emotional nature turns into madness as an inevitable response to the limitations of domestic life. On the other hand, this sociological scenario does not correspond to a neat, descriptive project. Rather, Cassavetes' choice of a mad woman protagonist, as well as his harsh picture of her domestic environment, undercuts a casual telos, turning this exploration of a contemporary social issue into a test case of the filmmaker's own romantic agenda.

"She's not crazy!" Nick Longhetti tells his buddy Eddie about his wife, Mabel. "She's unusual, but she's not crazy. She sews, she cooks, she washes the dishes, she takes care of the children." In the film, this qualified description establishes him as an all-too-human barometer of Mabel's oddities. It also verbally enunciates the question posed by the film: can a crazy woman advance both Cassavetes' agenda of imaginative expression *and* the broader issue of society's influences on women? Can his glorification of marginal eccentric types subsist once it is so clearly presented as the pernicious result of a patriarchal hierarchy?

The filmmaker has suggested that the film depicts men and women's irreconcilable differences, as well as the impossibility of women being able to sanely play their assigned role.[63] Nick, conceived by Cassavetes as a "real character . . . subject to the new moralities and the New Woman, and trying to do the best that he could,"[64] is the film's (and the filmmaker's) social spokesman. He is enthralled with her uniqueness. But his erratic demands on Mabel translate a male social optic, itself traversed by multiple pressures (his mother's, his fellow worker's, etc.). Nick's nuanced characterization counterpoints Mabel's portrait, and both seriously complicate any simplistic reading of who or what causes Mabel's behavior.

The film flows in waves of reactions. Nick breaks an unbreakable date with Mabel. After she hangs up the phone, she rises from the chair, and the tight camera framing loses her torso, but not her hands. One hand holds a cigarette, while the other, poised on her waist, is stilled in a frozen spasm. Bo Harwood's brilliant, sparse, jazz-styled score further dissociates this image of Mabel's body from the film's texture. Her disjointed reaction to being stood up is given a physical and cinematic correlate.

Afraid to confront her alone, Nick brings home a large group of hard-hat buddies. She responds to this imposition by playing her wifey role — but oddly. She greets his friends, but she does it by addressing each separately: "Hello, would you like some spaghetti?" Sitting at the end of the table, she asks each man his name and how he is doing. When one of the guys starts singing an aria, she stands up, approaches him amazed, looks at his face and then inside his mouth. How is it possible that such sound comes out of there? she seems to ask with her eyes. Nick looks on in embarrassment.

This movement between doing and watching, between performing and being aware of one's performance is present in other Cassavetes films. In *Faces*, Maria and her friends mate-dance around Chet. In *Opening Night*, as Maurice is embarrassed by her public breakdown, Myrtle tells him, in front of the audience, "Let's not ever forget this is only a play." And when Nick comes back home and goes wild on seeing Mabel's children's party running amok—we see his daughter dashing naked around the house and a horrified neighbor fleeing—Mabel sums up the director's philosophy: "You got embarrassed and made a jerk of yourself. I make a jerk of myself everyday." For each group of socially constrained spectators, Cassavetes creates at least one gloriously peculiar character who confronts their reluctant audiences—Sarah Lawson in *Love Streams,* Mr. Sophistication in *The Killing of a Chinese Bookie,* Mabel in *A Woman Under the Influence.*

What is the role fulfilled by such characters and their spectators? What is the meaning, for Cassavetes, of idiosyncrasy, and what happens once creativity dissolves into craziness, imagination into pain? In Cassavetes' work, eccentric characters are evidence of human complexity. Already when making *A Child Is Waiting,* the filmmaker tried to portray retarded kids as funny, as kids, instead of as "cases." With *A Woman Under the Influence* he faces a similar challenge. Mabel Longhetti is the ultimate example of the filmmaker's antireductive stance. She acts oddly, she has tics, she desperately tries to please, she becomes selfless, and then, in turn, her self escapes her, becoming too visible, too theatrical. By treading a precarious line between a medical and a social definition, between madness and difference, Cassavetes refuses to cast Mabel exclusively in one or the other camp. He stages both possibilities so fully that he preempts the possibility of pat diagnosis. We understand what triggers her reactions, and these times are only a little more extreme than normal ones. She has been stood up, she leaves the house and picks up a man; she feels cornered, she defends herself. But it is through Mabel's peculiarities that Cassavetes can state most clearly what he feels is at stake in individualized, personal expression. To aid a character whose unusualness is misunderstood, the filmmaker activates a panoply of corporeal language. Through Mabel, he renders social pressure and resistance visible. Mabel's nervous, quasi-hysterical body indicates the importance, for Cassavetes, of an alternate, nonrational language. If we are deaf to Mabel, the director and actress invent an idiosyncratic sign language. She cracks her fingers, raises her shoulders, mumbles, makes faces, mimics her mother-in-law and her husband. Twitches, mimicry, and noise signals introduce a preverbal, more truthful form of dialogue. Perhaps this dialogue informs her split self, maybe she is talking to herself, but most importantly for Cassavetes, these tics are proof of an untranslatable, authentic self.

Nevertheless, in a movement typical of Cassavetes' stories, the flights into self-expression, although clear signifiers of authenticity (or innocence), are subject to harsh reality checks. Though Mabel's gestures are both mysterious and spectacular, Cassavetes' insistence on her adaptation to a normal domestic routine is part of a most exacting realism, allowing as it does the perception of operative social forces.

When Mabel has just come home after having been committed to a mental institution for six months, Nick, in a characteristically inept gesture, organizes a party with his buddies from the water and power department as well as friends and recent acquaintances. His mother eventually disperses the guests so that only the family is on hand to celebrate Mabel's return. But as they sit around the dinner table, it is apparent that they are unsure how to deal with her. Trying to directly address their awkwardness Nick turns to Mabel and asks: "How was it at the hospital—bad, good, terrible?" And when Mabel softly resists by noting, "Everyone is here," Nick tells her how many people had been there before. Cassavetes multiplies the moments in which the perversity of family relations is at work. Mabel states that Nick and she want to go to bed together, and she asks everyone to leave. Her mother-in-law screeches that she came for a party! Mabel answers Nick's question by telling what happened to her in the hospital, but no one wants to listen. She tells a joke, but her mother-in-law says she has heard it before. She asks her father to stand up for her. The camera stays on her face as she murmurs, "That's not what I mean. Dad, will you stand up for me?" as her father in a shocking refusal to understand the nature of her request has literally stood up from his chair. Mabel has exhausted her entire range of polite, social behavior. It is only when she resorts to her madwoman pose—her body frozen in a swan dance poised on the edge of the sofa—that the family and friends finally "recognize" her and begin to leave the house.

When Mabel leaves the table and starts her stilted gestures, poised on the sofa, Cassavetes focuses on the general area by the door, on the broad entrance hall. We see the relatives' faces as they react to Mabel. Her blurred body, at times a hand, or bits of her dress intrude in the frame. Mabel's show is entirely framed by the preceding scene, and although one understands what caused her "attack," it is not entirely clear how much of this performance is a willed device to get rid of her guests. After they leave, Cassavetes refuses to ease the situation for the film's spectators. We watch as she remains in a trance, as Nick slaps her down onto the floor, as her kids uncomprehendingly try to defend her. The aftermath of this violence is baffling. Mabel stands up, and with Nick she puts the children to bed and washes the blood off her cut wrist. She comments that she must be really nuts, since she does not even know how this all started. When the couple moves the dinner table to make room for their pull-out bed in their convertible dining room-bedroom, they conclude the conversion of the spectacle of bumping pained bodies, the charged scenes of Mabel's breakdown, into a scene of collected domesticity. They pull the curtains of the glass-door partition of the bedroom, signifying the show's end.

Mabel's final integration into the household routine has far-reaching implications. It invites a causal reading (confirmed by the filmmaker in his statements) that blames her condition on a repressive patriarchy. Madness is, in this respect, simply the reverse of polite conformity. While this reading is possible, I am more interested in the aspect of contiguity—how can spectacle and routine coexist in the film as they do? Someone might act crazy, but in the next minute she discusses her difference and washes the dishes. For Cassavetes, in *A Woman Under the Influence* eccentricity is contiguous rather than antithetical to mundane and rational behavior.

The passage from Mabel's breakdown to normality seems part of an abrupt change of register rather than a character's sudden "cure." This change of register is signified by Nick's slap, which is intended to climax and terminate her hysterical pose, and it is in this act of violent contact that we can read Cassavetes' aesthetic creed.

The slap that thrusts Mabel from her pose on the sofa to the floor repeats a paradigmatic movement in Cassavetes' films—that of the regenerative fall. In *A Woman Under the Influence* the slap and the fall mark the potential convergence of (therefore also the needed distinction between) acting and being. In his study on gesture, semiotician Emilio Garroni uses the example of a fist raised in a threat as an example of a blurred performative moment. Is the threat already part of the beating or is it a mere sign? The continuity of natural movement has plagued semioticians with questions such as this one—when and where does meaning detach itself from nature? *Opening Night*'s slapping scene is one such semiotic crux. When Myrtle falls in anticipation of her stage slap, when she refuses to get up after falling, Cassavetes uses physical contact as transcending artifice.

In *Opening Night* Cassavetes searches for a deeper level of truth by separating effect from cause, by determining a fuller, more relevant reason for falling than just being hit. This reason must remain invisible. And it relates in Cassavetes' work to a very private motivation, an individual motor that works in fits and starts, as a constant—albeit slight—change in script. Cassavetes prefers not to explain or justify a character's behavior. Hence, if Mabel's shift from "hysteria" to "normality" seems to excessively confirm the male scenario of female hysteria solved by a well-intentioned manly slap, it also points to Cassavetes' desire to occlude explanation, to establish a purposeful blank spot in terms of character motivation.

Cassavetes' ideal of performance is best answered by brusque changes in register, for through such changes actors seem to confirm a continual existential exploration. In *A Woman Under the Influence*, however, he seems to raise the stakes of his social engagement when he suggests Mabel's sudden adaptation to domestic routine. Hence, it is through the very suddenness of this most conventional of dramatic devices—the slap and the fall—that Cassavetes refuses to explain Mabel's theatricality—her eruptions into the mode of spectacle—as well as her return to normality.

The complex transaction between a theatrical and a realist register in *A Woman Under the Influence* can also be explained, in part, by the film's production history. The script for the film came out of one of three plays written in 1971, and it was meant to be staged in successive nights with Gena Rowlands in the main role.[65] Several of Cassavetes' films were written initially as plays (*Faces, Love Streams*, written by Ted Allen and adapted for the screen by Allen and Cassavetes); indeed, the dramatic form is embedded in Cassavetes' cinema, placing the very notion of adaptation (from play to film) in question.

A Woman Under the Influence presents a number of exterior scenes involving Nick's blue-collar work in the water and power department, his beach outing with his kids after he commits his wife, and Mabel's wanderings through Los Angeles. But, true to the "stage version," most of it takes place inside the house, a fact that, according to Cassavetes, scared away potential investors.[66] The lack of movement and openness apparent in the script was addressed in practical ways. Though the house remained central to Cassavetes' sense of realistic detail, his use of telephoto lenses diminished the feeling of confinement—"the camera could be far away and the actors wouldn't be constricted by its proximity."[67]

After careful location scouting (they looked at about 150 houses in Los Angeles), Cassavetes found one that depicted "the kind of blue collar existence" that he had in mind.[68] Nevertheless, despite all the filmmaker's care in contextual setting, most of the film takes place in the dining room and the foyer, "basically from two angles,"[69] a structural and dramatic economy proper to a play. The camera's distance from the characters, as well as its fluidity, disguises the potential staginess of the scene. The repetition of Mabel's physical positioning in each set of scenes—at the head of the table and then cornered in the living room facing the entrance hall—is a clue to Cassavetes' dramatic sensibility. The broad entrance hall, which simultaneously separates and links private and public spaces, becomes a physical correlate of Mabel's permeability to external and internal influences. In one moment, the hall is the stage for Mabel's spectacular breakdowns (witnessed by friends and family); in another, the hall, like the convertible bedroom, allows for more mundane exchanges.

In 1977, in an essay entitled "Kitchen Without Kitch: Beyond the New

Wave I," Manny Farber and Patricia Patterson divided 1970s' films into two basic cinematic structures: dispersal and shallowboxed space.[70] Altman's *McCabe and Mrs. Miller*, Scorsese's *Mean Streets*, and Penn's *Alice's Restaurant* are all examples of "dispersed movies" where "the brusque, ragged movement . . . and ballad-like rhythm" are meant to suggest inconclusiveness and a mobile evasion of "big statements." On the other hand, Nagisa Oshima's *In the Realm of the Senses* (1976), Chantal Akerman's *Jeanne Dielman, 23 Quai du Commerce, 1080 Bruxelles* (1977), and Rainer Werner Fassbinder's *Katzelmacher* (1969) exemplify a minimalist sensibility, with the "low population image squared to the edges of the frame." Films in the second category often focus on a single, repetitive activity such as having sex, cooking, or gossiping. The repetitious setups of these films, as well as the nature of the events depicted, invokes a ritualized and—in a sense—theatrical space.

Cassavetes' position in this dispersed/boxed-in grid is singular. His visual style seems at first to work mostly to diffuse and minimize the radical theatricality permeating his cinema. His handheld camera registers a visual and performative volatility in a messy, noncomposed look. He alternates the clear, spatial display of long shots filmed in wide-angle lenses with telephoto shots whose dramatic center is minutely threatened by multiple layerings. Bodies, masses, and shadows obstruct and reshape the image, abstracting parts of it and commanding an even greater interest for what remains identifiable. Even when he centers on one character's performance, he devotes as much screen time to the other characters' reactions. On-screen, his main characters are often covered up by minor characters. For example, when Mabel's attack is extended to a pitch in which performance and pathology have become indistinguishable, Cassavetes suddenly changes the scene to a minor register.

After enacting shows of individual extremity, he (along with other filmmakers marked by 1960s' and 1970s' political disaffection) modulates the grand statement through a minor-key representation of the everyday.[71] The zoom lens, a trademark of Cassavetes and much of early-seventies' cinema, evinces an aesthetic impulse divided between engagement and distance.[72] An active zoom registers the pressure to follow the minute changes of reality; its shift of focus signifies the filmmaker's wandering eye. It is a realism split between a democratic depiction of central and peripheral action (Altman in *McCabe and Mrs. Miller*, Penn in *Alice's Restaurant*), and the expressive emphasis on character emotion (Scorsese in *Mean Streets*). The zoom also creates a modality of distance proper to the seventies' sensibility, its machinelike movement suggesting an ironic objectivity. This link between the optical qualities of the camera and ironic distance had been explored by Andy Warhol in the 1960s when he reversed the zoom's traditional valence of social interest (promoted in documentary film) into pure, random description. In *The Chelsea Girls* (1966) and *Four Stars* (1966), for instance, Warhol arbitrarily focused on different areas of the frame, suggesting that any surface was worthy of being looked at. In this form of record, the zoom operated as a willed digression, a programmatic indifference. In the negative register of Warhol, in the distanced, objective register of Altman,[73] or in the empathetic register of Cassavetes, the zoom identifies the scope of sixties' and seventies' realism.

With Cassavetes, the zoom as well as his use of the handheld camera make up a visual gauge of his emotional engagement and voracious attention. Cassavetes' composition moves back and forth between an obstructed image (a realist casualness), and a shot design in which focus is used expressively. His cinéma-vérité impulse—to record the integrity of performances from a distance—yields an image that is abstract, sensual, and intense. Far from signifying a random realism, Cassavetes' wide-angle, zoom, and telephoto shots create, with their extreme recession of perspective, or with their utter flattening of three dimensionality, a plastic space that simultaneously figures the director's and the characters' changeable sensibilities.

Cassavetes' films, however, cannot be described simply through his visual and aural style, an action-painterly principle whose abstract and random impetus shifts between realist and expressive valences. Along with the constant highs of performative excess, Cassavetes' theatricality persists in what seems an even more discrete strategy. His concentration on characters' reactions defines an intimist drama that resists dispersal. No matter how visually opened up these scenes might be, all of Cassavetes' scenes retain a relevant dramatic purpose. Cassavetes' visual scattering can thus be seen as a counterpoint to the clarity and ambition of his drama.

Cassavetes' cinema is contrapuntal. It creates an endless perceptual instability that corresponds to the nonjudgmental ethics forwarded in his work. He never explains a character, and judgment is scattered in reaction shots. His characters depend on the multiple audiences mobilized to watch, judge, and appreciate their quirkiness. Cassavetes' dispersed theater is formed by the creation of witnesses for his eccentric protagonists, and his cinema revels in the compassionate embrace of those kinds of character marginalized from Hollywood's narratives. At the same time, he fights a sentimentalizing strain by hurrying to place them on trial.[74]

The democratic, generous range of Cassavetes' cinema offers a revision (as well as a display) of spectatorial prejudice. Cassavetes has said that, "like Capra," he makes "films about the individual who asserts himself or herself in the face of a multitude."[75] Cassavetes' cinema rehearses a form of idealism as it addresses—with a vengeance—the notion of individual peculiarity.

In its specularization of madness, *A Woman Under the Influence* poses, in an extreme form, the central question in representing individuality: can individual expression be recognized without magnification, and can one avoid judgment once these traits are exaggerated? (Relevant here is Cassavetes' admiration for Carl Theodor Dreyer's work—*The Passion of Joan of Arc* (1928) in particular—and their common fondness for extended close-ups and reaction shots.)

Mabel's extremity (her madness) and her typicality (she is a mother and wife) parallel, respectively, her role as spectacle and her function in a patriarchal, domestic economy. The avoidance of both sentimentality and ironic distance makes for an emotionally wrenching film. As a male, albeit a well-intentioned one, Cassavetes is not interested in placing blame. In this case, however, the challenge posed by such a charged scenario demands pause. Cassavetes' confrontation of such a sensitive issue ups the stakes of his probationary tactics, making from this film perhaps the ultimate test case of the seventies' male-made woman's film. While other, similarly themed films such as *An Unmarried Woman* (1978) or *Looking for Mr. Goodbar* (1977) suggested a high degree of neurosis in women's desire for independence, Cassavetes tackles the broad issue of man-woman relationships by making Mabel a kookie, dependent creature. He emphasizes the stereotypical association of women and hysteria that feminists at the time were so intent on undoing. But, on the other hand, he gives Mabel center-stage in confrontations that place her illness in context. In a project intrinsically his own, Cassavetes depicts her "strangeness" as inseparable from her authenticity. And although this strategy far from validates his as a feminist project, the film does stand out in its unromantic approach to the forms of Mabel's co-optation; it is hardly comprehensible, albeit still believable, that her domestic routine can, and does, go on. The filmmaker's ferocious embrace of Mabel's oddity gives a purpose to his dramatic swings between spectacle and the everyday. This jagged and abrupt movement around elastic characters turns prejudice into compassion, abnormality back into an accepted, enriched version of reality. That this is also a film about a stifled woman's urge for independence seems to confirm Cassavetes' bold plunge into messy contradictions.

In Cassavetes, the hinge between trial and acceptance, playing a role and confessing, spectacle and routine, theatricality and realism is often eased through a repeated mantra—"I love you." In its variable nuance, "I love you" is a romantic modifier of the filmmaker's own Romantic views; the words tinge his idealism with bitterness, but they also, momentarily, operate in reverse. In the films, the phrase is said at any opportunity, to anyone, when one is embarrassed, tender, or even violent, making the change of gears always noticeable. When "I love you" is said, somehow it seems like a failure of the script, a gap the director is unsure of how to fill. And it is with this awkwardly repeated phrase, which does not quite fit, that Cassavetes says what he means by an "amateur filmmaker"—a true love of cinema is always unpolished and generously open. In short, though dramatically sound, it is always also experimental.

Notes

1 Throughout this essay I was helped by three major book studies on Cassavetes' work. The only two book-length treatments in the English language devoted to the filmmaker's work, Raymond Carney, *American Dreaming: The Films of John Cassavetes and the American Experience* (Berkeley: University of California Press, 1985), and Carney, *The Films of John Cassavetes* (Cambridge: Cambridge University Press, 1994), provide useful information on production history as well as valuable insights into Cassavetes' continuation of a romantic American ethos. In French, Thierry Jousse's *John Cassavetes* (Paris: Cahiers du Cinéma, 1989) is a particularly inspired evaluation of Cassavetes' themes and style. Epigraph quote taken from "A Director of Influence: John Cassavetes Interviewed by Gautam Dasgupta," *Film* 2, no. 26 (May 1975): 6.

2 Cassavetes, "The Director-Actor: A Talk with John Cassavetes," by Russell AuWerter, *Action: The Director's Guild of America* 5 (January / February 1970): 14.

3 Jonas Mekas is a filmmaker and ardent promoter of experimental and avant-garde film. For a comprehensive study of Mekas's share in 1960s' emancipatory ideology and cinema, see David James, "Introduction," and Paul Arthur, "Routines of Emancipation: Alternative Cinema in the Ideology and Politics of the Sixties," in *To Free the Cinema: Jonas Mekas and the New York Underground*, ed. David James (Princeton, N.J.: Princeton University Press, 1992), pp. 17–48.

4 Scorsese stated that along with *Citizen Kane*, Cassavetes' *Shadows* remained on the top of his list, even after he encountered the films of Truffaut, Godard, Chabrol, Antonioni, and others. When Scorsese moved to California, Cassavetes helped him in innumerable ways—for instance, urging him to move on to more personal projects such as *Mean Streets*. Martin Scorsese, "John Cassavetes, Mon Mentor," *Cahiers du Cinéma* 417 (March 1989): 17.

5 Larry Kardish, "Cassavetes, une aventure américaine," *Cahiers du Cinéma* 417 (March 1989): 25.

6 For obvious economic reasons, avant-garde filmmakers relied on friends and family as actors and crew. This practice resulted in a series of films displaying an autobiographical impulse. See P. Adams Sitney, "Autobiography in Avant Garde Film," in *The Avant-Garde Film: A Reader in Theory and Criticism,* ed. Sitney (New York: New York University Press, 1978).

7 Joseph Gelmis, *The Film Director as Super Star* (Garden City, N.Y.: Doubleday, 1970), p. 81.

8 Cassavetes also made *A Child Is Waiting* for United Artists in 1963, and he stepped in midway through production to finish *Big Trouble* (1985), a film starring Alan Arkin and Peter Falk, which had been half-filmed by Andrew Bergman when Cassavetes took over. The first film was disowned by Cassavetes since he had no say in its final editing, and *Big Trouble,* his last film, was an embarrassment for the director, who did not get the studio to agree to some of his projected changes.

9 Although Raymond Carney characterizes *Minnie and Moskowitz* as a screwball comedy (see *American Dreaming,* pp. 140–83) and one can rightly see *A Woman Under the Influence* as a woman's film, *The Killing of a Chinese Bookie* remains for me the only clear embrace of genre in Cassavetes' career. Several French critics have written beautifully on *Chinese Bookie.* See Noel Simsolo, "Note sur le Cinéma de John Cassavetes," *Cahiers du Cinéma* 288 (May 1978); Yann Lardeau, "Le Bal Des Vauriens," *Cahiers du Cinéma* 289 (June 1978); also see Jousse, *John Cassavetes.*

10 From publicity material on the film, cited in Carney, *American Dreaming,* p. 277.

11 Jousse sees this scene as shifting suspense into suspension. Jousse, *John Cassavetes,* p. 55.

12 Noel Simsolo uses this analogy to describe Cosmo and Mr. Sophistication's qualities in *The Killing of a Chinese Bookie* in "Note sur le Cinéma de John Cassavetes," p. 68.

13 Michael Ryan and Douglas Kellner mention Cassavetes as a true experimental filmmaker, but they do so in two dismissive lines. Their concern is the interface of politics and film in Hollywood film after 1980, and Cassavetes seems entirely irrelevant in this regard. See Michael Ryan and Douglas Kellner, *Camera Politica: The Politics and Ideology of Contemporary Hollywood Film* (Bloomington: Indiana University Press, 1988), p. 269.

14 As Carney states, "No American filmmaker trusted his audience more to fathom the seemingly bottomless obliquities of the unanalyzed, uneditorialized performances of his characters." Carney, "Complex Characters," *Film Comment* 25 (May-June 1989): 31.

15 Gelmis, *The Film Director as Super Star,* p. 81.

16 For a comprehensive account of Cassavetes' filmic career and his position vis-à-vis other late 1950s' cinemas, see Carney's *American Dreaming*, pp. 20–37.

17 Ibid., p. 23.

18 See Cassavetes, "Une manière de vivre," an interview with André S. Labarthe, *Cahiers du Cinéma* 205 (October 1968): 35.

19 See John Cassavetes, "Peut-etre n'y a-t-il pas vraiment d'Amerique peut etre seulement Frank Capra," *Positif* 392 (October 1993).

20 Cassavetes, "Une manière de vivre," p. 37. In another interview, Cassavetes adds that "people from the Army . . . and Shirley Clarke left some equipment for us. . . ." AuWerter, "The Director-Actor: A Talk with John Cassavetes," p. 12.

21 For an illuminating account of the period's film production and its links with the countercultural movement, read David James's indispensable chapter in *Allegories of Cinema: American Film in the Sixties* (Princeton, N.J.: Princeton University Press, 1989), pp. 85–165. James (p. 90) is particularly critical of *Shadows* use of the racist theme, and he sees the film's pretension of authenticity as being severely compromised by the casting of Lelia Goldoni, a white actress, for the role of a black woman.

22 Each film won, respectively, the first and second awards for best independent film. Their differing authenticity quotient was an issue hotly debated, and Parker Tyler's passionate article, "For *Shadows*, Against *Pull My Daisy*," suggests that *Shadows*' association with Frank's and Leslie's film, as well as with documentary film, is based on a "superficial technical kinship," and that *Shadows*' true value, its "casual directness" that "punctures . . . the skin of life, and as the bleeding goes on, vanishes before the outflow is stanched," should be seen as Cassavetes' immersion "not in public, but in human relationships." Tyler's article is in *Film Culture: An Anthology*, ed. P. Adams Sitney (London: Secker and Warburg, 1971), p. 112.

23 As opposed to the caricature of Beat culture presented in Hollywood studio products such as *The Subterraneans*, based on a Kerouac novella, or Roger Corman's *A Bucket of Blood*, these films' engagement with countercultural values was genuine.

24 For a detailed discussion of the Beat-Underground film connection, see James, *Allegories of Cinema*, pp. 96–97.

25 See Carney, *American Dreaming*, p. 26.

26 Cassavetes, "A Director of Influence," p. 6.

27 Gelmis, *The Film Director as Super Star*, p. 79.

28 For a complete account of Jonas Mekas's engagement with avant-garde film culture, see James, "Introduction," and Arthur, "Routines of Emancipation," in James, *To Free the Cinema*.

29 For the French New Wave position, see Godard's fierce attack on the preceding generation's film production, "Debarred Last Year from the Festival Truffaut Will Represent France at Cannes with *Les 400 Coups,"* *Godard on Godard* (New York: Da Capo Press, 1968), pp. 146–47. As Randal Johnson and Robert Stam note, the Brazilian politique des auteurs is politically committed. As Glauber Rocha explained, "if commercial cinema is the tradition, then auteur cinema is revolution. . . ." For an illuminating introduction, as well as historical manifestos of Brazilian Cinema Novo, see, Johnson and Stam, eds., *Brazilian Cinema* (Austin: University of Texas Press, 1982), pp. 64–71.

30 Jonas Mekas, "The First Statement of the New American Cinema Group," *Film Culture* 22–23 (Summer 1961); republished in *Film Culture: An Anthology,* ed. Sitney, pp. 79–83.

31 Cassavetes, "What's Wrong with Hollywood?" *Film Culture,* April 1959, p. 4.

32 See Carney, *American Dreaming,* pp. 34–35.

33 Cassavetes, "A Director of Influence," p. 5.

34 Cassavetes, "Derrière la caméra," *Cahiers du Cinéma,* May 1961, pp. 2–3.

35 Cassavetes, "A Director of Influence," p. 5.

36 He shot two additional scenes (the Museum of Modern Art visit by Ben and his friends, the seduction and after-sex scene between Lelia and Toni). The film was subsequently blown up to 35mm, but no commercial distributor was interested in this "commercial" product.

37 Cassavetes, "What's Wrong with Hollywood?" p. 4.

38 In one of the television episodes he directed ("Solomon") he displays a characteristic sense of space by using the entire expanse of the set for a terse choreography.

39 Cassavetes does all the handheld shooting himself. He likes "to use it where it wouldn't ordinarily be used . . . in an acting scene rather than in an action sequence—for fluidity, for intensity." Cassavetes, "A Woman Under the Influence," interview with Judith MacNally, *Filmmaker's Newsletter* 8 (January 1975): 25.

40 John Cassavetes, "Introduction" *Faces* (New York: New American Library, 1970), p. 7. *Shadows* was not immediately distributed in the United States. The film had to wait for a successful reception at European festivals and in London to achieve any notice beyond its splash on the avant-garde circuit. Cassavetes' artistic talent, allied with his ability to produce on a shoestring budget, led to an invitation to work for Paramount as a writer, producer, and director. He moved to Los Angeles, and in a short time wrote a screenplay and filmed *Too Late Blues,* which focused on a jazz musician holding to his dreams of musical independence. According to Cassavetes, the film was an artistic failure because he had no time to fully develop his script and filming process. After *Too Late Blues,* he directed *A Child Is Waiting* (1964), a film on retarded children that starred Judy Garland and Burt Lancaster. This second encounter with the Hollywood system was disastrous, culminating in Cassavetes' removal from the editing process, which was completed by Stanley Kramer, the film's producer.

41 Cassavetes, *Faces,* p. 7.

42 Ibid., p. 8.

43 Gelmis, *The Film Director as Super Star,* p. 87.

44 Al Ruban, "My Point of View," in Cassavetes, *Faces,* p. 13.

45 Cassavetes, "Une manière de vivre," p. 37.

46 Cassavetes, "A Woman Under the Influence," interview with McNally, p. 24.

47 In *I'm Almost Not Crazy,* one of the rare documents showing Cassavetes in the process of filming (*Love Streams*), we see him in private conversations with actress-wife Gena Rowlands and sitting down with his script girl as he dictates lines of dialogue.

48 The documentary made by Joe Lustig for the BBC is mentioned by Jousse in *John Cassavetes,* p. 29.

49 Gelmis, *The Film Director as Super Star,* p. 83.

50 As Sam Shaw, his producer for *Gloria,* notes in James Stevenson, "John Cassavetes: Film's Bad Boy," *American Film* 5 (January-February, 1980): 46.

51 See Cassavetes in Michel Ciment and Michel Henry, "Entretien avec John Cassavetes," *Cahiers du Cinéma* 289 (June 1978): 21.

52 Ibid.

53 Gelmis, *The Film Director as Super Star,* pp. 82–83.

54 Cassavetes, "A Woman Under the Influence," interview with McNally, p. 25.

55 Maria Forst going out with her friends is an example. Gelmis, *The Film Director as Super Star,* pp. 84–85.

56 Al Ruban, the cinematographer of *Faces,* experimented with black-and-white photography using multiple film stocks—Tri X reversal, Plus X reversal, Double X, or Four X—according to the nature of the scene to be filmed; Cassel also told anecdotes about the crew and actors taking turns with the camera. Seymour Cassel, "Tous les Acteurs Comme Des Stars," *Cahiers du Cinéma* 417 (March 1989): 20.

57 Cassavetes is quoted as saying, "I hate discipline, I despise it. If I walk on a quiet, polite set, I go crazy—I know there's something wrong because somebody has lessened himself in his own estimation and put either me or some actor above himself." "The Director-Actor: A Talk with John Cassavetes," p. 14.

58 Victor Turner, *The Ritual Process* (Chicago: Aldine, 1969); Arnold Van Gennep, *The Rites of Passage* (Chicago: University of Chicago Press, 1960).

59 In his invaluable essay on Jonas Mekas, Paul Arthur uses Victor Turner's concept of communitas to discuss Mekas's countercultural ethos. "Routines of Emancipation," pp. 42–46. Turner's notion of communitas, in particular its application to 1960s' culture, describes the general values embraced by Cassavetes. Turner states: "The hippie emphasis on spontaneity, immediacy, and 'existence' throws into relief one of the senses in which communitas contrasts with structure. Communitas is of the now; structure is rooted in the past and extends into the future through language, law and custom." *The Ritual Process,* p. 113.

60 Cassavetes, "*Playboy* Interview: John Cassavetes," *Playboy,* July 1971, p. 60.

61 See Carney, "Complex Characters," p. 31.

62 Parker Tyler, "For *Shadows,*" p. 112, remarked how the structure of *Shadows* resembled the subtle armature of Chekov's plays and short stories.

63 John Cassavetes, *Viva* (December 1974), p. 62.

64 "Cassavetes on Cassavetes," *Monthly Film Bulletin,* June 1978.

65 For a description of Cassavetes' unfinished plays, see Raymond Carney, "Unfinished Business," in *Film Comment* 25 (May-June 1989): 48–49.

66 Cassavetes, "A Woman Under the Influence," p. 25.

67 Ibid.

68 They wanted to avoid the kind of furniture covered in plastic, and they finally justified their location choice—a more upscale house with old furniture and old woodwork—as a hand-down to Nick. Ibid., p. 26.

69 Ibid., p. 25.

70 Patricia Patterson and Manny Farber, "Kitchen Without Kitsch: Beyond the New Wave I," *Film Comment* 13 (November-December 1977): 47–50.

71 Ibid., p. 47.

72 On the other hand, Cassavetes remarks that the overuse of certain techniques as "the telephoto lenses, the wide angle lenses, and the nuanced tone of soft lighting have contributed to films that all look alike." Cassavetes, "Peut-etre n'y a-t-il pas vraiment d'Amérique, peut-etre seulement Frank Capra," p. 54.

73 In *Nashville,* Altman uses the zoom lens as if the camera were a distracted onlooker amid a crowded reality.

74 He especially likes to undermine age stereotypes, constantly portraying older women who flaunt their flirtatious nature or physical desire (the mothers of the girlfriends in *The Killing of a Chinese Bookie,* and in *Love Streams* and *Faces*) or children's perverse adoption of adult tactics (in *Love Streams*).

75 Cassavetes, "A Director of Influence," p. 5.

Independent Features:

Hopes and Dreams

Chuck Kleinhans

Since awarding top honors to *sex, lies, and videotape* in 1989, the Cannes Festival has regularly highlighted U.S. independent films. With this boost from the most famous international film festival, a snowball effect ensued in public awareness of films produced outside Hollywood. At Cannes and other prestige festivals, each year brought forward new films and new directors and often new voices and visions that the mainstream had ignored, silenced, or pushed aside. African American and Asian American cinema expanded with films such as Reggie Hudlin's *House Party* (1990), Leslie Harris's *Just Another Girl on the I.R.T.* (1992), and Ang Lee's *The Wedding Banquet* (1994). More and more often, women directors seemed to break through, such as Allison Anders (*Gas Food Lodging*, 1992) and Mira Nair (*Mississippi Masala*, 1991) and gay and lesbian characters and stories multiplied in films by new directors such as Tom Kalin (*Swoon*, 1991) and Rose Troche (*Go Fish*, 1994). John Sayles and Spike Lee, once themselves upstart independents, are now well established.

By the mid-1990s the low-budget independent theatrical feature film gained enough consistent attention in the marketplace and public eye that such films were regularly reviewed across the media spectrum. As a prime showcase for independent features by new directors, the Sundance Film Festival became so well-known that *Vanity Fair*'s April 1996 cover could headline "Special Issue: Hollywood '96, From Sundance to Sunset." Becoming a household word and lending its name to a new cable TV channel highlighting independent films, Sundance stood for a new but well-publicized phenomenon: the low-budget, off-Hollywood film offering something distinctly different from the big studios' star-driven block-buster features.

With *Pulp Fiction*'s success at Cannes and at the box office in 1994, the entertainment press wrote another chapter of the American success myth with the story of how Quentin Tarantino went from video store clerk wannabe to big-time director, screenwriter, and celebrity. In 1996 book-stores displayed Robert Rodriguez's autobiographical book, *Rebel Without a Crew: Or How a 23-Year-Old Filmmaker with $7,000 Became a Hollywood Player*, detailing the making of *El Mariachi* (1992).

For many critics, hopeful filmmakers, aspiring writers and directors, and filmgoers looking for a film alternative to mainstream studio generic fare, the phenomenon of Sundance films, or off-Hollywood, or indy fea-tures promised a new alternative. But to what extent is that alternative truly different? To what extent does it challenge the status quo?

To answer these questions, we have to start with a broader context. We must first look at the dominant institution—Hollywood cinema—in order to understand the alternatives. Hollywood is more than just a place in Southern California where about two hundred feature films are made each year. It is also a financing system and a national and international distribution and exhibition enterprise. Its publicity, promotion, and mar-keting system crosses over into media celebrity, journalistic reviewing, advertising, and the marketing of associated products and images. Hol-lywood is intimately woven into popular television from stars promoting their new films on late-night shows hosted by David Letterman or Jay Leno through highlighted stories on *Entertainment Tonight* and politicians denouncing media sex and violence on the news, to recycled film refer-ences in *The Animaniacs*. Hollywood also exists as a linchpin of key video and broadcasting forms: the broadcast, cable, and videocassette (sales and rental) markets. It is the dominant force in international entertainment media. The rest of filmmaking exists below, beyond, subordinate to Hollywood.

"Independent," then, has to be understood as a relational term—independent in relation to the dominant system—rather than taken as indicating a practice that is totally free-standing and autonomous. Indi-vidual filmmakers who do not understand this relationship often end up frustrating and compromising their own efforts. In the United States, independent feature filmmaking (mostly dramatic fictional narratives, although a few are theatrical documentaries such as *Roger and Me*, 1989; and *Hoop Dreams*, 1994) always involves a tension between art and com-merce. For some media people, making an independent fiction feature is merely a first step toward a successful career within the dominant industry. The independent film is a calling card that allows Hollywood executives to see what a new director can do with a low-budget project so that she/he might be hired into a three-picture deal: probably assigned to a genre slot—horror, teen romance or comedy, neo-noir, action-adventure, homeboys/gangsta, etc. From the industry's point of view, contracting former independent directors gets the industry young talent that will work cheap, finish films on time and on budget, and satisfy the producers' specifications. Of course, some indie directors have insisted on remaining independent: Jim Jarmusch, for example (*Stranger Than Paradise*, 1984; *Down By Law*, 1986). And some take on larger-budget projects unsuccessfully, such as Gus Van Sant (*Even Cowgirls Get the Blues*, 1994). The other major routes into a directing career are (1) film schools, with the American Film Institute, the University of Southern California, New York University, and the University of California at Los Angeles being the best-known; (2) apprenticeship in the industry, including television, with some crossover from writing, acting, music, etc.; and (3) being born into the industry and having the family or business connections to jump-start a career.

From such a start, a young mediamaker hopes for frequent opportunities to direct, or write and direct, films that are both commercially successful (so as to keep on making films), creatively satisfying, and critically esteemed. The expectation is to get more and more creative control over one's projects. The underlying concept here is based on the critical idea of the auteur, which emphasizes the director. The auteur theory postulated that some directors within the studio system could use commercial films as creative, expressive vehicles. (Such a notion of authorship is opposed to the elitist high-culture dismissal of an industrial popular culture as inherently inartistic, and the dominant Hollywood view that stars and producers are the most important, and directors are interchangeable, and if too "creative"—like Orson Welles—not worth dealing with.)

Originally formulated following World War II by the French critic André Bazin and indirectly employed as an aesthetic justification for certain Hollywood films, the notion of the director as auteur was popularized in the United States by Andrew Sarris and others, validating figures such as an Alfred Hitchcock or a Douglas Sirk, who had been widely perceived in critical circles as talented craftsmen or simply skilled entertainers rather than true artists. In the 1970s, authorship seemed to fit a flock of new directors such as Martin Scorsese, Steven Spielberg, Francis Coppola, George Lucas, Brian De Palma, etc., who worked within the commercial system in general, but who also made some films that seemed personally important to them and that were highly regarded as cinematic art by the Academy of Motion Picture Arts and Sciences, by the public, and by critics: *Taxi Driver* (1975), *Jaws* (1975), *The Godfather I and II* (1972 and 1974), *Apocalypse Now* (1979), etc. And having established themselves, these directors have continued to create works that demonstrate their artistry and personal vision: *Schindler's List* (1993) and *Bram Stoker's Dracula* (1992), for example.

The upbeat positive side of the authorship myth validates those mainstream directors who combine personal vision with box-office success. But it downplays those directors who regularly produce mass market films, such as John Hughes (*Pretty in Pink,* 1986; *Planes, Trains and Automobiles,* 1987). And it forgets those directors who start strong but then stumble at the box office, such as Hal Ashby (*Harold and Maude,* 1971; *Shampoo,* 1975; *Being There,* 1979) and Peter Bogdanovich (*The Last Picture Show,* 1971; *What's Up, Doc?,* 1972; *Paper Moon,* 1973). For those directors written off as has-beens by studio executives, the myth provides the solace of validating their art, even though they failed at commerce. But it also can cut two ways for those permanently ensconced in Hollywood. Thus, Kevin Costner gets a large boost for directing his first-time-out pet project, *Dances with Wolves* (1990), but he loses big time when *Waterworld* (1995) is a critical and financial flop, even though he did not direct it.

The Filmmaking Aspiration

In the 1990s authorship and hype fueled the full-blown emergence of an earlier trend—the filmmaking aspiration. By this I mean that for a significant number of (mostly young) people, the desire to make films is a very strong motivation, even if they cannot successfully pursue it. Beyond that, filmmaking becomes a daydream entertained by many more, creating the infrastructure for attending to the activity. In other words, enough people care about the fantasy of becoming a filmmaker that they glom onto the legend of the young auteur breaking through—Spike Lee, the Hughes brothers, Quentin Tarantino, Robert Rodriguez, Allison Anders—and find it pleasurable. Just as an earlier generation of American intellectuals interested in narrative expression aspired to become novelists, by the end of the twentieth century the goal of becoming a screenwriter or screenwriter / director (or sometimes independent producer) was an important part of many young peoples' imaginations. The desire was fueled by examples of success—young people with little or no experience, training, or family advantage who managed to break through. In an otherwise devolving future for Gen Xers, new opportunities seemed to open up as film exhibition changed, and the expansion of cable, new telecommunications technologies, and delivery systems called for a dramatic increase in creative product.

The situation developed into something resembling a feeding frenzy with film magazines such as *Premiere, Movieline, Film Threat, Sight and Sound,* and *Moviemaker* highlighting new independent work—as in the obligatory reports from Sundance, Telluride, and other festivals—and purporting to provide inside information on the phenomenon and scene of Hollywood wannabees. (Why, we should ask, if the scene is so hot, are all these people occupied in writing trend-spotting journalism rather than being involved in production themselves?)

The expansion of media programs in higher education ranging from community colleges to research universities offered aspiring students the promise that they could train and qualify for positions in the industry. For those out of school, intense "insider" screenwriting workshops and weekend lecture series are packaged much like motivational lecture programs on how-to-make-a-million in foreclosed real estate, doing a mail-order business, or selling Amway or Herbalife from home. For yet others, an abundance of do-it-yourself books offered a cheap fix: Rick Schmidt, *Feature Filmmaking at Used-Car Prices: How to Write, Produce, Direct, Film, Edit, and Promote a Feature-Length Film for Less than $10,000,* John A. Russo, *How to Make Your Own Feature Movie for $10,000 or Less,* and Gregory Goodell, *Independent Feature Film Production: A Complete Guide from Concept to Distribution.*[1]

History

Independent production and diffusion (a term I will use to indicate distribution, including marketing and exhibition) was the norm when cinema was starting. But by the 1920s a dominant structure, the Hollywood studio system, was firmly in place, and all other cinematic expressions existed in some kind of relationship to it. Within Hollywood, stars Chaplin, Fairbanks, and Pickford formed United Artists in an effort to stand apart from the studios and maintain more control over their films' production and profits. At the same time, independence outside the studio system was embodied in figures such as Oscar Micheaux, an African

American entrepreneur, author, and filmmaker who produced about one dramatic film a year for exhibition in theaters serving black communities.[2] Similarly, in New York in the 1930s Yiddish films growing out of the Yiddish theater were regularly produced for a special subculture.[3]

In the 1930s the studio system created a new marginalized form, the B film. To attract Depression audiences, many theaters began the practice of running two feature films along with a newsreel, animated cartoon, and travelogue, making a full evening of entertainment. Within the major studios, production of these lower-budget films maintained investment in capital facilities, kept equipment operating and the screen talent and production personnel steadily occupied; such films also provided a second product for the theaters that the studios owned or effectively controlled. But the custom also created a space for minor studios, including Poverty Row outfits like Monogram, Mascot, and Republic, to maintain the industrial production of extremely low-budget films. This was the territory of genre films such as cheap horror pictures and westerns, with spare sets and costumes, minimal camera takes, and bare-bones overhead. Yet, within this framework, some directors were able to produce films that went beyond hackwork.[4]

Along with films, exploitation films "exploited" the marketplace. Ranging from sensation and pornography to niche specialization, exploitation films in the 1930s and 1940s derived from the carnival practice of maximum sensational publicity to get an audience on the basis of promising more than could be delivered. The horrors of drug addiction, the sensational aspects of prostitution and sexual disease, childbirth, and other hot topics were dressed up with a puritanical policing discourse.[5] *Reefer Madness* (1939), *Mom and Dad* (1947), and *Dust to Dust* (1944) are examples. This marginal trend continued in the postwar era when exhibition patterns changed again. As television took over much of the family entertainment sector in the 1950s, specialized markets emerged, such as the teen-oriented film (*The Wild One*, 1954; *Rock Around the Clock*, 1956), the drive-in film (*Caged Heat*, 1974; *Beach Party*, 1963), and later the blaxploitation film (*Shaft*, 1971; *Coffy*, 1975; *Superfly*, 1972). Again, despite the films' formulaic plots and low budgets, some directors were able to make interesting pictures and used the experience as a stepping-stone to the mainstream. Given the guarantee of available screens to show a B film, it was also possible for a driven eccentric such as Ed Wood to make some features (*Glen or Glenda*, 1953; and *Plan 9 from Outer Space*, 1959) with minimal bankrolling and incredibly cheap production.

With the emergence of an art house exhibition system in the postwar period, it was possible for some filmmakers to make expressive films intended as cinematic art as well as entertainment that could make at least a modest profit. Such social problem dramas as *The Quiet One* (1948) and *Lost Boundaries* (1949) are examples. The art house audience enjoyed (mostly European) dramatic features that addressed a presumably more serious, more educated audience, with such Italian neorealist films as *The Bicycle Thief* (1949) and *Bitter Rice* (1949), and such British comedies as *Kind Hearts and Coronets* (1950) or *The Lavender Hill Mob* (1950). The creation of a clearly defined art house audience, in turn, encouraged further production of this type.[6]

In the 1960s a new generation of independent filmmakers appeared who constituted an American New Wave. Often identified as the New American cinema, the movement ranged from short, visually complex experimental works through cinéma-vérité documentaries to unique dramatic features. Some of these films attained widespread distribution, most notably documentaries such as *The Endless Summer* (1966) and *Point of Order!* (1964), and some intended for television broadcast such as Frederick Wiseman's *Hospital* (1969) and *Basic Training* (1971), or performance documentaries that made theatrical debuts such as Bert Stern's *Jazz on a Summer's Day* (1958) or the Maysles brothers' *Gimme Shelter* (1970). On the art-house front, a wide range of critically esteemed, though not always commercially successful dramatic films appeared in the 1960s: Cassavetes' *Shadows* (1958), *Faces* (1968), and *Husbands* (1970); Shirley Clarke's *The Connection* (1961) and *The Cool World* (1963); Michael Roemer's *Nothing But a Man* (1963); Frank Perry's *David and Lisa* (1962); and Joseph Strick's *The Balcony* (1963). Because the Hollywood system in the 1960s and 1970s was unable to adapt quickly and effectively to a changing culture, baby boom demographics, and economic changes, instability created more opportunities for new directors who could occupy the in-between places during a time of upheaval and drastic change. The 1980s saw more stability in the mainstream production and distribution, but also the concerted efforts of new institutions such as the Sundance Institute and the Independent Feature Project promoting the development of off-Hollywood product.[7]

While these films varied from the dominant Hollywood *product,* they still worked within the logic of the dominant *system,* occupying the relatively freer areas of the margins and in-betweens of the conventional industry. What is important to understand is that with changes in finance, in production process, in distribution, in exhibition, that is, in the material forces of production, and in the cultural environment, and the audience and its own cultural and historical development, different elements can combine with others and as a result new possibilities emerge. Part of the phenomenon of independence is also due to the fact that very large industries such as Hollywood by their very nature are not totally rationalized in all respects. Rather, they have, in the course of their evolution, changes that present new (often temporary) options, gaps that can be filled. The corollary of this thesis is that much smaller cinema systems, such as those in various European countries, can in fact be managed in much more consistent ways and often are much less open to new options.

Independence Today

Dreams of independence, dreams by independents are often the stuff of fantasy and illusion because no matter how low-budget one goes in production, it is the dominant capitalist system that defines the basic structures. One of the most enduring structures of the present era is the basic economic truth that for Hollywood, *money is made on hits.* Since the breakup of the vertically integrated studio system after World War II, with ownership extending from finance and studio operations to local theaters, people who finance film production are playing a speculative game of producing a number of films, many of which will fail to make back their costs, in hopes that one will become the megahit. Today the dominant film form is "high-concept," in which stars and genres mesh to produce a new product that already is familiar and that can easily and obviously be marketed through TV advertising.[8] A high-concept film can be pitched in one sentence, something repeatedly well parodied in Robert Altman's comic satire *The Player* (1992). It can be promoted in one image: John Travolta dancing the Hustle in *Saturday Night Fever* (1977), the costumed figure of *Batman* (1979), blowing up the White House in *Independence Day* (1996). These films depend on the already known: many are based on pop culture comics and television shows, the generically familiar, well-known stars, and a signature style (high production values, stupendous special effects, and attractive audiovisual design).

High-concept filmmaking aims at exploiting a specific marketing strategy, one that has developed with changing institutions. First, the release pattern for films slated to be blockbusters relies on the interacting synergy of multiplex theaters, saturation TV (and some print) advertising, star / celebrity, and merchandising. From the early 1970s on, the older pattern of staggered theatrical release (in which films ran at certain prime theaters for long runs, followed by gradual release to neighborhood, smalltown, and drive-in markets over time) has changed to the current norm of nationwide saturation presentation at multiplexes, which can open a film in several rooms at once to maximize attendance and then shift down to smaller spaces as the run wears on and the audience shrinks. This national breakout is highlighted by saturation advertising and extensive hyping of the new film immediately before release. Late-night TV talk show hosts Letterman and Leno are simply the most obvious cases; their guest lists are trimmed exactly to the next weekend's releases. With ancillary tie-ins such as Lion King and Pocahontas T-shirts and McDonald's meals and Space Jam toys, anticipation builds before a film appears, guaranteeing a presold desire to see the film and to continue the circulation of the films' image after viewing.

Conventional thinking in the industry assumes that any new film has at most a three-week window of opportunity to become a hit. Most films today "make it or break it" in the three days of their first weekend in release. Since active moviegoers can see only one or two films a weekend, and Hollywood is releasing more than two hundred a year, on any one weekend only one or two new releases will be able to gather an exceptionally large audience. If a film does not fly the first weekend, it *might* be possible to recoup something with a change in promotion and advertising by the second and third weekends, especially if the film can exploit positive word of mouth.

Release patterns also are governed by demographic considerations. Holiday weekends supply important launch points and often highlight a kids-oriented feature such as *Toy Story* (1995) or *101 Dalmatians* (1996). Black-themed and Latino-themed works premiere in urban centers with some hope for a suburban crossover. But many neighborhoods have too few screens for potential audiences. When former basketball star Magic Johnson in the mid-1990s opened a multiplex in south central Los Angeles and set attendance records, industry "experts" were surprised.

Similarly, demographics influence the successful release of films aimed at women and girls. The underserved female audience takes longer to attract to films for numerous reasons. Women and girls typically make less money than men—compare the pay for baby-sitting and yard work. They have greater family obligations, have less disposable income to spend on entertainment, and in heterosexual dating the male conventionally pays and usually chooses the films to be seen. But word of mouth and attending events in a group are more characteristic of females. Thus, some women's films have built more slowly (such as *Thelma and Louise*, 1991) than overnight male action blockbusters, but they have "long legs" and can run many weeks in one of the theaters of a multiplex, often developing repeat viewings. *Little Women* (1994), *Sense and Sensibility* (1995), and *Like Water for Chocolate* (1992) are cases in point.

Some Numbers

The major studios released 212 movies in 1995. John Pierson estimates that about four hundred independent features were produced that year.[9] That number is debatable because it includes not just extremely low-budget films by new screenwriters and directors, but also some carefully crafted features for cable and direct-to-video, which are relatively low-risk bets (examples would be John Dahl's *Red Rock West* (1993) or *The Last Seduction* (1995) and Zalman King's *Wild Orchid* (1990) and other erotic thrillers) because they can be easily sold or even presold. Cable or direct-to-video films require virtually no promotion or advertising. But even discounting those parts of the low-budget independent market, we still have well above two-hundred films every year that compete for attention in the Independent Feature Film Market (the annual fall showcase in New York for completed and in-progress works) and for festival slots at Sundance, Telluride, New York, Toronto, San Francisco, and other major festivals that are important showcases for independent films. Some of these films do get festival screenings, but they do not get distribution deals. Some get distribution deals, but they do not succeed at the box office.

In the United States there are about 26,500 movie screens in about 10,000 houses. Half of those screens are effectively controlled by the major Hollywood studios. The largest art-house chain, Landmark, owns about 120 theaters, but for the past few years it has been owned by the financially troubled Goldwyn Company, and the chain may be sold off. It is possible that new owners could judge Landmark less profitable as an art house chain and turn to the mainstream product. Thus, the largest and most stable set of venues for independent features could disappear overnight, leaving indy distributors with costly product with no outlets, and producers and directors without a future.

Essentially, Hollywood—the dominant capitalist film entertainment system—has increasingly come to use low-budget independent films as an inexpensive, low-risk source for an increasingly differentiated market and as a kind of minor league training ground for new talent—mostly younger hopefuls who, if they succeed, will then be promoted to the majors. Rather than investing its own money in initial production, the industry sponsors—by purchasing distribution rights in some cases, or by option-ing a filmmaker's next film—a highly speculative system in which about three hundred independent films a year are winnowed down to about thirty that are released nationally, and about ten that are profitable or at least come close to returning their investment. However, the dominant feature of the system remains the blockbuster model. Upper-level Holly-wood executives are concerned exclusively with highly speculative proj-ects (in search of the prospective megahit), not with films that offer small or modest returns. Industry executives do not make their careers by having a long string of films that return a modest profit; rather, they succeed by having big hits that return massively on investment. Holly-wood is governed by a highly speculative mode of capitalism.[10]

With this understanding of the dominant system as a background, we can better understand the actual situation for independents. First of all, financing production for independents is significantly different than for major studio projects. The majors finance a project up front and monitor progress to make sure that budgets are being met. In contrast, indepen-dents almost always finance production from a variety of sources in a series of stages before distribution is secured. Money is cobbled together from the filmmaker, family and friends, speculative investors, and in some cases institutions such as the Corporation for Public Broadcasting or, in the past, the National Endowment for the Arts or the American Film Institute.

Much of the initial labor for scriptwriters and directors in particular involves deferred payment. Production personnel and screen talent also may work on a deferred basis. Often the crew is nonunion, or sometimes a union crew, under a special arrangement, works for lower or deferred salaries. However, if a film's crew is unionized (which guarantees a certain level of professional skill and experience), certain standards for meals and accommodations must be met, which makes much location shooting extremely expensive. It would be impossible, for example, to do a road movie involving different scenes in widely dispersed places on a limited budget. Casting scenes with many extras, ensemble acting, and numerous open-air scenes that depend on consistent weather also present economic problems. Of course, these limitations can be turned to a distinct advan-tage by an imaginative and talented director. Thus, a small cast and limited interior locations are exploited to great effect in such independent films as Jim Jarmusch's *Stranger Than Paradise,* Steven Soderbergh's *sex, lies, and videotape,* Hal Hartley's *Trust* (1991), Allison Anders's *Gas Food Lodging* (1992), Spike Lee's *She's Gotta Have It* (1986), Wayne Wang's *Dim*

Sum—A Little Bit of Heart (1984), and even Quentin Tarantino's *Reservoir Dogs* (1991).

One much-hyped aspect of independent features—working on bare-bones budgets—is extremely misleading to the uninformed. It makes good copy to say someone shot their first feature for $100,000 or $25,000 or even $7,000, but the realities are very different. For example, Robert Rodriguez gained a lot of attention for the claim that he made his first feature *El Mariachi* for $7,000 in out-of-pocket expenses. But this disguises many details. First of all, that figure covered only the initial outlay for film stock and processing. Rodriguez was "discovered" while in the editing stage of his film (actually while making a videotape rough version), and he then was signed to a deal that allowed him to finish editing and pay for the blow-up to 35mm, the cost of optical effects (such as fades and dissolves), the remixing of all sound including extensive Foley (sound stage) work, payment for music rights, and so forth, to bring the film to completion. Beyond those payments was the distributor's cost of prints, advertising, negotiating distribution and exhibition deals, and residual sales abroad as well as cable TV and videocassette sales. The actual cost of the film was hundreds of thousands of dollars, perhaps several million dollars, not a few thousand. The myth of *El Mariachi* also omits that Rodriguez borrowed rather than rented his equipment, made a kind of film that required no skilled physical acting or dialogue delivery or synchronized sound recording. With a minuscule cast and crew, and by editing his rough cut at cable access on videotape, living and working at home and thus having no office or overhead expenses, Rodriguez executed the film's production and postproduction with remarkable efficiency.

The legend of *El Mariachi* also omits the fact that Rodriguez had made and edited little videotape "movies" at home; he actually had ten years and thousands of hours of shooting and editing experience before starting on this project, which was initially intended to be a Mexican action film— a genre popular and well-established in Mexico and the American Southwest—with typically very low production values. That was all Rodriguez was aiming for when he was picked up by U.S. producers who thought *El Mariachi* seemed similar to the newly discovered films of Hong Kong director John Woo (who had influenced Tarantino and others). So part of Rodriguez's success is due to being in the right spot at the right time, but his ingenious and aggressive improvisation and years of experience were crucial to his film's being catapulted into successful distribution.

Unlike the major studio projects that are guaranteed financing at all stages from script and star acquisition through production to distribution and advertising, independents face new financial hurdles at each stage. Getting a film to a completed video edit, or securing one print for festival submission, becomes a filmmaker's central goal in pursuit of a distribution deal. Such contacts can range from selling the film outright to attempting to arrange a percentage of the gross or net profits after distribution expenses.[11] (However, the naïve should remember that given Hollywood's peculiar accounting practices, a film such as *Forrest Gump* (1994)—in 1997, among the highest-grossing films in Hollywood history—still has not returned any net profits.)

Taking a fully completed film from festival screening to full-scale exhibition is an exceptionally expensive venture that ties up considerable capital. Opening a film nationally involves making costly prints, an expense paid by the distributor. In the case of a high-concept film intended to be a blockbuster, hundreds of prints are needed for simultaneous national release to theaters. And some multiplexes may show several prints of the same film on several screens at the same time, so that a different print is needed for each projection room.

In the 1990s "average" Hollywood films have had projection costs in the $30 million range, added to an average standard of $10–15 million for advertising. Given such industry practice, it is easy to see that however low-budget an independent film is in its production costs, it is still massively expensive to compete for attention in the same market. In addition, mainstream films have the advantage of star power to gain promotional attention from popular magazines such as *People,* celebrity news shows (for example, "Entertainment Tonight") and talk shows on TV, while independents must rely on promotion to a smaller audience mostly through specialty publications.

The dominant system also sets the terms for distribution / exhibition economics. In general, distributors of major hits can drive a harder bargain with exhibitors, who expect to make a profit on popcorn and concessions as well as on volume of patronage. But the reverse obtains for small-market films—exhibitors expect to get a larger percentage of the gate for themselves because of the smaller audience. As businessmen, distributors are in the position of advancing money for the expenses of prints and advertising against expected returns from exhibition, and they want to pay off those expenses first, before sharing anything with the filmmaker. Reasonably enough, they expect to make back their money and a profit before sharing it.

There are a relatively small number of theaters that regularly screen independent features. So, although some films may break through into e mainstream world of multiplexes, as *sex, lies, and videotape* did, most of indy features are shown at art-house theaters, such as the 120 or so run he Landmark chain in Yuppie and Gen X urban neighborhoods and college towns that have well-established, loyal audiences. There is some room for slotting independents into multiplexes, especially when there is not enough successful mainstream product to go around. But multiplexes do not provide a stable and predictable environment for independents, who can be easily squeezed out if a studio film might make more money on the same screen.

The filmmaker may see money returned from foreign sales, sales to cable TV, and videocassette sales, and this may be a significant part of the revenue return for some filmmakers. For example, veteran independent director John Sayles (*Matewan,* 1987; *Lianna,* 1983; *Lone Star,* 1996) has observed:

[Video] is not the ideal way to see a movie. But we're aware, just like any Hollywood director is aware, that more than half the people who see your movie are not going to see it on the big screen. That's just the way it works these days. But it's one of the reasons independent film has flourished. We financed two of our movies through home video pre-sales. If video did not exist, or cable, there'd be a lot fewer independent filmmakers running around.[12]

However, for an unknown director, this area is much less reliable. Foreign sales are a better bet for films such as martial arts, action, and erotic thrillers that can be followed with minimal attention to dialogue / subtitles. The "little personal film" stressing dialogue, character drama, and acting is a hard sell in virtually every theatrical and ancillary market.

Critical Investment

While occupying an economic and aesthetic position at the edge of the mainstream commercial cinema, independent films represent a large critical investment for some people. These are the kinds of films that cinephiles, media teachers and students, and many liberal intellectuals like to watch (at their local art house), or on PBS, or on a specialty cable channel, or, in small college towns without an art house in the form of rent-by-mail from Chicago's Facets Multimedia (the largest distributor of such films on video) or some other mail-order distributor so that they can overcome the cultural deprivation of the heartland. These are films like *Art for Teachers of Children, The Devil Never Sleeps, Safe* (all 1995), or *Clean, Shaven* (1993) that were shown at the 1996 annual meeting of the Society for Cinema Studies, the professional organization for film teachers. They are films that some members of SCS give papers on: *Clerks* (1994), *Hoop Dreams, Liquid Sky* (1982), *Speaking Parts* (1989), *Household Saints* (1993), and (arguably) *Pulp Fiction* (1994). These are the films most film teachers like to teach and often (implicitly or explicitly) hold up as models for aspiring students who want to make films. They are the films that teachers and critics might even daydream about making themselves.

Film critics and teachers are, after all, the ones who tell the legendary story of the *Cahiers du Cinéma* critics (Jean-Luc Godard, Alain Resnais, François Truffaut, Jacques Rivette, etc.) becoming the French New Wave—the early 1960s' creative outburst of new French cinema. They are the ones who discuss Orson Welles's career in Hollywood as a tragedy. They are the ones who are most attracted to legends of author / directors who succeed on their own terms within the system such as Robert Altman, John Cassavetes, Woody Allen, or Spike Lee.

This critical investment is grounded in something worth validating: the artistically accomplished dramatic feature that speaks to (and sometimes with, and sometimes for) an audience that wants entertainment and enlightenment through a film that seems to express an artist's vision. These are the films that cinephiles, critics, scholars, teachers, journalists, curators, preservationists, and intellectuals validate. Scholars do not have any influence on box-office success, but they are the ones who construct the canon of works, write the histories, and argue the moral, social, and artistic value of those films that are produced. For example, Charles Burnett's *Killer of Sheep,* a film that never had a commercial release, is on the National Film Registry of the Library of Congress because film scholars validated it.[13] And Oscar Micheaux's work is receiving the reevaluation it deserves because of SCS members.

The concern for low-budget independent films has been a vital part of identity and constituency developments in film studies. Feminist, gay / lesbian / queer, African American, Latino, Asian American critics and scholars have all focused attention on significant feature-length films and created the context of intellectual discussion necessary for coalescing and developing an ongoing audience awareness as well as a critical discourse among makers, exhibitors, critics, and the public. The prime example would be Marlon Riggs's *Tongues Untied* (1989), which not only spoke from, for, and to a black gay male experience, but also, in fact, helped bring community and public political awareness into being. The video, which rapidly circulated in both public forums and through private copies passed hand-to-hand, functioned as a starting point for discussion and organization of a nationwide constituency brought to self-awareness through the vehicle of the tape. At times, some such films occupy an importance within critical discourse far out of proportion to their actual aesthetic or political value. For example, *She Must Be Seeing Things* (Sheila McLaughlin, 1988) was vastly overvalued by some lesbian-feminist critics simply because it was the first narrative feature of its kind depicting the complexity of lesbian love relations and fantasies and thus allowed certain critical and political issues to be discussed.[14] Today, with more lesbian dramatic features in existence, that breakthrough film seems pale by comparison. Lizzie Borden's *Born in Flames* (1983) represents a similar example in terms of feminist discussions.[15]

However, the considerable intellectual (and emotional) investment that some people have in these films should not blind us to some *other* investments that people have. I mean financial monetary investments that people—individuals and corporations—make in order to make money in the film industry. In other words, analysis of a personal libidinal and social economy must be balanced in relation to a political economy. Independent features can be viewed as either/both a cultural phenomenon or/and an economic one. For some, making an independent film is simply a stepping-stone to the big time of Hollywood. For example, Edward Burns made the dramatically conventional feature, *The Brothers McMullen*, for about $100,000, showed a two-hour rough cut at the Independent Feature Film Market in 1994, and sold the picture to Fox Searchlight for a bit more than $1 million. The film went on to gross over $10.2 million domestic, hitting 367 screens at its widest release.[16] For others, personal expression and/or social-political statement may be more important than economic rewards. Thus, the field always contains diverse motivations.

Independence alone does not confer political, social, or aesthetic value. But there is a long-standing connection between *some* independent feature films and political advocacy or minority cultural expression. The left-wing projects *Native Land* (directed by Paul Strand, 1942) and *Salt of the Earth* (Herbert J. Biberman, 1954), made by artists blacklisted during the McCarthy era, provide significant examples. Similarly, films such as Lionel Rogosin's *Come Back to Africa* (1960, set in the black townships of South Africa), Roemer's *Nothing But a Man,* and Robert M. Young's *The Ballad of Gregorio Cortez* (1982) demonstrate the possibility of making politically acute and dramatically effective independent films. But many of the lessons learned by independents are sobering. Critic Michele Wallace describes the tension between optimism and realism for African Americans:

The Spike Lee phenomenon began [in 1986] with the mistaken assumption, on his part and everybody else's, that blacks making their "own" films would improve the quality of black representation in the public sphere. . . . We can now see that the notion of blacks making their "own" films presupposed the existence of a monolithic black community, unified enough to possess a common ideology, ethics, morality, and culture, sufficient to override such competing and divisive interests as class, gender, sexuality, age, and education. Also implicit in this formulation of blacks having their "own" films was the nagging question as to whether such representations would somehow make black people's lives better overall. Regardless of whether a film has any value as art, it can, if it chooses, closely mirror or reflect the problems and inequalities of its society. People make the mistake of thinking that a film can therefore also correct inequalities. This is because we, as a culture, are still trying to figure out what representation fully means in still new and exponentially expanding forms: what such forms can and can't do, what we should and should not ask of them.[17]

Some Lessons of Independent Low-Budget Film

1 Low-budget independent features do not make huge net profits. A gross of $2 million to $6 million (considered an excellent return) can easily be eaten up by expenses and shares. Independent filmmaking is not a dependable way to make a living.

2 The net that is received by the original director / producer is almost never enough to finance production of another film. For the director, it is at best the start of a track record to make another film (which usually will fit much more within the Hollywood system). Spike Lee is a good example. Although moderately budgeted and generally successful, his films have had to seek funding from scratch each time out.

3 The films that are the most successful within this particular system tend to be those with a clearly marked niche market. For example, John Pierson characterized the audience for Julie Dash's *Daughters of the Dust* (1991) as "black women who read Toni Morrison novels," and the marketing of the film carefully developed that audience.[18] The film *Crumb* (1995) was a sure thing with the legion of the adult comic artist's fans and then picked up some critical acclaim. And *Go Fish* and *The Incredible True Story of Two Girls in Love* (1995) were eagerly viewed by the unserved young lesbian market.

4 But when a niche is identified, Hollywood then steps in with its own vehicle. The success of independent gay features shows that you can sell a

33 The new, independent African American cinema: the Hughes brothers'
Dead Presidents (Hollywood Pictures, 1995).

Philadelphia (1993). *Daughters of the Dust* demonstrated a market potential
filled by *Waiting to Exhale* (1995). And in almost every case, the Hollywood
copycat is less imaginative, less politically committed, and less interesting
than the original. But the copycat then sets the terms for the continuation
of the niche. So, while independents can initially gain a leg up through
constituency markets, once those markets are identified, independents
must compete with the mainstream industry's attempt to exploit the
same market. Thus, the Gen X hit *Slacker* (1991) opens the space for
the star-powered *Reality Bites*. And Richard Linklater goes from the unpolished *Slacker* to *Dazed and Confused* (1993), nice enough as a teenage
comedy, but no more than that, and a good deal less than, say, the most
successful independent teen picture of all time, *American Graffiti* (1973).

5 Conditions of financing, distribution, and exhibition are changing rapidly. The current deregulation of telecommunications and the rapidly
changing delivery technologies promise further and greater change. Last
year's strategy may be totally wrong for next year's situation. For example, current reports indicate that the direct-to-video film niche may
be oversaturated.[19] And there is a clear distributor and critical backlash
against the glut of no-talent films being produced by opportunists for the

independent feature market. As the number of films made increases, the percentage of worthy ones has declined.

Future Prospects

The pressure of the marketplace is decisive in shaping the continuing work of independents. Thus, any further analysis must recognize:

1 Some new directors are simply interested in making a calling card film to enter the mainstream and thus use low-budget filmmaking as an alternative to (expensive) film school. When we reflect that going to USC, NYU, or UCLA costs between $90,000 and $120,000 for four years (plus costs of filmmaking), there is a certain logic for an eighteen-year-old to simply hit up mom and dad for enough money to make a feature.

2 To continue to make auteur cinema within the present system is increasingly tenuous. In the 1980s, NEA, NEH, AFI, and foundation grants and coproductions with Channel 4 in England and German television helped out many young filmmakers. Today those opportunities seem to be gone forever.

3 Appealing to or exploiting constituency markets is risky business. If successful, independents risk competition from the mainstream industry for the same market. Identification of a constituency market may hurt the very institutions that initially created the phenomenon. For example, B. Ruby Rich's trumpeting of a New Queer Cinema of feature films in the *Village Voice* and *Sight and Sound* a few years ago[20] convinced filmmakers that they should *not* show their features at gay / lesbian / queer festivals because it would disrupt the later marketing plan. Thus, the identity festivals that created a market and demand and that had highlighted features in premieres, opening nights, and festival closings and awards were subsequently cut off. And some filmmakers have shown their bad faith and bad politics by then arguing that they do not want to be "ghettoized" in such festivals that gave them their first recognition and encouragement.

4 Many alternative distributors are not really independent. New Line is owned by Turner Broadcasting, which is now owned by Time Warner. Miramax is currently carrying fifty or more independent features in initial release or some stage of prerelease, which means they have more product under their control than some studios release in a year. Miramax, now a minimajor, is a subsidiary of the mammoth Disney conglomerate.

We should not be naïve about the actual constraints on "independent" film. But we should not be cynical or despairing either. Independence is not just a state of mind; it is a set of potentials that can only be realized in a real world situation with real economic institutions and constraints. Knowing what those are gives the filmmaker a much better chance of being successful in aesthetic terms as well as economic ones. And for the critic and viewer of independent work, knowing the field provides an important context for assessing specific films. The legendary stories of filmmaking make for interesting reading and daydreaming. But in the long run, an informed audience is a better one for ensuring the lasting position of the independent sector.

Notes

1 Gregory Goodell, *Independent Feature Film Production: A Complete Guide from Concept to Distribution* (New York: St. Martin's Press, 1982; John A. Russo, *How to Make Your Own Feature Movie for $10,000 or Less* (New York: Barclay / Zinn, 1994); Rick Schmidt, *Feature Filmmaking at Used-Car Prices: How to Write, Produce, Direct, Film, Edit, and Promote a Feature-Length Film for Less than $10,000*, rev. ed. (New York: Penguin Books, 1995).

2 Jane Gaines, Pearl Bowser, and Charles Musser, eds., *Oscar Micheaux and His Circle* (forthcoming); Ronald Green, *Straight Lick with a Crooked Stick: The African-American Middle-Class Cinema of Oscar Micheaux* (forthcoming).

3 J. Hoberman, *Bridge of Light: Yiddish Film Between Two Worlds* (New York: Schocken Books, 1991).

4 Todd McCarthy and Charles Flynn, eds., *Kings of the Bs: Working Within the Hollywood System: An Anthology of Film History and Criticism* (New York: Dutton, 1975); Don Miller, *"B" Movies* (New York: Curtis Books, 1973).

5 David F. Friedman, with Don De Nevi, *A Youth in Babylon: Confessions of a Trash-Film King* (Buffalo, N.Y.: Prometheus, 1990); Eric Schaefer, *"Bold! Daring! Shocking! True!": A History of Exploitation Films, 1919–1959* (forthcoming).

6 Barbara Wilinsky, "An Alternative Mode of Film Exhibition: The Rise of Art Houses After World War II," unpublished Ph.D. dissertation, Northwestern University, 1997.

7 For the 1980s, see the collected case studies in David Rosen, with Peter Hamilton, *Off-Hollywood: The Making and Marketing of Independent Films* (New York: Grove Weidenfield, 1990).

8 Justin Wyatt, *High Concept: Movies and Marketing in Hollywood* (Austin: University of Texas Press, 1994).

9 John Pierson, *Spike, Mike, Slackers and Dykes: A Guided Tour Across a Decade of American Independent Cinema* (New York: Hyperion, 1995), p. 204.

10 The Begelman affair provides a classic example. A studio executive was caught red-handed stealing money from a prominent actor's studio account, but he eventually kept his position because he was well-connected and otherwise considered successful in the industry. See David McClintick, *Indecent Exposure: A True Story of Hollywood and Wall Street* (New York: William Morrow, 1982).

11 For a full discussion of distribution economics, see Pierson, *Spike, Mike, Slackers and Dykes.*

12 John Sayles, quoted from interview with Michael Janusonis, *Providence* (R.I.) *Journal Bulletin,* July 19, 1996, quoted by Jim Marsden on H-Net List for Scholarly Studies and Uses of Media (July 20, 1996).

13 Full disclosure: I gave the first scs paper on Burnett's film, "Charles Burnett's *Killer of Sheep,*" Society for Cinema Studies, UCLA, June 1982, on a panel I organized called "Independent Black, Chicano, and Asian Filmmaking in Los Angeles." The film society as a whole was instrumental in establishing the registry, along with the Academy of Motion Picture Arts and Sciences, and other industry and library interests. The registry marks films which have aesthetic and historical importance and which the relevant institutions attempt to preserve as part of our film heritage.

14 Teresa de Laurentis, "Guerrilla in the Midst: Women's Filmmaking in the 80's," *Screen* 31 (Spring, 1990): 6–25.

15 Teresa de Laurentis, "Rethinking Women's Cinema: Aesthetics and Feminist Theory," in *Multiple Voices in Feminist Film Criticism,* ed. Diane Carson, Linda Dittmar, and Janice R. Welsch (Minneapolis: University of Minnesota Press, 1994), pp. 140–61.

16 See "1995's Sundance Domestic Box Office Chart," *Filmmaker,* Winter 1996, p. 37.

17 Michele Wallace, "Doin' the Right Thing," *Village Voice,* Film Special, May 21, 1996, pp. 10, 12, 14. Quote on p. 10.

18 Pierson, *Spike, Mike, Slackers and Dykes,* p. 208.

19 Richard Corliss, "There's Gold in That There Schlock," *Time,* August 26, 1996, pp. 55–56.

20 B. Ruby Rich, "New Queer Cinema," *Sight and Sound,* May 1992, pp. 31–34.

A Circus of Dreams and Lies:

The Black Film Wave at Middle Age

Ed Guerrero

The circus proves to be a useful and poetic metaphor for black "progress" in the film industry since the end of the civil rights movement. With Jesse Jackson and other social critics and activists once again protesting Hollywood's exclusion of African Americans,[1] that effort toward progress seems to have come full circle to the issues at the frustrated beginnings of African American open mass resistance to white supremacy in the 1950s and 1960s. In spite of some measured progress evinced by a sprinkling of new black actors, directors, and producers, and because of very few changes in Hollywood's executive offices, the 1990s' black film wave has hit an interesting, stalemated, plateau, one that belies its auspicious take-off in the mid-1980s. Then, young, insurgent filmmakers like Spike Lee with *She's Gotta Have It* (1986), Robert Townsend with *Hollywood Shuffle* (1987), and Julie Dash with *Daughters of the Dust* (1991), all playing variations on the theme of "guerrilla financing," helped crack the discriminatory insider networks and investment ceilings that kept African Americans shut out of the production end of the movie business after the collapse of the Blaxploitation period in the mid-1970s. While varied discussions and definitions of "black cinema" and its much debated independent and mainstream tendencies will always be important issues, this essay will primarily focus on the tangled workings of *race* in Hollywood and those feature films made by or about black people that enjoy a popular commercial audience. As well, I will rummage through and unpack some of the symptomatic expressions of *blackness* and *whiteness* on the big screen.

With the new black movie boom now slouching toward middle age, one can expect the industry to release twenty or so black-cast, black-focused, or black-directed mainstream films a year. Beyond the industry's persistent color problems, its hackneyed short-term, profit-focused scenarios and stereotypes, the potential for a fully developed black cinema, featuring an interesting range of new films and directions, cultural trends, and rising star personas, still lingers, albeit as a long shot. For one thing, the 1990s' black film wave has been amplified and sustained by the emergence of a new generation of black film scholars and critics who have defined and shaped black cinema's circus maximus.[2] But also, black films and filmmakers are enjoying an expanded audience, a heterogeneous mix of crossover, youth, and art-house moviegoers.

It is this variety of expressions, activities, and innovative directions in black filmmaking and its critical discourse that stands out as the most promising development in popular black cinema production and consumption to date. However, as with all circuses—cinematic, metaphoric, or otherwise—we must contend with a downside. As Federico Fellini so recurrently and brilliantly revealed in his work, the circus and cinema are closely related cultural forms. Both are built on the suspension of disbelief and rely on large doses of spectacle. Both forms also mask a certain hard reality with the veil of aesthetic pleasure, illusion, and trickery. In the case of Hollywood's ongoing, lucrative flirtation with the sign of *blackness*, the tricks have been habitual and predictable as the film industry persists in skewing its representations of African Americans while containing much of black film production within the limitations of its most profitable and expedient marketing strategies and narrative formulas.

Accordingly, then, the most important circumstance informing the production of all *blackness* on the big screen, and the primary hoax or illusion in the American collective psyche, continues to be the social construction and contestation of *race* and how we as a nation negotiate our turbulent, multivalent racial definitions and power relations. The symptoms of our delusional and troubled racial condition endure, from the sensational Susan Smith double murder case with its traditional evocation of the black bogeyman,[3] to the acrimonious black / white split in public opinion over the verdict in the O. J. Simpson murder case, to the emancipatory upsurge of 1996's Million Man March, to a wave of black church burnings throughout the South, or the rise of militant militias across the land. Feeding the tense racial climate in the mid-1990s, the affirmative action backlash metastasized with the passage of California's anti-immigrant Proposition 187. Also by mid-decade, there arose a racially polarized national electorate, accompanied in print by the pseudo-scientific claims of black intellectual and genetic inferiority in *The Bell Curve* or the laughable assertion in *The End of Racism* that white supremacy and racism are no longer social malignancies.[4]

"The Devil's greatest trick was to convince the world that he doesn't exist," goes the most memorable line in the hit film noir thriller *The Usual Suspects* (1995). And so it goes when confronting the elusive, resilient demon of racism. In a pattern that has endured at least since the end of World War II, *race* continues to smolder unconsciously in the social psyche and then, with dismal regularity, leap into public focus and media exploitation with the revelation of some racial crime or competition over shrinking resources. Ralph Ellison's trope of "invisibility," which, really, maps the dialectic of visibility / invisibility, in part explains this phenomenon. Blacks are highly visible in society as suspects, victims, or noble exceptions to the rule, while simultaneously they are invisible when it comes to the recognition of their individuality, their basic humanity and rights. Unfortunately, the cycle has degenerated into vulgar, reflexive habit that begins with a state of racial denial that passes for the social norm. This so-called normative state is disrupted by some racial issue or "incident," followed by feigned shock and innocence, then anger, then repression and a return to a primal and uneasy state of collective denial. It should come as no surprise, then, that with the end of the Cold War, a nation that has defined and unified itself with the threat of an external, foreign enemy for more than fifty years now increasingly finds *race,* in all of its themes, coded forms, and endless variations, such a charged and divisive issue.

Consequently, our chronic racial condition multiply informs how we play out *blackness,* from the most evident stereotypes to the most encrypted racial arguments and assumptions in commercial cinema and the media. Making the point in a perhaps unconscious way, the hit noir comedy *Get Shorty* (1995) depicts the exclusion and marginality that Hollywood routinely inflicts on African Americans in its portrayal of two underworld figures, John Travolta as a mob loan collector and Delroy Lindo as an upper-level black L.A. drug dealer, who are competing to turn a hot film script into a movie. As the plot unfolds, the Lindo character is eliminated while the Travolta character prevails and goes on to produce the film. Besides Hollywood's usual death wish for Lindo as a stereotypical black loser, what *Get Shorty* reveals about the inner workings of the film industry is exactly how and why African Americans are excluded from its highest decision-making levels. Using a series of clever deceptions, mixed with some intensive networking, Chili Palmer (Travolta), manages to sign a big-name star, Michael Weir (Danny DeVito), to his script and move it into production. Thus, *Get Shorty* explores the same Hollywood business culture that made Robert Altman's *The Player* (1992) and the art-house film *Swimming with Sharks* (1995) so potently satirical. All of these films self-reflexively depict how the executive levels of the movie industry work as a "relationship business."[5] Whatever superficial liberal concessions Hollywood makes to racial equality on TV talk shows, or in its award ceremonies, when it comes to greenlighting film scripts, building stars' careers, or investing large amounts of capital for short-term profit, moviemaking is an overwhelmingly white male insider's game. And unfortunately, at the highest levels in the movie industry, *blackness* bears little "relationship" to "business."

In one sense, contemporary Hollywood has broken with its openly white supremacist origins. Mostly gone are the obvious caricatures: the stern Mammies, bug-eyed Sambos, grunting Tontos, and pidgin-speaking Charlie Chans of the past. Yet, off the screen in the executive suites, when it comes to money and power, control over decision-making in production and distribution remains firmly a white monopoly. But then again, one also could argue that crude racial slander has outlived its usefulness and been supplanted by the more insidious and potent subtleties of racial coding, implication, symbolism, and co-optation. It is in this more subtle sense that Hollywood's racial ideology remains intact and almost thoroughly pervades the content of its films. Take, for instance, the sustained racial allegory in the science fiction hit *Stargate* (1994), which deploys a recuperative strategy[6] that subtly undermines the popular black claim to African American origins in classical Egyptian civilization by portraying the Sun God Ra (Jaye Davidson) as the epitome of black, decadent, hermaphroditic evil. In one of the more fantastic rationalizations of the history of slavery, an expedition of white-led earthlings shows up on an alien planet to destroy Ra and liberate an enslaved population from this wicked black slavemaster deity.

Some of the conflicting ideological tensions between mainstream and black independent perspectives become clear when one contrasts Hollywood's *Stargate* with the independent, black-cast *Space Is the Place* (1972). Exactly inverting *Stargate*'s perspective, in *Space Is the Place* jazz musician Sun Ra, playing the Sun God Ra, comes to planet Earth to liberate its population with the blues-jazz harmonies of his advanced musical technology. But clearly, blues and jazz as the mark of *blackness* in mainstream cinema will, mostly, continue to be the locus of an exotic, oppositional pathology or evil, as drug- and alcohol-addicted geniuses stumble through

34 The jive-talking, blues-jazz spook, Oogie Boogie, in
The Nightmare Before Christmas. Photo by Joel Fletcher
(Touchstone Pictures, 1993).

tired master narratives under the care of white buddies, as in *Round Midnight* (1986) or *Bird* (1988). Or as with the otherwise innovative animated feature *The Nightmare Before Christmas* (1993), in which, yet again, the epitome of primordial evil looms as a jive-talking blues-jazz spook, the "Boogey-woogey Man."

Blackness has always been in vogue in the mainstream culture industry, but it has been relentlessly framed and controlled by an overdetermining sense of white hegemony. Consequently, the biracial buddy formula has found ample expression in the 1990s, continuing to be dominant cinema's most standard and profitable strategy for representing African Americans on the commercial screen.[7] Among the rising black talent of the decade, Samuel L. Jackson, Morgan Freeman, Angela Bassett, Denzel Washington, Will Smith, Laurence Fishburne, and Wesley Snipes have all found themselves in the protective custody of buddy movies that achieved varying degrees of commercial success. The box-office hit *Pulp Fiction* (1995), starring crime partners John Travolta and Samuel L. Jackson, marked one of the revealing, ironic variations on the buddy theme, with Jackson's character and performance carrying much of the film. Articulating that transgressive sense of outlaw blackness that Hollywood finds so alluring, Jackson plays an L.A. hit man who by the end of the flick goes through a profound inner transformation.

The ironies of *Pulp Fiction* are at least twofold, the first irony dealing with the way the script exploits and fetishizes the word "nigger," liberally and gratuitously invoking it throughout the dialogue. Here, one of the grand absurdities of American identity applies. For everyone fantasizes about being black in a hip, cultural sense, even to the point of using "niggah" in the familiar, insider way that some black people do. However, no one wants to live the social disadvantage and inequality of actually *being* a nigger, as only black people are forced to do in America. This is the great hidden energy that animates the Jackson and Travolta performances, the painfully tangled, historic interplay between the black man and his white buddy, from Jim and Huck Finn, to Rocky and Apollo Creed, on into infinity. *Pulp Fiction's* second irony is more important, simply because it is about Hollywood power relations—that is to say, how the political economy of stardom works when it comes to *race*. Samuel Jackson was the dramatic force and moral center of the film, but it was John Travolta's career that was revived by the film and took off. With credit to his talent and effort, Travolta has moved, once again, into the adoring, national spotlight, to star in a series of popular films including *Get Shorty, Phenomenon* (1996), and *Broken Arrow* (1996). Conversely, the equally talented and persevering Jackson has moved laterally through the endless purgatory of yet another biracial buddy movie, *Die Hard with a Vengeance* (1995), the supporting lead in *A Time to Kill* (1996), and the lead in the pedestrian, black-made comedy *The Great White Hype* (1996).[8] Morgan Freeman also exemplifies how the buddy system traps and contains black talent, as he moved from *Driving Miss Daisy* (1989), to prison buddy of Tim Robbins in *The Shawshank Redemption* (1994), to driving Mr. Brad Pitt in his star vehicle, the complex psychological thriller *Seven* (1995).

By far the most interesting buddy variation of the 1990s has to be *Strange Days* (1995), a film so thematically radical and creative in terms of race, gender, and romance that it is inevitably located in an uncomfortably near, authoritarian future Los Angeles. *Strange Days* ends on a powerful note of love and miscegenation between the black female and white male leads, Angela Bassett and Ralph Fiennes. In part because it premiered at the moment of the O. J. Simpson verdict and held its disturbing focus on *race*, the moviegoing audience stayed away from the film. On the other hand, *Money Train* (1995), starring buddies (in this case, foster brothers) Wesley Snipes and Woody Harelson, extended the crossover refrain of conscience liberalism that the pair struck in *White Men Can't Jump* (1992) and made out at the box office. With the grand success of Will Smith and Jeff Goldblum bonding to fight alien *otherness* in the megahit *Independence Day* (1996), or Arnold Schwarzenegger rescuing Vanessa Williams in *Eraser* (1996), it is a safe bet that the biracial buddy formula in all of its variations is destined for a long run in Hollywood.

35a and b Whiteness naturalized and revered: Disney's *Toy Story*
(Walt Disney Company, 1995).

In contrast to the endless ways that dominant cinema subordinates and plays *blackness* in its films, it would be productive here to mention some of the ways that Hollywood constructs the naturalized, privileged sign of *whiteness*. It is easy enough to realize that in a long trajectory of historical contexts, representations, and films, *blackness* has been trapped in expressions of the primitive, the physical body, violence, and eros, while *whiteness* has been associated with civilization and the refinements and gifts of the mind and intellect.[9] Certainly the films *Powder* (1995), about a Caucasian albino gifted with psychic powers, and *Phenomenon* (1996), about an average white guy who is suddenly anointed with the gift of superintelligence continue this traditional black-body, white-mind binary. But perhaps most interesting for this discussion we should explore how one of the leading, box-office hits of the 1995–96 season constructs the naturalized, sovereign sign of *whiteness*. In spite of all its cutting-edge technology, the computer-animated blockbuster *Toy Story* (1995) follows a traditional strategy of omitting *blackness* from its simulacral world, underscoring one of the most powerful but subtle ways that *whiteness* is naturalized and revered.[10] Yet if there are no black subjects in the film per se, because no film can ever completely escape the *race* context of the society whose values it mediates, *Toy Story*, at least by analogy, recognizes the pain of a "double consciousness" in the way that Buzz Lightyear must come to realize that he is not human, but rather "just a toy."

Toy Story's characters and scenes are rendered in the clean, flawless, "look" of new toys, pristine images completely free of anchoring in material reality. What is most interesting about this visual style is the crisp, untanned white-pink skin, and standardized middle-America voices of the leading characters,[11] who in complexion, sound, and appearance come to represent a purely simulated, primally innocent *cyber-whiteness*, suggesting the possibility of some future homogeneous, psychic state free of the social tensions of constantly taking into account a sense of an oppositional *blackness*. With complete demographic heterogeneity encroaching on the social horizon, such a world can exist only in the total constructedness of cyber-animation. Yet, however innocent their makers allege animation to be, its ideological effects are pervasive. Cartoons and animated features most deeply influence the habits and perceptions of Hollywood's youngest and therefore longest-term consumers: children. And *Toy Story*, with its idealized white world, like *Home Alone* (1990) before it, turned out to be one of Hollywood's biggest blockbusters.

There is, however, some good news in all of this. Because the moviegoing audience is vast, heterogeneous, and fragmented along many lines—including the abiding differences of class, race, gender, education, religion—and because consumer tastes and trends are perpetually agitated and shifting as a result of the way the market economy functions, no studio executive can predict exactly which big entertainment investments, no matter how calculated or formulaic, will be box-office winners. So Hollywood's domination of commercial cinema can never be seamless or absolute. Also contributing to the insecure, sometimes contradictory nature of Hollywood's hegemony, black filmmaking and black participation in mainstream cinema tend to evoke a variety of oppositional currents, resistant practices, and insurgent moments against the norms and conventions of the dominant cinema system. Generally speaking, these currents of opposition arise simply because black filmmaking mediates the experiences of a marginalized, oppressed people. Thus, narratives about blacks or the black world tend to be set in more desperate, contested circumstances, or they tend to voice resistant social critiques, often in new, emergent vocabularies and cultural styles.

In response to the challenge of an oppositional black creative discourse, one of the most subtle and powerful features of the dominant cinema system has been its ability not so much to eradicate independent and countercultural impulses, but rather to co-opt and contain them. Hollywood has been highly proficient at buying up and then marketing the signs of *blackness,* emptied of their social and political meanings. In the entertainment and movie businesses, the interplay of resistance and co-optation is a continual push-pull dynamic. This, in part, explains the commercial success of films like Ice Cube's ribald, insider black comedy *Friday* (1995), or the music phenomenon of gangsta rappers decrying the desperate realities of ghetto life to their audience, which is in large part white youth safely tucked away in the suburbs. (As rapper Dr. Dre succinctly puts it: "We're marketing black culture to white people.")[12] Under these complex conditions, then, black-focused filmmakers are struggling to maintain a diverse middle-market niche by drawing the attention and limited investment of industry executives and the mixed patronage of a sizable consumer audience.

Thus, a key factor informing low-budget to mid-budget black filmmaking aimed at popular circulation comes down to the need to increasingly innovate and diversify the subject matter and the genres of black-focused product. The exploration of a shifting range of ideas and subjects in anticipation of an ever-restless moviegoing audiences' boredom with worn-out styles and genres is foremost an economic necessity. But, since black people are themselves a diversely heterogeneous social formation that cannot adequately be defined by any one style, class, or identity, this is also a cultural imperative. So as the end of the decade, century, and millennium approaches, the most obvious genre to exhaust itself is the ghetto-action-gangsta flick. After reaching the zenith of box-office success with *Boyz N the Hood* (1992), and a sort of perfection of representational violence in the Hughes brothers' hit *Menace II Society* (1993), or a more socially conscious tone in the Oliver Stone-sponsored *South Central* (1994), or Spike Lee's *Clockers* (1995), or a range of expression in a dozen or so genre fillers like *Trespass* (1992), *Sugar Hill* (1994), *New Jersey Drive* (1995), the popularity of these films has dwindled. The best of these movies manifest signs of a more sophisticated noir style, which I will return to. But reminiscent of the collapse of the Blaxploitation boom, critics and audience members, black and white alike, once again have tired of seeing

African American life narrowly portrayed in terms of ghetto violence, adventure, and pathology.[13]

On the horizon, the fresh, imaginative work of filmmakers such as Rusty Cundieff, David Johnson, Darnell Martin, Leslie Harris, and Carl Franklin suggests creative directions and solutions to stylistic dead ends and genre traps like the homeboy-ghetto-action flick. Such new impulses in black cinema frequently show up in low-budget or independent features that do not always enjoy success at the box office. Often, these films find their largest audience on the second bounce, at the video store. Such has been the case with Rusty Cundieff's relentlessly funny and cutting satire of the rap, hip-hop music business, *Fear of a Black Hat* (1994), which was made for less than $1 million. Taking up the mock documentary style of *This Is Spinal Tap* (1984) with its critique of heavy metal bands, *Fear of a Black Hat* follows the escapades of "Niggaz with Hats" in a brilliant parody of the cultural politics and style of the rap group N.W.A. *Fear of a Black Hat* outdoes the more expensive, Nelson George-produced rap comedy *CB4* (1993), which suffers from being too close to the material it is trying to satirize to be more than erratically funny. By contrast, what makes director-star Cundieff's humor so potent is his ability, as rap star "Ice Cold," to mock the way that some rappers deftly gloss over or excuse their blatant misogyny or violence with the most ludicrous rationalizations. Ice Cold, for instance, glibly tries to explain why the group's hit, "Booty Juice," is uplifting for women, or that their *Kill Whitey* album has no social implications but refers specifically to a former manager named Whitey. Unfortunately, both *CB4* and *Fear of a Black Hat* slept at the box office, in part because of the speed at which rap culture evolves. By the time they were released, their object of parody, the swiftly mutating rap scene, had moved light years past either films' gags and allusions. However, Cundieff's humor and his mocking style with its distinct political edge and subversive currents suggest a potent new direction in the development of black comedy.

Cundieff refined his talent for casting his work in specific political or social contexts without lapsing into editorial propaganda in his second feature, the Spike Lee-produced *Tales from the Hood* (1995). Working in an unevenly explored genre for black filmmaking, the horror flick, and taking off on the style of the *Tales from the Crypt* films and TV series, *Tales from the Hood* consists of a round of stories told to three homeboys by a spooky mortician. If, as critics Michael Ryan and Douglas Kellner, among others, argue,[14] the great dread in the horror film is the return of those repressed energies in the form of the monster that society cannot openly deal with, then the monsters that animate *Tales from the Hood* are markedly political and express the great terrors of African American life: police brutality, lynching, racism, the catastrophic effects of social inequality. In one of the most original and stunning sequences in the horror genre, a gothic scientist (Rosaland Cash) tries to deprogram the violent rage of a black youth by exposing him to a sustained, gruesome photo montage of real lynchings, thus replaying the long historical nightmare of genocidal violence inflicted upon black people. This brutal vignette clearly makes the point that the homicidal rage that young black men inflict on each other merely carries on the work of more traditional racist forces in the society.

Tales ends with that perfect register of cultural humor and irony that marks the insider perspective of a black filmmaker with an ear for the subtle nuances of 'hood vernacular. By the end of the round of tales, predictably, the three young gang-bangers realize that they are actually dead and that their host, played brilliantly by Clarence Williams III, is not a mortician at all, but really the Devil. In a close-up, the undertaker's face morphs into that of Satan, who articulates his deep mastery of worldly wickedness by turning the homeboys' obscene argot back on them, as he announces their fate. In a controlled sardonic voice, the Devil closes the film with the line "Welcome to Hell, *motherfuckers,*" as the camera pulls back on the homeboys dancing in a whirlwind of flames. It is exactly this kind of sly counter-current linguistic detail that distinguishes *Tales from the Hood* from the comparatively more mainstream black horror flick, the Wes Craven-directed star vehicle for Eddie Murphy and Angela Bassett, *Vampire in Brooklyn* (1995).

Another low-budget, Spike Lee–produced feature that mediates complex expressions of opposition to the Hollywood norm is *Drop Squad* (1994), directed by David Johnson. Following up on the concerns of a string of early 1990s' black-made, or black-focused, mainstream films like *Jungle Fever, Livin' Large,* and *Strictly Business* that explore the contradictions and anxieties of black middle-class identity and "buppie" assimilation into white business culture, *Drop Squad* pushes these identity issues a step further by questioning what constitutes a stable, "authentic" *black* self and who gets to determine precisely what that "self" might be. Adopting the atmospherics of a movie about the French Resistance during World War II, *Drop Squad* depicts the clandestine operations of an underground black organization that abducts and deprograms sellout "buppies." When Bruford Jamison, Jr. (Eriq La Salle), is kidnapped, or "dropped," the dauntingly complex problem of black identity formation in the heterogeneous post-Black Nationalist 1990s is rigorously interrogated and satirized.

36 Black identity in the 1990s, diverse and in process: David Johnson's *Drop Squad* (Universal, 1995).

Drop Squad assumes an open or unfinalized stance on nineties' black identity in that Bruford's deprogramming ends on a painful, ambiguous note. While Bruford's neoconservative, Uncle Tom consciousness is the source of the film's funniest moments and best satire, in the Drop Squad's frustrated attempts to deprogram him, their tactics turn increasingly violent and fascistic. Bruford comes to see the error of his buppie ways, but he justifiably feels injured and wronged by the deprogramming process itself. Finally, the Drop Squad's two leaders—Rocky (Vondie Curtis-Hall) and Garvey (Ving Rhames)—are forced to recognize that the moment for 1960s-style shock therapy and consciousness-raising has passed. They split over tactics and decide to fulfill the mission of salvaging black identity and culture by going their separate ways. Contrary to the dominant Hollywood style that provides pat solutions to an assenting consumer audience, *Drop Squad* dares to leave the audience with more questions and uncertainties than answers. The film's ending is open and unresolved, quite simply because black social identity in the mid-1990s itself is diverse, in process, and so unresolved.

Also drawing upon dissident perspectives or modes of expression, Mario Van Peebles's *Panther,* and the Hughes brothers' *Dead Presidents* (both 1995) exemplify that curious mix of oppositional tendencies with grudging concessions to mainstream Hollywood style often found in popularly consumed black films. In that it offers a creative, revisionist reading of history, Van Peebles's *Panther* fits well into the trajectory of political works by more established directors: Spike Lee with *Malcolm X* (1992), Oliver Stone with *JFK* (1991) and *Nixon* (1995), or Costa-Gavras with *Missing* (1982). In the case of *Panther,* its relevance as black, oppositional cinema was confirmed by how instantly and vehemently the film was attacked by the right wing of the media, and how quickly it disappeared from theater circulation after its release.[15] Set in Oakland, *Panther* sympathetically depicts the founding of the Black Panther Party in 1966 and explores the grievances of the black community, the police terror, ghettoization, and social injustice that brought the party into being. *Panther* ends, à la Hollywood, with both a climactic shootout (with nefarious, government-controlled drug dealers) and a sentimental denouement in which a stoplight that the community had agitated for in the beginning of the film is finally installed. Following up on an idea raised by Spike Lee's *Malcolm X,* perhaps, *Panther*'s most significant contribution to black cinema will be to suggest that one can tell a politically focused story from a collective black point of view in the style of the heroic, individualistic, Hollywood action-adventure narrative.

With historical hindsight, *Panther* stops at a prudent historical moment in that the film ends with Huey Newton at the height of his powers, before his decline and eclipse in the seventies and eighties. The Hughes brothers' second film, *Dead Presidents* (1995), is located in the same politicized, Vietnam era as *Panther,* and it deals with many of the same injustices, discontents, and tensions. *Dead Presidents,* the title alluding to those faces on paper money, tells the coming-of-age story of a black youth, Anthony (Lorenz Tate), as he journeys from a working-class background in the Bronx, through Vietnam combat, and back to the old neighborhood and the economic and social dead end that dominant society has carefully prepared for so many of its returning black veterans. Anthony's pent-up frustrations culminate in one of the most visually fantastic heists in recent cinema. If *Dead Presidents* is canonized for nothing else, it will be for the brilliantly imagined image of black bandits in whiteface stealing the government's worn-out, about-to-be-burned money, or for the close-up shots of the faces on that money burning in the film's opening credit sequence.

The politics of *Dead Presidents* can be found in the way that the film subtly charts the slow economic and social decline inflicted on the black community after the formal, legal gains of the civil rights movement, as well as the resulting epistemic break between black generations. Youth that once settled disputes with fisticuffs now ice each other with guns.

Brothers that would have held down steady jobs or gone to college now stand idle and junked-out on ghetto street corners. Yet, however ambitious their intentions, Allen and Albert Hughes have created a rambling, panoramic narrative that tries to cover and say too much—from the Bronx to Vietnam and back again. This is the film's minor weakness, but it is also the filmmakers' great strength, the fact that the Hughes brothers are willing to experiment with form, to push the material beyond normative boundaries of what would have routinely been edited down by the dominant film industry to an apolitical, adventure-caper flick.

To date, the most glaring example of Hollywood's cultural hegemony, and the way its system of discrimination intensifies according to the number of excluded categories one fits into, has to be the token to nonexistent participation of black women in mainstream filmmaking. If black male filmmakers have been marginalized, even with the "new black movie boom" producing, at most, a dozen or so new black male directors, then black women are, to employ the phrase of legal scholar Derrik Bell, the "faces at the bottom of the well."[16] Since Julie Dash's breakthrough with *Daughters of the Dust* (1991), only two black women, Leslie Harris and Darnell Martin, have each managed to make and release one mainstream theatrical feature film. This is not to say that many other black women have not made outstanding feature films, including the late Kathleen Collins's *Losing Ground* (1982), Ayoka Chinzera's *Alma's Rainbow* (1992), and Camille Billops's *KKK Botique Ain't Just Rednecks* (1993). A number of varied explanations—from black women's narratives being too "soft" and not conforming to Hollywood marketing formulas or genres, to black women not being able to "fit in" and negotiate in what is overwhelmingly a male business—have surfaced to account for the lack of black women's films in theatrical distribution.[17] As a result, the major share of black women's feature filmmaking has been declared "independent" and channeled into the relatively more obscure circuit of museums, universities, and film festivals.

Contesting the inertia and exclusionary games of the industry, then, Leslie Harris's gender-focused *Just Another Girl on the I.R.T.* (1993), takes an unsentimental look at the enormous difficulty of making one's way out of the lower-class / working-class projects, especially if one is black, female, and young. *Just Another Girl* opens with a precocious high school student, Chantel (Ariyan Johnson), telling the audience of her plans to skip her senior year and go directly to college and then medical school so that she can become a doctor. The painful irony of Chantel's words becomes clear as the film contrasts her ambitions with the limitations imposed by her

37 Ariyan Johnson biding her time in Leslie
Harris's unsentimental teenpic, *Just Another Girl
on the I.R.T.* (Miramax, 1992).

cramped, impoverished environment. Things are further complicated
when she hits one of life's speed bumps. In that accidental teenage manner that afflicts so many urban youth, she becomes pregnant and has to
put her imagined future on hold.

   Leslie Harris's directorial touch is subtly intelligent and noneditorial; it
points out that talk is cheap, and it is always a struggle to make one's
dreams tangible. Related to gender, it is interesting to note that *Just
Another Girl* was released at the same time as *Menace II Society*, and that
while both films focused on the same generational and locational politics,
they performed very differently at the box office. *Menace*, which was made
for New Line Cinema for $2.5 million, was released to more screens and
better press coverage and went on to become a hit, grossing $30 million,
while *Just Another Girl*, made for $500,000 did modestly well at the box
office and went into the comparable obscurity of "film society" screenings and the shelf in the video store.[18] While it would be reductive to

argue that these two differently financed, made, and positioned dramatic features should have performed at the same box-office level, it is worth noting that their gender and genre orientations had much to do with how they were marketed and consumed.

Darnell Martin's *I Like It Like That* (1994), mediating many of the same gender concerns as *Just Another Girl,* explores the rigors and complications of pursuing one's ambitions and dreams in spite of the barriers set so relentlessly against lower-class black/Latino women. Lisette (Lauren Velez) struggles to realize a business career in the record industry and feed her three children, while her rakish husband languishes in jail for petty theft. Besides depicting the usual hassles and limitations of the 'hood environment, Martin's film brilliantly articulates the concept of *mestizaje* (cross-cultural hybridity and heterogeneity). Bringing to life a creative mix of tensions and dialogue that shifts between the tragic and comic, Lisette is biracial black, her husband is Latino, and her brother is a transvestite. At one hilarious point, Lisette's mother-in-law (Rita Moreno), who considers her family to be of "pure Castilian blood," sarcastically comments while dressing her grandchildren, "I don't know how to comb nappy hair." Martin's talent resides in her ability to handle a fast-paced, character-focused drama with moments of comedic punctuation. Her work was backed by Columbia Pictures with a $5.5 million budget, and the film had a modest box-office success.[19] Wisely, both Leslie Harris and Darnell Martin have confronted one of the main problems contributing to the underdevelopment of black women filmmakers. Both have managed to sustain their momentum in the business by quickly moving on to their next projects, with Martin working on a horror movie about a mixed race family called *Listening to the Dead,* and Harris working on a feature about a black woman executive in the music industry called *Royalties, Rhythm and Blues.*

The issue of black women's film and filmmaking has been further highlighted, and complicated, by the release of the mega-hit screen adaptation of Terry McMillan's novel, *Waiting to Exhale* (1996). Although directed by a black man, Forest Whitaker, in many ways *Waiting to Exhale,* fulfills the promise and aspirations set in motion by earlier films made about, or by black women, especially the mainstream *The Color Purple* (1985) and the independent *Daughters of the Dust.* Black women went to see *Waiting to Exhale,* en masse, held discussion groups and parties after shows, and sparked an ongoing dialog-debate in the media. *Waiting to Exhale*'s enthusiastic reception has evinced how, as black film critic Jacqueline Bobo argues in her work, black women are accustomed to "reading through the text" as in the case of *The Color Purple,* in order to extract their own resistant, countercurrent meanings.[20] Accordingly, because so few mainstream features are made from a woman's point of view, let alone a black woman's, over the years, tremendous, pent-up demand to see their subjectivities on the big screen has steadily built up among black women as a segmented audience. These issues, partly, account for the film's success among black women, and the film's crossing over to women in general.

Like many other successful, black feature films from Blaxploitation on, *Waiting to Exhale* became a major hit as much for its music, featuring women singers from Whitney Houston to Aretha Franklin and Patti LaBelle, as for its gendered subjectivity and the appeal of its stars, Whitney Houston, Angela Bassett, Wesley Snipes, and Gregory Hines. Consequently, *Waiting to Exhale's* thematic, dramatic soundtrack has gone platinum as a CD. Add a phenomenal $45 million gross in the first seventeen days of exhibition on a production cost of a mere $15 million, and profits running at $70 million for an overall profit-to-cost ratio of 5-to-1 after twelve weeks, and one has every studio VIP's fantasy of a hit black film. There is, of course, another, more critical take on all of this. *Waiting to Exhale* is every studio mogul's dream because black films are held to an unequal standard compared to commercial films made by white male filmmakers. In 1997 the average cost of a mainstream white film was about $30 million, with the industry demanding a profit-to-cost ratio of 3-to-1. At the same time, the cost of a black feature was held well under $20 million, while executives used a much higher profit-to-cost ratio, 5-to-1, in order to consider a black film a success. When it comes to Hollywood economics, the system is markedly skewed. When one considers the meager capital that Hollywood puts into black films, it becomes clear that the bottom of white people's expectations forms the sky of black people's aspirations, as in so many other areas of American life.

Beyond those black men who were critical of what they perceived as their routine devaluation in yet another dominant cinema vehicle, *Waiting to Exhale's* phenomenal success has sparked a lively critical debate. While many women writers have praised the film throughout the media, other women, black and white, have been insightfully critical. In an episodic, soap-opera narrative style that carries considerable gender appeal, *Waiting to Exhale* follows the romantic misadventures, entanglements, and triumphs of four successfully career-oriented, middle-class black "girlfriends." The film celebrates the idea that black women can find true love and "exhale." These women, as well, form deep, insightful relationships among themselves without the domination and validation of men. Critic Karen DeWitt rightly observes in a *New York Times* article that *Waiting to Exhale* has tapped into the tremendous pent-up demand of black women for their equitable representation in cinema, and that the film's instincts about black women's issues and lives have been more than confirmed by the fervent way it has been consumed by them. DeWitt, along with studio executives, notes that black women found the film a realistic portrayal of their concerns as well as entertaining. But curiously, she goes on to say that the film "has nothing to do with racism, interactions with whites, or ghetto life," and that while black men have their Million Man March, *Waiting to Exhale* "becomes the female equivalent."[21]

Immediately after DeWitt's survey of the positive social impacts of *Waiting to Exhale,* black writer and social critic bell hooks responded in a *New York Times* op-ed piece wryly titled "Save Your Breath, Sisters." In her essay, hooks notes that the film is practically empty of any feminist consciousness or politics. Moreover, hooks comments on Hollywood's co-optative strategies, saying that "no doubt it helps crossover appeal to set up a stereotypically racist and sexist conflict between white women and black women competing to see who will win the man in the end." As for the comparison of *Waiting to Exhale* to the Million Man March, hooks perceptively cautions black women that "we are being told and we are telling ourselves that black men need political action and black women need a movie."[22] While the desire of black women to see themselves honestly depicted on the big screen is palpable and valid, one must always remember that, here, the situation is intensely ironic. For Hollywood, through its relentless discrimination against black women, has created this "desire" in the first place.

Co-optation is Hollywood's strongest suit. So once again, as was done with *Lady Sings the Blues* (1972) and *Mahogany* (1972), the film industry has colonized the very desire that it is responsible for inflicting on black women, and then it has sold it back to them as packaged "entertainment." In her *Sight and Sound* review, critic Amanda Lipman puts it most eloquently, saying that *Waiting to Exhale* "raises difficult, thorny issues, then strangles them with the velvet glove of liberalism."[23] The true and direct expression of black women's subjectivities in commercial cinema is going to have to emerge out of the political consciousness and filmmaking of, first, black women, but also black people as a collectivity, beyond the ideological filters and games of Hollywood's co-optative liberalism. Put simply, the real promise of black women's filmmaking resides in the work and imaginations of black women directors themselves, perhaps always working against the grain of Hollywood.

Another interesting development now under way in popular black cinema has to do with the subtle genre shift away from the 'hood action flick and toward an emerging black-made, or black-focused, film noir genre. Certainly African Americans have been well-acquainted with the noir world since their arrival in America. W. E. B. Du Bois's notion of "double consciousness," literally having to think and see "double," within the constraints of an oppressive, racist system, that is, having to read the official version as well as its corrupt underside, has been a primary black survival tool as well as a motive force in much of black cultural production. If the classic style of film noir depicts the decaying urban world of crime and violence fed by corrupt, indifferent social institutions, then, from the resistance and protest of the slave narratives, through the crime fiction of Richard Wright, Chester Himes, and Donald Goines, or the ghetto action of Blaxploitation, or the gangsta poetics of Snoop Doggy Dogg and Coolio, black people have honed and perfected a wicked, penetrating vision of America's noir world that can only be described as the funkier side of noir.

The principal difference between the noir world of dominant cinema and the funky noir world of African Americans resides in the difference between the perception of crime as the act of a deviant individual and crime as survival, as the informal resistance of a subject people against a racially unjust system. When a white person turns to crime, he or she turns against a system that is designed to defend and support dominant white society. Conversely, while black people are fully aware that they are disproportionately the victims of crime and want criminals held accountable for their acts, most blacks, including those leading "the life" as they call the funky noir experience, also view crime more in a socioeconomic context.[24] They feel that they have been trapped by a racist system that was set against them from the beginning. "I know it's rough, but it's the only game the Man left us to play," as Eddie tells Priest in *Superfly* (1972). In the noir world, whites feel that they have failed the system, while, conversely, blacks feel that the system has failed them.[25] This sense of funky noir accounts for an implicit yearning for racial justice in many films or the ambivalent endings that negate or expose the hypocrisy of the system. It is not surprising, then, that one can see the persistent developing momentum of this funky noir genre over a long trajectory of black films from *Shaft* (1971) and *Superfly* on through *A Rage in Harlem* (1991),

38a and b Funky noir: Carl Franklin's *Devil in a Blue Dress.*
Photo by Bruce W. Talamon (Tri-Star Pictures, 1995).

Deep Cover and *One False Move* (both 1992), *Menace II Society* (1993), to *Devil in a Blue Dress, New Jersey Drive,* and *Clockers* (all 1995).

The case of *Devil in a Blue Dress* proves instructive in exploring the funky noir style and, more importantly, in demonstrating how the success of a black film is always, in some way, defined and overdetermined by the social construction and contestation of *race.* Directed by Carl Franklin and adapted by him from Walter Mosley's successful novel, *Devil in a Blue Dress* follows the adventures of part-time detective Easy Rawlins (Denzel Washington) as he tracks the whereabouts of mystery lady Daphnie Monet (Jennifer Beals), who is known to occasionally stray from the white world in search of the forbidden pleasures of soul food, jazz, and "dark meat." The funkier side of noir reveals itself in *Devil in a Blue Dress*'s subtle cultural details and gestures, its racial ironies, its character sketches and impressions, as Rawlins follows Monet through the soul food cafés, juke joints, and smoky bars of L.A.'s black world, circa 1948.

Marking a funky noir detail, instead of the cheap suit favored by the typical film noir detective, Rawlins wears casual clothes that suggest his predicament as a laid-off aircraft worker. Just as the cheap suit marks the social distance between the white private eye and his wealthy clientele, Rawlins's attire, slacks with an undershirt, signify his black, working-class status and his powerlessness relative to the forces he is up against. With mortgage payments due on his modest house, Rawlins is an ordinary brother trying to unravel a mystery while being acutely aware that he is meddling in the corrupt business of powerful, contending white forces: the city government's ruling elite, the police department, and organized crime. Denzel Washington's characterization shows that he understands the psychology and appeal of Easy Rawlins, a black man painfully squeezed by every conceivable power relation and economic circumstance of a racially unjust society. He plays Rawlins with social vulnerability and caution mixed with an underlying toughness that gradually builds into a cunning, assertive rage against injustice. "They thought I was some kinda new fool . . . and I guess I was," Rawlins says, recognizing the doubleness and danger of his situation.

The principal, tangled mystery of *Devil in a Blue Dress* is caught up in a nagging question: how could a film based on a hit novel, directed by an outstanding filmmaker, a film generally praised by critics and featuring the star power of Denzel Washington, be such a flop with the mass audience? The answer to the mystery of *Devil* at the box office alludes to the answer to the central mystery of the film's narrative. The key to both mysteries is found in the multiple expressions of that enduring social construct, the great American obsession and catastrophe: *race*. The persistent irony driving all of Mosley's novels, and certainly Franklin's film, has to do with the many ways that Easy Rawlins's double consciousness as a black man empowers him as an occasional private eye. The fact that he is on this case at all is because he is black and can move and detect things in those circles and in that world forever closed to whites. The ultimate joke of *Devil* is that double consciousness facilitates double vision. Rawlins can see deeply into both black and white worlds and find out what powerful whites want to know, but also what they do not want to know, or at least what they want to deny or hide.

As the plot evolves, we find out that the elusive Daphnie Monet is engaged to one of L.A.'s leading mayoral candidates, Todd Carter (Terry Kinney). In an attempt to protect herself in a counter-blackmail scheme,

Daphnie disappears into the black world where Rawlins finds her. Then, in a spectacular gunfight, he rescues her from the gangsters who originally retained him. At this point the mystery of the *Devil in a Blue Dress* is revealed: Daphnie is not white at all, but rather a "tragic mulatto" passing for white. Miscegenation being the potent, socially charged act that it was in 1948 (and still is), Daphnie's upcoming marriage into one of the city's elite families definitely can never happen. In a final act of double consciousness, confirming his true powers as a black private eye, Rawlins functions as go-between, negotiating the breakup of the miscegenational liaison. For his trouble, Rawlins receives a generous settlement for all concerned, as the rich white man, Carter, retreats back across the color line into his world. All is happily resolved as noir becomes a bright L.A. morning, the foreboding ghetto turning into the African American community.

As for that other mystery, *Devil in a Blue Dress*, like *Strange Days*, had the misfortune to be released at the time of the O. J. Simpson verdict. And that, in my opinion, was the kiss of death at the box office: guilt by association. The film clearly did not cross over. Racial divisions and tensions were running high, and angry, post-verdict whites were in no mood for a tour of L.A.'s black world, 1948, 1999, or otherwise. But *Devil* also failed to attract the youth audience, black or white, and this group accounts for much of the black film market. While black cinema's audience is restless for new experimentation, genres, and narratives, and the 'hood-homebody-action flick is brain-dead, filmgoing youth did not make the leap to consuming this excellent cinematic translation of a brilliant novel. This lack of response has much to do with contemporary youth's focus on electronic rather than print culture. But the film's disappointing box office also relates to the difference in social visions between now and the time of the film, 1948, between an inherent contemporary pessimism, as opposed to the basic optimism of post-World War II America. *Devil* ends on exactly the right historical note, but a note perhaps strange to the contemporary audience. Easy Rawlins, the homeowner, walks down the sunny, modest street of his friendly community, as superficial order—and separation between the races—for the moment, at least, is restored.

Our discussion concludes at an interesting social moment that exemplifies so many of the issues and problems related to the way in which the film industry deals with the sign of *blackness:* the annual Academy Award ceremonies. While the 1996 event was produced by Quincy Jones, hosted by Whoopi Goldberg, and featured walk-ons by a number of black celebrities, it was protested from the sidewalk outside the auditorium by

Jesse Jackson and roundly denounced in *People* for its complete "blackout" of African American nominations to its award categories. Beyond the pertinent issue of why black people should even seek validation from such a racially corrupt institution, or the utter banality of the year's event (with a film about a pig contending for best picture), we should note the disparity between appearance and reality. For Hollywood is always the circus of dreams and lies as it exploits the image of *blackness*, while relentlessly keeping African Americans and other people of color marginalized and locked out of its mode of production and executive process, and therefore out of the largest share of its power and profits. In other words, the reality of Hollywood's racism resides in the disparity between Quincy, Whoopi, et al. as images on the stage, and the dismal figures for black participation in the film industry's guilds and unions, figures that have not changed since the late 1960s.[26]

On a final, cautiously optimistic note, I will speculate that there is still room for expansion in the shaky, but ongoing, black film wave. The display of Eddie Murphy's talents in his comeback hit, *The Nutty Professor* (1996), or the release of the prolific Spike Lee's *Girl 6* (1996) are relevant here. And certainly there have always been a few courageous innovations in the way that Hollywood reads *race,* like Robert De Niro's excellent *A Bronx Tale* (1993) or the Harry Belafonte and John Travolta acting collaboration, *White Man's Burden* (1995). But cause for whatever limited hope there is pales against the cultural apartheid expressed in Hollywood's images, films, and genres, or the racist and sexist employment practices structuring all levels of the industry. Rather, because the moviegoing audience is a perpetually shifting social formation that is increasing in its heterogeneity, and a fair number of "blockbuster" investments regularly fail, and because black people overconsume film in proportion to their numbers, the industry has committed itself to making a steady quantity of small-to-medium-budget black features as insurance against white, big picture flops.[27] Beyond the great industry motivator of greed for ever-larger profit margins that, for now at least, will keep open a narrow window of opportunity allowing low-to-mid-budget black films to survive and, hopefully, gain an increasing portion of the diversifying consumer audience, one must always keep in mind, when considering whether or not Hollywood can change, that the industry's racial attitudes cannot be separated from the broader social context of America's racial condition. In a way, then, it all comes back to the resistance practices and oppositional currents of black filmmakers, critics, and the potential for an evolving social consciousness of the audience. For if Hollywood can never free itself from, or change, the dictates and images of American racial protocol, then black filmmaking, in all of its many aspects is destined to struggle against, represent, and mediate the fundamental condition of black people in America.

Notes

1 Pam Lambert, "Hollywood Blackout," *People,* March 18, 1996, pp. 42–52.

2 Kevin Hagopian, "Black Cinema Studies: Shadows and Acts," *Journal of Communication* 45 (Summer 1995): 177–85. Some of the influential new authors and works mapping the emergent area of black cinema are James Snead, *White Screens, Black Images;* Ed Guerrero, *Framing Blackness;* Manthia Diawara, ed., *Black American Cinema;* Mark Reid, *Redefining Black Film;* Jesse Rhines, *Black Film, White Money;* Nelson George, *Blackface;* Thomas Cripps, *Making Movies Black;* and Ella Shohat and Robert Stam, *Unthinking Eurocentrism.*

3 Don Terry, "Woman's False Charge Revives Hurt for Blacks," *New York Times,* November 6, 1994, pp. 12, 32; Kevin A. Ross and Bruce Levitt, "Drowning Case Embitters U.S. Race Relations," *New York Times,* November 11, 1994, pp. A18, A30.

4 Tom Wicker, *Tragic Failure: Racial Integration in America* (New York: William Morrow, 1996); Derrick Bell, *Gospel Choirs: Psalms of Survival for an Alien Land Called Home* (New York: Basic Books, 1996); Earl Ofari Hutchinson, *Beyond O. J.* (Los Angeles: Middle Passage Press, 1996).

5 Dennis Greene, "Tragically Hip." *Cineaste* 20 (October 1994): 28–29.

6 Guerrero, *Framing Blackness: The African American Image in Film* (Philadelphia: Temple University Press, 1993), pp. 113–17.

7 Ed Guerrero, "The Black Image in Protective Custody: Hollywood's Biracial Buddy Films of the Eighties," in *Black American Cinema,* ed. Manthia Diawara (New York: Routledge, 1993), pp. 237–46.

8 Ellen Holly, "Waiting for a Black Heathcliff," *New York Times,* March 26, 1996, p. A19. Holly makes the insightful observation that Hollywood puts most black male actors' careers on hold, deploying them as character actors until they are safely past their youth and prime. Consequently, black men cannot become the high-voltage sexualized movie stars that white men can.

9 Shohat and Stam, chap. 4, "Tropes of Empire," *Unthinking Eurocentrism: Multiculturalism and the Media* (New York: Routledge, 1994); Richard Dyer, *Heavenly Bodies: Film Stars and Society* (New York: St. Martin's Press, 1986), pp. 67–139.

10 Shohat and Stam, *Unthinking Eurocentrism,* pp. 223–30; Snead, *White Screens / Black Images* (New York: Routledge, 1994), pp. 6–7.

11 It is relevant to note here that the voices of the two lead toys were provided by white comedy stars Tom Hanks and Tim Allen, in contrast to the distinctly *black voice* of Levi Stubbs in *Little Shop of Horrors* (1986). See Guerrero, *Framing Blackness*, pp. 56, 118.

12 John Leland, "Rap and Race," *Newsweek*, June 29, 1992, p. 48.

13 Bernard Weinraub, "Black Film Makers Are Looking Beyond Ghetto Violence," *New York Times*, September 11, 1995, pp. C11, C14; Guerrero, chap. 3, "The Rise and Fall of Blaxploitation," in *Framing Blackness*.

14 Douglas Kellner and Michael Ryan, *Camera Politica* (Bloomington: Indiana University Press, 1990), p. 169; Michael Paul Rogin, *Ronald Reagan, the Movie, and Other Episodes in Political Demonology* (Berkeley: University of California Press, 1987), p. 237.

15 "*Panther*: An Interview with Mario Van Peebles," *Tikkun* 10, (July-August 1995): 20–24, 78.

16 Derrick Bell, *Faces at the Bottom of the Well* (New York: Basic Books, 1992).

17 Rhines, *Black Film, White Money* (New Brunswick, N.J.: Rutgers University Press, 1996), pp. 125–42.

18 Amy Taubin, "Girl in the Hood," *Sight and Sound*, August 1993, pp. 16–17.

19 Jan Hoffman, "Mom Always Said, Don't Take the First $2 Million Offer," *New York Times*, October 9, 1994, p. 28H.

20 Jacqueline Bobo, *Black Women as Cultural Readers* (New York: Columbia University Press, 1995).

21 Karen DeWitt, "For Black Women a Movie Stirs Breathless Excitement," *New York Times*, December 31, 1995, pp. 1, 25.

22 bell hooks, "Save Your Breath, Sisters," *New York Times*, January 7, 1996, p. E19.

23 Amanda Lipman, "*Waiting to Exhale*," *Sight and Sound*, February 1996, p. 56.

24 Andrew Hacker, chap. 11, "Crime: The Role Race Plays," in *Two Nations: Black and White, Separate, Hostile, Unequal* (New York: Charles Scribner's Sons, 1992); Wicker, chap. 9, "Throwing Away the Key," *Tragic Failure*.

25 Manthia Diawara, "Noir by Noirs: Toward a New Realism in Black Cinema," *African American Review*, Winter 1993, pp. 525–38. Also see H. Bruce Franklin, *Prison Literature in America: The Victim as Criminal and Artist* (New York: Oxford University Press, 1989), p. xv.

26 In one instance black representation has actually declined since the late 1960s when black membership in Hollywood's craft unions stood at barely 6 percent. Today, Local 44, which includes set decorators and property masters, has less than 2 percent black membership. Compare the numbers in Dan Knapp, "An Assessment of the Status of Hollywood Blacks," *Los Angeles Times*, September 28, 1969; "Black Craftsmen's Talents Untapped on Entertainment Scene," *Los Angeles Times*, October 5, 1969; Pam Lambert, "Hollywood Blackout," *People*, March 18, 1996, pp. 42–52.

27 Rhines, *Black Film, White Money*, p. 216; Bernard Weinraub, "Dismay Over Big-Budget Flops," *New York Times*, October 17, 1995, pp. C13, C18.

Culture as Fiction:

The Ethnographic Impulse in the Films of

Peggy Ahwesh, Su Friedrich, and Leslie Thornton

Catherine Russell

*I call ethnography a meditative vehicle because we come to it neither as to a map
of knowledge nor as a guide to action, not even for entertainment. We come to it
as the start of a different kind of journey.*   Stephen Tyler

Over the last ten to fifteen years, experimental and documentary filmmakers have been encroaching on each others' terrain. Documentary, for
example, has begun to place its own authority in question, and in doing
so, has embraced many techniques associated with experimental film.
While this borrowing of techniques and blurring of boundaries has enriched film culture in many ways, it is a tendency that can also be traced to
larger developments in the cultural role of media. Reality TV and media
events such as the Rodney King and O. J. Simpson trials have demonstrated that "reality" cannot be taken for granted, that truth is how and
where you make it.[1]

In the late 1980s and early 1990s a number of documentaries were
produced in which the filmmaker's personality and / or perspective thoroughly informed the depiction of reality. Michael Moore and Ross McElwee placed themselves centrally in their films *Roger and Me* (1989) and
Sherman's March (1985). Errol Morris constructed his own version of a
crime in *The Thin Blue Line* (1988) and succeeded in having a wrongly
accused man released from Death Row. These films achieved a modicum
of success among mainstream audiences and might be seen as an inter-
mediate zone between tendencies in popular culture and parallel developments in experimental filmmaking.

In films like Moore's and McElwee's, the filmmakers' obsession with inadequacy constitutes what one critic has described as "an aesthetics of failure."[2] The personality of the filmmaker in these instances often compensates for the breakdown of representation. As Linda Williams has argued, even in postmodern cinema, truth "still operates powerfully as the receding horizon of the documentary tradition."[3] In films further out on the margins of the mainstream, outside that tradition, neither authenticity nor documentary truth are so easily reinstated. In the void of documentary veracity, in an apparent acceptance of a breakdown of realist aesthetics, a new cinematic language has evolved. It is a style that draws on "fictive" strategies of representation and is concerned with cultural observation. It is a film form that does not come out of nowhere, though; it draws heavily on the history of experimental filmmaking as well as documentary and narrative practices.

The term "antidocumentary," first coined by the Dutch filmmaker Johan van der Keuken, has been used to refer to a kind of filmmaking that rejects the truth claims of conventional documentary practice. Subjective or poetic documentary in fact has a very long history, including the Soviet filmmaker Dziga Vertov, the World War II British filmmaker Humphrey Jennings, and the Left Bank New Wave filmmakers Chris Marker and Alain Resnais. In the United States, however, we need to turn to a cinema that is more readily labeled "experimental" for such an example of a "subjective documentary" project, i.e., the films of Andy Warhol and Jonas Mekas in the 1960s.

Warhol and Mekas were very different filmmakers, but both were interested in developing film languages that could convey something about the microcultures in which they lived and worked—two overlapping pockets of the New York art world of the 1960s.[4] For Warhol, this meant using cinema like a machine through which his actor friends were transformed into cultural commodities; for Mekas, it meant using cinema as a romantic form of expression in which his filmmaker friends were depicted as the poets of a new world. It was not "documentary" with which these filmmakers were concerned, but rather a new means of representing culture in which people and art could be fused in new forms of cultural production that lay resolutely outside the film industry and all that it represented.

These labels of "experimental" and "documentary," and even "nonfiction" which is often used to link them, are clearly limited. If they are inadequate to categorize the films of the sixties, they are even less appropriate to contemporary filmmaking. A more useful term, one that cuts across these formal definitions, is that of ethnography. Taken in its broadest sense, ethnography refers to the documentation of culture and has come to incorporate a poetics of observation.

Since the 1950s, ethnographic films have consistently adopted reflexive techniques to inscribe the relation between filmmaker and subject(s) within the film. References in the text to the means of production—shots of camera equipment, voice-over comments, dialogue with ethnographic subjects about the making of the film—have for a long time served as attempts to correct the imbalance of power between those who are making the film and those who are being studied.[5] More recently, a great deal of writing and experimental practice has been devoted to rethinking ethnography within a postcolonial framework. This means not only theorizing ways of filming, but going beyond the idea of "other cultures."

The representation of cultural difference, cultural history, and cultural transformation needs to be carried out from a decentered perspective in which "us and them" is no longer the governing paradigm: we are all each other's other. The most prominent proponent of this new approach to ethnography is the Vietnamese-American filmmaker / theorist Trinh Minh-ha, who argues that a new way of making films meaningful is required.[6] Her claim in *Reassemblage* (1983) that she intends not to "speak about" but "speak nearby" is a radical subversion of the totalizing structures of realism that ethnographic film, as a form of visual knowledge, tends to assume. Her ethnographic style allows glimpses into African cultures, while refusing to allow those glimpses to add up to a total, seamless picture.

At the same time as these changes have been happening in documentary and ethnographic filmmaking, experimental filmmakers have moved closer to documentary. That is to say, they tend to be more concerned with cultural representation: they are "political" without necessarily being didactic or polemical. One of the lessons of ethnographic film is that any culture can be objectified, including one's own. Many filmmakers have turned to ethnography as a means of examining tendencies within American culture that impinge on psychological profiles. Personal filmmaking has, for many, become an examination of the ways that identity is constructed in culture. The films of the direct cinema movement of the 1960s are early examples of this tendency. In the work of D. A. Pennebaker and the Maysles brothers, an observational style of filmmaking was employed to analyze individuals, both celebrities (like Bob Dylan in *Don't Look Back*, 1967) and "ordinary people" (like the Bible salesmen in *Salesman*, 1968). It is the poetic quality of this filmmaking, as much as its observational style, that qualifies it as ethnography. Individuals, perceived as "social actors,"[7] become sites of alienation within a complex and often cruel American society.

One can trace ethnographic tendencies in the American avant-garde as far back as Maya Deren, who went to Haiti in the 1940s to film voudou ceremonies.[8] In her book on Haitian possession rituals, she notes that, as an artist, she felt she had an affinity with the natives and may have been able to relate to them better than would a professional anthropologist.[9] While this may smack of a certain romantic naïveté, there is some truth to the sense of marginality that experimental filmmakers share with the many different groups of people who are marginalized by mainstream Western commercial culture.

Mekas and Warhol, Shirley Clarke, Beat photographer Robert Frank, and even experimental filmmaker Stan Brakhage deployed imagery of people and culture in innovative ways throughout the 1960s. However, experimental filmmakers in the 1970s and 1980s played down the documentation of culture and shifted their attention to a critique of the various available languages of cinematic representation. Structural and feminist filmmakers, in very different ways, carried out a radical critique of narrative in search of a rarefied cinematic space, free of bourgeois and patriarchal structures of meaning. In the last ten years, as the avant-garde has diversified into a range of different practices and media, ethnography has emerged as a return to the rich poetic imagery of an earlier avant-garde.

Three Filmmakers

In June 1989 Peggy Ahwesh, Su Friedrich, and Leslie Thornton were among a group of seventy-six filmmakers who signed an open letter contesting the "official history" of experimental film promoted by the International Experimental Film Congress held that month in Toronto.[10] This letter, in the vitriolic language of a manifesto, pointed to the need for a new critical vocabulary adequate to an avant-garde that had moved beyond formal experimentation and personal expression into the messy business of cultural politics. It is important to recognize the interdependence of critical discourse and avant-garde practice, and this letter pointed to the lag between the two that emerged late in the 1980s. Since that time ethnography has developed as a poetics of visual culture that can provide the necessary terms of a renewed experimental film language.

Ethnography in its most progressive sense refers to a level of specificity and detail that remains autonomous from the generalizing conceptualizations of anthropological knowledge. The significance of the quotidian and the everyday is also a key strategy of feminist discourse, but it would be a mistake to characterize the ethnographic impulse of recent experimental filmmaking as "simply" a feminist strategy. Although I intend to focus on three women filmmakers, an ethnographic tendency is equally inscribed in the work of their contemporaries Steve Fagin, Craig Baldwin, Roddy Bogawa, Isaac Julien, John Akomfrah, Tracey Moffatt, Sadie Benning, Abigail Child, and many other men and women besides those discussed here. These filmmakers share not only the ethnographic effects, but also many of the specific techniques to be discussed.

Thornton, Friedrich, and Ahwesh stand out for me because they share a number of techniques that, taken together, delineate a new kind of experimental filmmaking that is at once fully conscious of the avant-gardes that have come before and is committed to "the social" and its politics of representation. Their films may or may not have a feminist "agenda" or a feminist "aesthetic," but they definitely emerge from and are addressed to a gendered cultural milieu. From feminism, these filmmakers have developed a critique of authenticity, authority, and mastery. In contrast to the "aesthetics of failure" evident in the autobiographical documentaries of McElwee and Moore, these filmmakers portray American culture as a site of transformation. They position themselves on the brink of a future that still has possibilities for social renewal.

As reality itself has become a contested terrain, Ahwesh, Friedrich, and Thornton negotiate access to history and culture through fictional devices, that is, through the artifice of filmmaking. Instead of cinematic codes of "realism," they have found ways of accessing and depicting "the real" as history, memory, and the body. In their films, ethnography is mobilized as a cinematic language that is able to articulate "culture" as "fiction." They create narrative spaces where difference, family, otherness, and desire are put into play, and where authorship is a deeply embedded cultural practice, as opposed to being simply "personal." As ethnography, their work is profoundly unscientific, and it consistently challenges all forms of objective representation. As such, it runs parallel to a larger rethinking of ethnography that is taking place across a spectrum of disciplines and media.

Ahwesh, Friedrich, and Thornton, all based in New York, have been making films since the early 1980s and are still very active. Like many contemporary filmmakers, they have made videos and move fairly easily between the two media. Video provides a different aesthetic that they have each occasionally exploited as an extension of their experimental practice. The low-grade image, along with its ease of access and economy, make video a valuable tool, especially for documentary and ethnographic material, but it has not replaced film, by any means, in any of their oeuvres. I have chosen to focus here on a few films and tapes by each film-maker that are most indicative of the ethnographic aspects of their work.

The term "culture as fiction" comes from a piece that Leslie Thornton wrote about her film-video epic *Peggy and Fred in Hell* (1984–1992).[11] To represent culture as fiction is in one sense to make a false documentary about people; *Peggy and Fred* is only a "fiction" in the broadest sense of the word. *Peggy and Fred* was made in several installments on both film and video. The entire cycle was designed to be projected simultaneously on video monitors set up in front of a film screen. The work has actors and sets and fragments of stories, but it is far from a conventional narrative. The performances are all slightly detached, and everything is constantly interrupted by borrowed music, stolen images, and all kinds of extraneous material that clutters the space in which Peggy and Fred subsist. They are children in a postapocalyptic dystopia. "There are no other people in the world," writes Thornton. "Something has happened to them, but Peggy and Fred are unconcerned. And since the only other people they ever see are on TV, they figure that people are watching, learning from, or ignoring them as well. This constitutes their idea of the Social."[12]

If ethnographic film is populated by "social actors," Thornton has extracted this concept and flayed it open. Peggy and Fred, played by Donald and Janis Reading (aged six and eight at the beginning of the production in 1984, and fourteen and sixteen by its end), perform for the camera a kind of garbled mimicry of popular culture. They take on roles that are fleeting and barely articulated, all the while performing as "themselves," as American children. The settings that Thornton has placed them in, and her montage of disparate imagery and sounds, are carefully constructed heaps of debris and ruins that evoke an imploded decadent society that has annihilated itself and left behind its waste. From their environment and from their half-directed, half-improvised performances, the children seem to be making—being—something new. Thornton, likewise, works toward the creation of a language of representation built on the ruins of the old, used, burned-out cinematic forms, and it is precisely in the ethnographic presences of the children that transformation and renewal are made possible.

39 Outside the church in Su Friedrich's *First Comes Love* (Su Friedrich, 1991).

Thornton's videotape *There was an Unseen Cloud Moving* (1983) is a travelogue and a biography that circles around its subject, Isabelle Eberhardt, refusing to pin her down. Eberhardt was a woman who traveled through northern Africa in the nineteenth century masquerading as a man. Thornton's version of her story is assembled from fragments of images, sounds, and voices; various performers indicate the value and significance of Eberhardt for contemporary women. The colonial context of the story is developed by way of quotations from other films made in and about Arabic-African culture, as Thornton engages with the mythology of Eberhardt's adventures.

Su Friedrich's most "ethnographic" film to date is *First Comes Love* (1991). It features black-and-white footage of the church weddings of four different couples, accompanied by a soundtrack of familiar American popular music about love and romance. The wedding imagery is interrupted only once for a long list of 149 countries, including the United States, in which single-sex marriages are not recognized. At the end of the 22-minute film, a title announces that Denmark had recently legalized homosexual marriage. The exclusion of gays and lesbians from the institution of marriage elsewhere in the world is stated in terms that are at once romantic and ethnographic. The itemized list of countries is a bald statement of fact. The weddings are documents of a heterosexual culture, and because Friedrich mixes footage of four different weddings in which the same activities, poses, and practices are enacted, they are clearly ritualistic.

Although Friedrich herself is not "present" in the film, her gaze lurks on the margins of the wedding parties. Her own desire is registered in an end credit that dedicates the film: "For Cathy." Along with other North American gays and lesbians, Friedrich is part of a shared musical culture obsessed with "love" that is coded as incomplete without the seal of approval available only to heterosexual couples. The objectivity of the film balances a delicate ambivalence between a thwarted desire to be part of this ritual and a critique of its ritualistic emptiness.

In two other very different films, Friedrich uses ethnographic techniques to transform even more explicitly autobiographical material into cultural and social documents. In *The Ties that Bind* (1984) she interviews her mother about her years in Nazi Germany, extracting a highly emotional monologue on the noncomplicit German citizen whose world literally crumbled around her. As a portrait of an "ordinary" woman in extraordinary times, the film invokes not only ethnographic codes, but also those of the woman's film. Here, as in other women's films, history is domesticated into an emotional tale of struggle and tears, and it is mapped onto the relation between mother and daughter.

Sink or Swim (1990), Friedrich's cinematic analysis of her ambivalent relationship with her father, works very differently. In contrast to the empathy that is achieved with her mother in *The Ties that Bind,* this film is characterized by a great tension between sound and image. Its narration, spoken by a young girl, refers to "a girl" and her father in the third person. The image track features many girls and fathers—Friedrich's own home movies mixed with found footage and other original material shot by Friedrich.

Friedrich routinely incorporates footage of anonymous people into her work in order to explore the parameters of her identity—as a woman and as a lesbian. In contrast to conventions of autobiographical filmmaking, she refrains from "personal expression" as a key to identity. Instead, she finds herself struggling with social codes and cultural constructions. Cinematically, she relies on observational documentary techniques to represent herself as a witness even to her own childhood.

Peggy Ahwesh uses cinéma-vérité techniques, but not to the ends for which they were developed thirty to forty years ago. She intervenes and *stages* "reality" before the camera, and she juxtaposes different scenes, "interviews," and "confessions" for dialectical and associative effects. Yet, despite this apparent playfulness, the vérité shooting style inscribes an indexical access to a historical "real," rediscovering the spontaneity that originally accrued to this form of ethnography. Ahwesh's *From Romance to*

Ritual (1985) and *Martina's Playhouse* (1989) are both "about" sexuality, but in each case the perspective is "from below"—that is, from a level of detail, experience, anecdote, and incident, much of it performed and therefore "inauthentic," brought together in a random inversion of social-science discourse.

From Romance to Ritual (1985) consists principally of performance pieces by Renate Walker, Margie Strosser, and ten-year-old Mandy Ahwesh. An accumulation or amalgam of stories, bodies, and images, the tape is an oblique ethnography of women's sexual culture. Whether the stories or the histories are "true" or not is far less important than the ways that the sections of the tape play off against each other. The implied relationship between "prehistoric" cultures with their dancing virgins and contemporary scenes of prepubescent girls constitutes a form of comparative ethnography. Insofar as both groups are "lost" to the filmmaker, and to us, "after sex," the film invokes a form of anthropological desire fully consistent with "the salvage paradigm"—the "desire to rescue 'authenticity' out of destructive historical change."[13] This desire is registered cinematically through Ahwesh's low-end production techniques that refer, obliquely, to an "authentic" reality that lies just outside or before the film.

Although most of Ahwesh's films feature her own friends and family, *Strange Weather*, a videotape codirected with Margie Strosser, offers a different strategy. On one level, the tape is an ethnographic study of a group of crack-addicted teenagers in Florida. The teen "actors" in the tape perform as young people (three of them white, one black, all middle-class dropouts), who they may or may not be, or may have been, themselves. The documentary shooting style and narrative codes work against any "performance" or "theatrical" cues, giving the work an aura of authenticity. The intercut TV announcements of an impending hurricane stand in for the filmmakers' commentary on this microculture of decadent America.

Each of these filmmakers is preoccupied with children, not as the cuddly innocents of Hollywood, but as prehistoric versions of themselves. Of *Peggy and Fred*, Thornton has said, "children are not quite us and not quite other. They are our others. They are becoming us. Or they are becoming other. They are at a dangerous point."[14] Thornton echoes an anthropological fascination with "the primitive" as always bound by the limits of modernity, and the need to save an image, at least, of a culture before it "vanishes" (which is to say, before it becomes "us"). Otherness is in many ways the product of ethnography, although it is often disguised as its subject. In the work of these filmmakers, otherness is a necessary fiction, produced within the fissures of American culture, as a means of making it strange, in order to see it differently.

The Free-Floating Gaze

A handheld camera is what gives the works of Ahwesh, Friedrich, and Thornton their most distinctive style. When shooting people talking or performing, the framing is often "too close": it lops off a head, it wanders away from the person being filmed, or it refuses to follow people as they walk out of a shot. Meanwhile, the person behind the camera is hard to position, as camera movements are never directly tied to the filmmaker's movements. Instead, they take on a life of their own, literally "floating" over the field of vision. For example, Friedrich's interview with her mother in *The Ties that Bind* consists largely of close-ups of Lore Friedrich's hands and feet. The camera occasionally wanders to her face, but the image is never in synch with Lore's voice-over monologue. Thornton's footage of Peggy and Fred is at times so close to the children that we cannot tell immediately who is being filmed. Ahwesh is more inclined to speak from her camera-person vantage point and always shoots in synch, but the roving gaze of her camera is equally unsettling and disorienting.

The visual aesthetic of "unfixedness" is crucial to these filmmakers' critique of authenticity and authority. Compared to the fixed frame of Warhol's early films and the structural filmmaking that came later, for these filmmakers the cinematic apparatus no longer signifies "control" or "mastery." As David James notes of Warhol's films: "The camera is a presence in whose regard and against whose silence the sitter must construct himself. As it makes performance inevitable, it constitutes being as performance."[15] In the films of Ahwesh, Friedrich, and Thornton the camera is less a "presence" than a microscope or telescope searching for its subject. The effect is a kind of flirtatious game between the camera person and her subject in which even performance is subverted by a lack of "correct" framing.

Bruce Baillie's *Valentin de las sierras* (1967) provides an experimental model for extreme close-up ethnography, but whereas he blacks out the frame around his peephole camera, these three filmmakers are careful to avoid such voyeuristic structures. Because the takes are so long, because so much is given over to the pro-filmic in Ahwesh, Friedrich, and Thornton's films, the free-floating gaze does not evoke the notion of the visionary artist (as it does in Stan Brakhage's films).[16]

In Ahwesh, Friedrich, and Thornton's films people frequently tell stories to the camera that may or may not be true and which may or may not be autobiographical. Testimonials, often in the form of voice-over mono-logues, take the form of storytelling rather than authorial discourses. Storytelling, according to Walter Benjamin, is a narrative form anchored in the history of its telling, a form that opens out centrifugally, as opposed to the novelistic insistence on closure.[17] It is a narrative form that is anchored in the experience of the storyteller, and, unlike "information," the more common form of documentary voice, it is the discourse of a traveler or wise elder: "The most extraordinary things, marvellous things, are related with the greatest accuracy, but the psychological connection of the events is not forced on the reader. It is left up to him [sic] to interpret things the way he understands them, and thus the narrative achieves an amplitude that information lacks."[18]

From the cacophony of details, ambiguities, and elliptical editing that make up these films, a number of engaging narratives emerge. The direct address to the viewer may be seductive, but we are always offered these stories as "symptomatic," as dream tales that are surface talk, without depth. Stories are language and are told, in several instances, not by people, but by typewriters banging them out letter by letter (Sink or Swim, Peggy and Fred).[19] Other techniques of distanciation put the stories at one remove, like the omniscient child narrator of Sink or Swim.

Of Peggy and Fred, Linda Pekham says: "Fred has eliminated thought. He transmits speech directly, which, like the excesses of Hell, faces the act of total indiscriminate recall. Peggy has no direct speech at all. Her thoughts accumulate in a confused noise outside her head."[20] The stories themselves are not deeply meaningful but are resonant with the act of being told, of being spun out of nothing and referring back simply to the fact of their telling.

The "talking cure"—the Freudian therapeutic technique—is inverted in these films insofar as speech, especially in the form of monologue, creates a sense of dis-ease and doubt, distending the tissue of language beyond the language of film and video. In the same way that "acting" is transposed into "performance," narrativity is replaced by storytelling. While the per-formances produce a doubling of body and character, storytelling also in-corporates a sense of doubleness. A split between the telling and the told marks many of these monologues, and with the exception perhaps of The Ties that Bind, there is no guarantee that the story is in fact based in experience. The films are extraordinarily discursive, weaving complex cultural webs of desires and identities that challenge the constitution of a coherent speaking subject (one who speaks "the truth" about himself or herself).

For example, in Thornton's videotape *There was an Unseen Cloud Moving* seven different actresses play Isabelle Eberhardt dressed in variations on the Muslim cloak and veil. Several of them simply sit in front of the camera and speak to it, performing themselves performing Eberhardt, reclaiming her for a camera that participates fully in their cultural anarchy. In *Sink or Swim* the young girl's narration literally takes the form of a fairy tale.

Anecdotal stories reside at the heart of these films, but they are never "of" the film. Even in *The Ties that Bind,* Lore Friedrich's life is made up of a series of stories, a series of anecdotes broken up by her daughter's editing and scrawled questions on the screen. Joel Fineman writes that the anecdote "lets history happen by virtue of the way it introduces an opening into the teleological, and therefore timeless, narration of beginning, middle and end. The anecdote produces the effect of the real, the occurrence of contingency. . . ."[21] As ethnographic traces of culture, storytelling binds narrative to experience, regardless of the veracity of the stories that are told. Ahwesh's performers are often flamboyant and dramatic, but they always retain a certain cultural integrity as people that underwrites their performances and grounds their stories in a history, if only the history of the making of the film.

Sound vs. Image

Both Thornton and Friedrich shoot their material silently and edit it with soundtracks made up of rich compilations of music, voice-overs, and sound effects. Ahwesh usually shoots in synch sound, but she often includes ambient music or sound that threatens to drown out the performers. All three directors use the relations between sound and image for dramatic, ironic, and other effects that variously challenge the realist potential of the film medium. The dialectical effects that are produced by the tension between sound and image are crucial to their politics of representation.

Friedrich's *Sink or Swim* consists of a series of voice-over anecdotes that are "illustrated" or "accompanied" by images, creating a dynamic interplay between two levels of discourse that occasionally converge. Shots of water, of a girl and father skating, of children swimming, occasionally "match" points in the narration where these things are mentioned. Sometimes the relation between sound and image is poetic or ironic, as when shots of female bodybuilders accompany a story of Atalanta, a Greek goddess, heroine of a favorite bedtime story of "the girl's." The most ambivalent convergence is the relation between the speaking voice (that of a young girl's)—"the girl" that she refers to—and the many shots of different young girls, all of which are in turn ambiguously related to the filmmaker herself.

Young girls in confirmation dresses are all that we see while we are told a terrifying story of the father punishing the girl and her sister by holding their heads underwater in the bathtub. Occasionally, though, sound and image match much more closely. Toward the end of *Sink or Swim,* as the narrator tells about the strained relationship between the girl (now a woman) and her father, we see shots of Friedrich herself in the bath, at the park, and then typing the words that we hear being spoken.

The convergences of *Sink or Swim* are stunning precisely because they figure the displacement between sound and image as an impossibility of representation, a perpetual gap between image and reality that is never, except momentarily, bridged. If the shots of African American children swimming do not match the story of the girl's first swimming lesson literally, they do match it figuratively. Like so much of the material in *Sink or Swim,* these shots are of anonymous people, and therefore the girls are "simply" young black girls swimming in a pool. When we finally see Su Friedrich herself, identified as the writer, the slippages of identification are *almost* corrected. But we still are not entirely sure, especially the viewer who does not recognize the filmmaker. The difficulty of self-representation becomes that of cinematic representation. "Identity" becomes dispersed across a cultural spectrum of "positions" and discourses. Although the film is autobiographically based, it takes on a broader significance as a story about patriarchy, girls and fathers, family dramas and American culture.

In another section of the film the girl writes a letter to her father (in silence, the letter is banged out on a typewriter). She describes the German song that her mother played repeatedly after he left them. She says, "It's so strange to have such an ecstatic melody accompany those tragic lyrics. But maybe that's what makes it so powerful: it captures perfectly the conflict between memory and the present." This conflict is precisely the dynamic of the film. The narration, a very literary, stylized form of storytelling, belongs to the present, while the home movie and other footage belong to memory and the evocation of memory.

In fact, all voice-over narration embodies this historical difference that most documentaries try to cover up and mask. In Ahwesh, Friedrich, and Thornton's films, there is no necessary correspondence between voice and image, but once they are firmly pried apart, the excesses of the image track are returned to history and ethnographic specificity. The storytelling, for its part, is rendered as a discursive attempt to make the randomness of the everyday into something meaningful.

Thornton's *There was an Unseen Cloud Moving* gives further evidence of the role of correspondences in this style of filmmaking. At one point in the tape, Thornton relates some details of Isabelle Eberhardt's upbringing. Under her casual, informal voice-over, shots of plants, houses, dishes, and people "evoke" Eberhardt's world without necessarily actually representing it. Even the portraits—still photos blown up to screen size—may or may not actually be of Isabelle. Symbolically, the image illustrates the soundtrack and the story, but at the same time it is something in and of itself. It points to another space and time of which it is the indexical fragment. As in *Sink or Swim*, the image track is consistently in excess of the film. What makes these films so thoroughly ethnographic is that even the most symbolic imagery is also "literal."

The sound and image tracks in Friedrich's *First Comes Love* are radically different. The wedding imagery is shot silent, forcing the details of behavior to bear the weight of meaning. Friedrich's camera comes in very close to examine the way people touch each other and smile at each other; she pans over the fabric of dresses, the movements of small children among adults. Meanwhile, the music of Janis Joplin, Willie Nelson, Bonnie Raitt, and James Brown provides a cultural background, lifting the particular to the level of the general (an ethnographic convention) while retaining the difference between the two levels. The ironic juxtaposition of James Brown's "Sex Machine" with a couple preparing their formal pose in front of a church is a very open irony. It is neither exactly condescending nor flattering, but it poses the question: what does this song have to do with this picture?

Home Movies

Like many of the techniques listed here, the incorporation of home-movie footage is not unique to this group of films. Nor is the approximation of a home-movie aesthetic. Avant-garde filmmakers have always had an affinity with the low-end anti-industrial qualities of the home movie. Stan

Brakhage, Jonas Mekas, George Kuchar, and Sadie Benning have all developed versions of the home movie into multilayered, highly personal aesthetics.

While super-8 film was once the privileged gauge of the home movie, video has since taken over that role and has become the medium of choice for documentation of family rituals and daily life. A number of filmmakers have taken up Fisher-Price's pixel-vision video camera designed for children. Its low-definition image is a direct challenge to the production values associated with cinematic realism, and its shallow depth of focus restricts the image to an extremely localized, close-up field of reference. Of the three filmmakers under discussion, Peggy Ahwesh works most closely with the home-movie format. She favors color super-8 film, and she also uses pixel-vision.

By eliminating the tripod, the free-floating gaze is like that of the amateur camera operator, if somewhat more deliberately misplaced. In the final versions of her films Ahwesh often includes her performer's questions before and after a take—"Are you ready?" "Was that OK?" "Is that good enough?"—as well as comments about her own relationship with the performer. Jennifer in *Martina's Playhouse* even tries, halfheartedly, to seduce Ahwesh behind the camera. She finally admits she does not have a chance with Ahwesh, and that if it were not for the camera that she has loaned to the filmmaker, she would not even have been invited to the filming. When Friedrich performs in Thornton's *Unseen Cloud,* she is clearly distracted by the amount of time they have left to shoot before they have to be somewhere else.

Ahwesh often has people perform in domestic spaces—bedrooms, kitchens, living rooms. This contributes to the aura of authenticity and, ironically, to the overall sense of playacting. *Martina's Playhouse,* for example, is shot entirely in such spaces, making the home (various New York apartments where the film is shot) a series of performance spaces. When six-year-old Martina and her mother, Margie Strosser, reverse their roles, the effect of substitution and displacement is that much stronger because of the home-movie framework of their performances. The quotations from Lacanian psychoanalysis concerning the child's desire for the mother take a new resonance when the child, Martina, pretends to nurse her mother on their own sofa, in their own home.

The homes of Ahwesh's performers in *Martina's Playhouse* and *From Romance to Ritual* are cluttered, slightly "bohemian" settings with telephones, television sets, and stereos constantly interrupting or accompanying the "action." The familiar iconography of the home itself becomes discursive, a subversion of the visual language of TV sitcoms. Thornton's *Peggy and Fred* series constitutes a kind of anarchic domestic space, an implosion of 1950s' family culture. With the score to Roman Polanski's *The Tenant* (1976) appended, *Peggy and Fred and Pete* becomes a terrorized home movie.

The home-movie aesthetic contributes two key effects to these women's films. Formally, it constitutes a challenge to the aesthetics of mastery implicit in more high-tech film forms. Secondly, it offers an ethnographic specificity of the once-only that defines the home movie. The informality of the home-movie aesthetic enables these filmmakers to perform their ethnography surreptitiously, "at home."

Found Footage

Recycling "found" images implies a profound sense of the already seen, the already happened, as well as a certain failure of the new and a collapse of history. It is not surprising but nonetheless significant that we encounter found footage of apocalyptic and violent events in so many experimental films today. One sees it in the work of Ahwesh, Friedrich, and Thornton as well as in Craig Baldwin's *Tribulation 99* (1991) and (its originary moment) in Bruce Conner's *A Movie* (1958).

In the Peggy and Fred series, found footage is an integral part of the environment-cum-mediascape that the children inhabit; *Peggy and Fred in Kansas* ends with a house falling off a cliff. Ahwesh's *From Romance to Ritual* contains a single image of the demolition of several high-rise buildings; *Strange Weather* is punctuated by television footage of an impending hurricane threatening to devastate the film's depiction of a decadent, degraded, drug-addicted Florida subculture. Friedrich's *The Ties that Bind* includes footage of the aftermath of bombing in German towns as well as film of military aircraft and canons.

Ahwesh, Friedrich, and Thornton make extensive use of industrial films, TV imagery, feature films and newsreels, and early cinema. It may be argued that the close affinity between the avant-garde and early cinema constitutes a cinematic version of "the salvage paradigm." Many experimental filmmakers, such as Hollis Frampton, Ken Jacobs, and Ernie Gehr, have appropriated fragments of early cinema, harking back to a preindustrial, preinstitutionalized, even prehistoric film culture.[22]

While these filmmakers may be equally eager to find themselves in relation to such a history, their use of early cinema is somewhat less reverent than their predecessors'. Rather than fetishize that history through re-photography, they tend to use it as one element of a many-textured montage structure. Combined with other forms of archival material, the palimpsest of found footage places "the real" at one remove, and it is this detachment that provokes the repeated use of destructive, violent, and apocalyptic found footage. The loss of the real signifies a loss of history. The burden of accumulated public memory becomes a cultural garbage heap through which the filmmaker has picked.

Despite the suggestion of "loss," found footage also can provide a critical distance that can be valuable to experimental ethnography. As the relationship between the filmmaker and her subject is made indirect, the subject arrives "already filmed." Ahwesh, Friedrich, and Thornton take this effect one step further by making their own footage often *look* as if it were "found." Thornton has suggested that "in the future the position of the archival footage may not be so much with quotation marks around it."[23] Indeed, throughout the Peggy and Fred series and in Friedrich's two family films, the difference between archival and original footage is hard to detect. Shooting in black and white in their off-center ways, Friedrich and Thornton imitate the style of the found footage and create an effect that is akin to the surrealist objets trouvés. Just as the surrealists bridged the gap between art and ethnography by bringing artifacts into the art gallery, these filmmakers evoke the "shock of the new" with extensive collage-narratives that juxtapose a diversity of imagery and sounds.

Two examples may clarify the way that history is revitalized in these films' incorporation of "used" imagery. Thornton's *Unseen Cloud* concludes with the final scene of a documentary introduced earlier in the tape, "The Moslem World—Part One: Lands of the Camel." As a line of teenage girls marches over the dunes toward and past the camera, the male narrator says, "Now these girls, brave and real, not mere shadows on the sands, they too sense the desert glory that it is our additional privilege to *see.*" The segment is not manipulated in any way by Thornton, and yet, as the conclusion to a film about Isabelle Eberhardt, it is transformed into a text on and about colonialism and gender. The girls, dressed in school uniforms, are clearly Middle Eastern colonial subjects, directed to enact this curious scene within an apparatus of visual and ideological oppression. As the conclusions of the two films converge with "The End," the found footage becomes a text of cruelty.

In *The Ties that Bind*, we hear Lore Friedrich say, "I would *not* say Heil Hitler," as we see an early film (ca. 1900) of a girl holding an American flag, dressed in a stars-and-stripes outfit, dancing a can-can. The sound-image relation is ironic, but like the Middle Eastern example, the irony betrays the filmmaker's politicized sensibility as she draws an oblique parallel between her mother's experience and another context of (American) nationalism. Instead of being cut off from history through the use of archival material, Thornton and Friedrich use found footage to engage with history. They en-gender the footage, meaning not only that they underline inscribed codes of gender, but they analyze its production of knowledge.

Piecemeal Narrative

One of the most controversial issues in avant-garde cinema is narrative. In the 1960s and 1970s narrative was dismissed as "bourgeois," as American experimental filmmakers distinguished themselves even from European art cinema, which remained preoccupied with narrative forms. In the early 1980s "new narrative"—feature-length avant-garde films—hoped to make experimental films more accessible and more widely distributed. New narrative films such as Lizzie Borden's *Born in Flames* (1983) and Betty Gordon's *Variety* demanded very different structures of financing and production methods and often a complete departure from the artisanal, personal filmmaking associated with experimental cinema.[24]

Ahwesh, Friedrich, and Thornton have not (yet) ventured into feature-length filmmaking. Nevertheless, their short films do have a narrativity that is best described as piecemeal. Narrative returns partially in the form of storytelling, as described above, but also in the form of assemblage, or montage construction. Documentaries often are structured out of a series of scenes, events, interviews, and archival footage assembled by the filmmaker in the editing room. The result is a narrative coherence that is different from the psychological narrative space developed by the editing codes of dramatic realism.

Unlike the conventions of documentary, these filmmakers are as likely to foreground effects of juxtaposition as those of continuity in their editing. They also combine "fiction," in the form of performances, with found footage, landscape, and other imagery. Friedrich's *Sink or Swim,* for example, is structured as a series of autonomous fragments, each named in reverse alphabetical order, from "Zygote" to "Athena / Atalanta / Aphrodite." Each section corresponds to a story told by the child narrator and a montage of imagery. The diversity of material of which the film is

40 Fred caught in mid-shout in Leslie Thornton's *Peggy and Fred in Hell* (Leslie Thornton, 1984).

constructed has no internal logic; yet the piecemeal effect is by no means random.

The *Peggy and Fred in Hell* series is among the most densely structured of these works. *Peggy and Fred in Kansas* begins with this order of fragments:

1. A lightning storm / landscape image. Peggy and Fred's voices and the ambient sound of their set is heard over images 1 to 4.
2. An underground tunnel (archival).
3. An industrial fire.
4. A chemical storage site.
5. Peggy and Fred play a "scene" in a domestic set that involves a telephone and a table. They talk about a kidnapping and a murder.
6. A military aircraft with a male voice-over speaking in code.
7. Peggy eating in close-up.
8. Fred sitting in a chair pretending to be an astronaut.
9. Peggy plays with a doll; Fred gets up and says he has to "get the milk."
10. Shots of the parts of an unidentified mechanical apparatus, with continued sounds of Peggy and Fred playing.
11. An intertitle: "Later."
12. Fred throws things around the set.
13. Peggy and Fred play a "scene" in which Fred speaks and sings into a microphone.

In this nine-minute excerpt the inserted fragments of industrial and landscape imagery create a narrative environment for Peggy and Fred. The overtones of disaster, decay, and technoculture provide a context for the children's activities. Through juxtaposition and collage, an apocalyptic sensibility is developed out of an assembly of details in which the children's bodies become sites of survival. The intertitle "Later" is typical of Thornton's flirtation with narrative codes of temporality; it is borrowed along with all the other found footage and draws attention to the artifice of narrativity.

Ahwesh and Strosser's tape *Strange Weather* is constructed as a series of disparate performances and "scenes." Characters reappear, and a loose narrative can be discerned from the various fragments, but the impression is one of slices of life or glimpses into a cultural milieu. It is impossible to determine whether these are "real people" or "actors," or real people acting like real people. In fact, they are actors, but the fragmentation of the film into pieces is instrumental in preventing the film from slipping into illusionism. The piecemeal narrative keeps the question of "truth" in circulation and prevents the film from adopting a moral stance regarding its subjects / characters, who remain at a distance that the filmmakers do not try too hard to bridge.

The aesthetic of fragmentation pertains as much to the montage within scenes as to the structures of the films as wholes. The jump-cut is the modus operandi of these filmmakers, and yet it functions as much more than a reflexive device. In filming people it becomes a means of analyzing behavior. Cutting on gestures, breaking up movements and dissecting bodies—in concert with the free-floating gaze—is a key means by which people (actors and nonactors) become objects of study. Each performance is made into a text this way, even the voyeuristic surveillance footage that Friedrich occasionally uses in *Sink or Swim*.

Piecemeal narrative structures enable these filmmakers to hold "objective realism" at bay, even while they engage with other documentary codes. On the level of the segment, and on that of the whole, fragmentation pits reality against itself. Nothing is natural, everything is fictional, because everything within scenes is broken down and reconstructed; the films themselves openly bear their signs of construction and assembly. The lack of narrative closure of any of these films leaves them open to history, and as ethnographies open to cultural change and transformation. A very high level of craft unites these three filmmakers, for whom editing—and sound / image counterpoint—is a process of "writing" culture.

Overtly manipulative techniques like bleaching the image (*Martina's Play-house*), scratching words on the celluloid (*The Ties that Bind*), or reversing a sequence (*Peggy and Fred*) seem on the surface to be anathema to the ethnographic priority of objectivity. After all, even in the most reflexive of documentaries, documentary codes aim to preserve the integrity of the real. These filmmakers at times work directly on the celluloid, making "their mark" on the film in a material way by scratching or coloring the film stock. The ethnographic effect of this manipulation is to forestall questions of ethical manipulation. Given the filmmaker's work on the image, the film becomes entirely theirs, and everything in it is thus subject to their manipulation. It further alludes to a kind of cruelty (toward those captured on film) inherent in "realist" cinema and renders the image a tangible, textural object-in-the-world.

Work directly on the image problematizes notions of authenticity and realism. But such a problematic is essential to this new, experimental ethnography. Trinh Minh-ha argues:

The real world: so real that the Real becomes the one basic referent—the pure, concrete, fixed, visible, all-too-visible. The result is the elaboration of a whole aesthetic of objectivity and the development of comprehensive technologies of truth capable of promoting what is right and what is wrong in the world, and by extension, what is "honest" and what is "manipulative" in documentary.[25]

The ethical questions predicated by an "aesthetic of objectivity" are made irrelevant in the absence of claims to objectivity. But objectivity is not simply opposed by "subjective filmmaking." As these filmmakers reveal, subjectivity is a means to express not only one's inner self, but one's cultural self as well. It is important to realize that "meaning" in these films is not "simply" phenomenological or experiential, but it is produced by the lived body, which is, in all instances, a cultural body.

Video can potentially inscribe a mediated "screen" between the viewer and the world filmed through the digitized image of pixel-vision or another low-grade variation on the medium. Thornton sometimes borrows video imagery with the time code still intact. Ahwesh's use of pixel-vision in *Strange Weather* is exemplary of a transformed observational cinema. Because of the short focal range of the Fisher Price camera, the tape is composed primarily of close-ups. Although the conception of the tape is very close to a neorealist conceit of dramatizing reality, the aesthetics are

41 A girl and her toys: Peggy Ahwesh's *Martina's Playhouse* (Peggy Ahwesh, 1989).

radically different. Missing the visual context of backgrounds, sets, or the mise-en-scène of environment, the level of detail is microscopic, organized around the bodies of the actor-subjects.[26] The viewer may be drawn into a false belief in the veracity of the performances, but at the same time the digitized image maintains a sense of doubleness and a distance from the performers. Ahwesh is not claiming any honesty in her aesthetics, and she is thus absolved from "manipulation."

Conclusion: Marginality

Ahwesh, Friedrich, and Thornton are white American women. Their intervention into ethnography is not to speak from the position of "the other," not even as women. Indeed, "feminism" is relegated in their work to yet another master discourse. The otherness of non-Western, non-white, and gay cultures figures in their films as discursive positions and cultural possibilities, and never as authentic sites of identity. Marginality thus figures more appropriately as a site from which they speak aesthetically and politically. "It is essential to imagine a marginality," Thornton writes, "to perceive an edge from which to work."[27]

The "edge" that these filmmakers have found is on the boundary between experimental, documentary, and fiction filmmaking. These women filmmakers force documentary and fictional materials and processes together like land masses, and what emerges is an experimental ethnography, an examination of "culture" from the perspective of "art" that inverts and reinvents the conventions of cultural representation. Trinh Minh-ha has called for such a foregrounding of all art in ethnography,[28] and yet where her own work falls into the trap of aestheticization and stylization, these filmmakers have recourse to a long inventory of avant-garde techniques and histories, the deployment of which constitutes a more critical and rewarding form of cultural intervention.

Found footage, edited into these densely structured films, incorporates an ongoing commentary on image culture. Not only is personal expression thoroughly mediated, it is also constructed historically. The process of "growing up" is depicted as one of growing into image culture, of a negotiation with visual languages and learning how to manage them. If anonymity is the privileged ethnographic identity, even the people in original footage—like "Peggy," "Fred," and Lore Friedrich, Martina and Mandy Ahwesh become ethnographic subjects. Ahwesh, Friedrich, and Thornton's films are, in fact, densely populated. A tape such as *Strange Weather* brings experimental visual techniques into a virtually neorealist terrain. Cultural observation becomes a form of cultural invention through the deployment of fictional strategies where they are least expected.

The kind of experimental ethnography that has been described here tends more toward the postmodern than the modern. If modernist ethnography entails a critique of realism and objectivity, these filmmakers take it one step further. Besides working on the language of representation, they work directly on the pro-filmic. That which is in front of the camera is as fictional as the film it ends up in. Their work also points to something outside discourse, something prior to it, accessible only as allegory. The performances of social actors, the assortment of recycled imagery, and the many stories that are told, all are grounded in "the real" and have the aura of ethnography. And yet reality in these films is never natural; it is always cultural, always already at one remove. As a form of ethnography, these films are indeed evocative rather than representational. Our desires and fantasies are engaged to "depart from the commonsense world only in order to reconfirm it and return to it renewed and mindful of our renewal."[29]

In the final monologue of *Strange Weather* the storyteller is a blonde girl wearing a bra and cutoff shorts. She tells about a particular night of drinking and crack-smoking. As she speaks, she drinks a beer and the camera wanders to her crotch: "one thing he said to me was 'Florida is cosmetic.' And I thought about that and I thought about that whole night and everything that had happened and I was, like, he's right because behind those potted plants and behind those big double doors, nobody from the outside world really knows what's going on. They can't see through it." Culture is likewise not transparent. If the task of experimental ethnography is to represent culture differently, the new avant-garde of Ahwesh, Friedrich, and Thornton is a leading example.

Notes

This research has been funded by a grant from the Social Sciences and Humanities Research Council of Canada.

1 The relation between new forms of documentary film and developments in popular culture and TV culture have been extensively developed by Bill Nichols in *Blurred Boundaries: Questions of Meaning in Contemporary Culture* (Bloomington: Indiana University Press, 1994).

2 Paul Arthur, "Jargons of Authenticity (Three American Moments)," in *Theorizing Documentary*, ed. Michael Renov (New York: Routledge, 1993), p. 132.

3 Linda Williams, "Mirrors Without Memories: Truth, History and the New Documentary," *Film Quarterly*, Spring 1993, p. 11.

4 Warhol and Mekas were part of a much-larger underground cultural scene that was crossed by the Beat poets, emergent gay, black, and feminist voices, and various other expressions of the 1960s' counterculture. See David James, *Allegories of Cinema: American Film in the Sixties* (Princeton, N.J.: Princeton University Press, 1989), for a more thorough analysis of the films of this period.

5 Key filmmakers in this tradition are Jean Rouch, David and Judith McDougall, and John Marshall.

6 Trinh T. Minh-ha, "The Totalizing Quest for Meaning," *When the Moon Waxes Red* (New York: Routledge, 1991), pp. 29–52.

7 Bill Nichols uses the term "social actor" to refer to the individuals who appear in documentary films: "Those whom we observe are seldom trained or coaxed in their behaviour. I use 'social actor' to stress the degree to which individuals represent themselves to others; this can be construed as a performance." *Representing Reality: Issues and Concepts in Documentary* (Bloomington: Indiana University Press, 1991), p. 42.

8 Unfortunately, Deren was not able to complete her film *Divine Horsemen* before her death in 1961, although it has been edited and released posthumously by Charyl Ito.

9 Maya Deren, *Divine Horseman: The Living Gods of Haiti* (New York: Documentext, 1953), p. 7.

10 The "Open Letter" is reproduced in Paul Arthur, "No More Causes? The International Experimental Film Congress," *The Independent* 12 (October 1989): 24.

11 Thornton, "Culture as Fiction," *Unsound*, p. 30.

12 Thornton, "We Ground Things, Now, on a Moving Earth," *Motion Picture* 3 (Winter 1989–90): 13.

13 "The salvage paradigm" refers to the anthropological commitment to "disappearing" cultures in which "authenticity" exists just before the present and outside the industrialized, urban world. As an ideology, it keeps the idea of non-Western cultures in a position of marginality with respect to "modern" culture. James Clifford, "Of Other Peoples: Beyond the 'Salvage' Paradigm," in *Discussions in Contemporary Culture*, no. 1, ed. Hal Foster (Seattle: Bay Press / Dia Art Foundation, 1987), p. 121.

14 Thornton, "We Ground Things, Now," p. 15.

15 James, *Allegories of Cinema*, p. 69.

16 The "profilmic" refers to the reality that is before the camera; the term is usually used in reference to the ongoing events that a documentary filmmaker records.

17 "The novel reaches an end which is more proper to it, in a stricter sense, than to any story. Actually there is no story for which the question as to how it continued would not be legitimate. The novelist, on the other hand, cannot hope to take the smallest step beyond that limit at which he invites the reader to a divinatory realization of the meaning of life by writing 'Finis.' " Walter Benjamin, "The Storyteller," *Illuminations*, trans. Harry Zohn (New York: Schocken Books, 1969), p. 100.

18 Ibid., p. 89.

19 The typewriter image is a technique that many filmmakers use, including Jonas Mekas in *Lost Lost Lost* and Vanalyne Green in *A Spy in the House That Ruth Built*. It is a valuable technique for mediating a personal discourse and demonstrating the way that voice-over is in fact the reading of a script. Like the use of diaries and letters, it often alludes to an intermediary stage of writing that takes place between personal experience and filmmaking.

20 Linda Peckham, "Total Indiscriminate Recall: Peggy and Fred in Hell," *Motion Picture* 3 (Winter 1989–90): 17.

21 Ivone Margulies, "After the Fall: Peggy Ahwesh's Vérité," *Motion Picture* 3 (Winter 1989–90): 31, quoting Joel Fineman, "The History of the Anecdote: Fiction and Fiction," in H. Aram Veeser, ed., *The New Historicism* (New York: Routledge, 1989), p. 61.

22 Frampton's films include *Gloria* (1979) and *Public Domain* (1972); Ken Jacobs, *Tom Tom the Piper's Son* (1969); Ernie Gehr's *Eureka* (1974) and *History!* (1970). See Bart Testa, *Back and Forth: Early Cinema and the Avant-Garde*

(Toronto: Art Gallery of Ontario, 1992), and Noel Burch, "Primitivism and the Avant-Gardes: A Dialectical Approach," in *Narrative, Apparatus, Ideology,* ed. Philip Rosen (New York: Columbia University Press, 1986), pp. 483–506.

23 Quoted in Bill Wees, *Recycled Images: The Art and Politics of Found Footage Films* (New York: Anthology Film Archives, 1993), p. 97.

24 Paul Arthur, "The Last of the Last Machines?: Avant-Garde Film Since 1966," *Millennium Film Journal* nos. 16 / 17 / 18 (Fall–Winter 1986–87), p. 81.

25 Minh-ha, *When the Moon Waxes Red,* p. 33.

26 Another videomaker who has worked extensively in pixel-vision is Sadie Benning. She uses her bedroom as the site of many of her tapes, and her work is intensely personal, based in her experience as a young lesbian. In her case, the short range and low definition of the image is exploited for its sense of privacy and intimacy, and she has developed a kind of ethnography of the young girl's room.

27 Leslie Thornton and Trinh T. Minh-ha, "If Upon Leaving What We Have to Say: A Conversation Piece," in *Discourses: Conversations in Postmodern Art and Culture,* ed. Russell Ferguson, William Geander, and Marcia Tucker (New York: New Museum of Contemporary Art / MIT Press, 1990), p. 50.

28 Minh-ha, *When the Moon Waxes Red,* pp. 61–62.

29 Stephen A. Tyler, "Post-Modern Ethnography: From Document of the Occult to Occult Document," in *Writing Culture: The Poetics and Politics of Ethnography,* ed. James Clifford and George E. Marcus (Berkeley: University of California Press, 1986), p. 134.

Selective

Bibliography

Adair, Gilbert. *Hollywood's Vietnam: From "The Green Berets" to "Full Metal Jacket."* London: Heinemann, 1989.

Affron, Charles. *Cinema and Sentiment*. Chicago: University of Chicago Press, 1982.

Allen, Richard. *Projecting Illusion: Film, Film Theory, and the Impression of Reality.* Cambridge: Cambridge University Press, 1994.

Allen, Robert C., and Douglas Gomery. *Film History: Theory and Practice.* New York: Knopf, 1985.

Anderegg, Michael, ed. *Inventing Vietnam: The War in Film and Television.* Philadelphia: Temple University Press, 1991.

Arkoff, Samuel. *Flying Through Hollywood By the Seat of My Pants.* New York: Birch Lane, 1992.

Arrighi, Giovanni. *The Long Twentieth Century: Money, Power and the Origins of Our Times.* New York: Verso, 1994.

Auster, Albert, and Leonard Quart. *How the War Was Remembered: Hollywood and Vietnam.* New York: Prager, 1988.

Bach, Stephen. *Final Cut.* New York: William Morrow, 1985.

Baker, Fred, ed. *The Movie People.* New York: Douglas, 1972.

Balio, Tino. *Hollywood in the Age of Television.* Cambridge: Unwin Hyman, 1990.

Bart, Peter. *Fade Out.* New York: William Morrow, 1990.

Battock, Gregory, ed. *The New American Cinema: A Critical Anthology.* New York: Dutton, 1967.

Belton, John. *Widescreen Cinema.* Cambridge, Mass.: Harvard University Press, 1992.

Biederman, Donald, et al. *Law and Business of the Entertainment Industries.* Westport, Conn.: Praeger, 1996.

Bobo, Jacqueline. *Black Women as Cultural Readers.* New York: Columbia University Press, 1995.

Britton, Andrew. "Blissing Out: The Politics of Reaganite Entertainment." *Movie*, no. 22 (1984).

Bukatman, Scott. *Terminal Identity: The Virtual Subject in Postmodern Science Fiction*. Durham, N.C.: Duke University Press, 1993.

Burrough, Bryan. "The Siege of Paramount." *Vanity Fair*, February 1994.

Cagin, Seth, and Philip Dray. *Born to Be Wild: Hollywood and the Sixties Generation*. Boca Raton, Fla.: Coyote Press, 1994.

Carney, Ray. *The Films of John Cassavetes: Pragmatism, Modernism, and the Movies*. Cambridge: Cambridge University Press, 1994.

Carroll, Noel. "Address to the Heathens," *October* 23 (1982): 89–163.

Carson, Diane, Linda Dittmar, and Janice Welsch, eds. *Multiple Voices in Feminist Film Criticism*. Minneapolis: University of Minnesota Press, 1994.

Caughie, John. *Theories of Authorship*. London: Routledge, 1981.

Caughie, John, and Annette Kuhn, eds. *The Sexual Subject: A Screen Reader on Sexuality*. London: Routledge, 1992.

Chion, Michel. *David Lynch*. London: British Film Institute, 1996.

Chown, Jeffrey. *Hollywood Auteur: Francis Coppola*. New York: UMI Research Press, 1981.

Ciment, Michael. *Kubrick*. Translated by Gilbert Adair. New York: Holt, Rinehart and Winston, 1982.

Clifford, James, and George Marcus, eds. *Writing Culture: The Poetics and Politics of Ethnography*. Berkeley: University of California Press, 1986.

Clover, Carol. *Men, Women, and Chainsaws: Gender in the Modern Horror Film*. Princeton, N.J.: Princeton University Press, 1992.

Cohan, Steven, and Ina Rae Hark, eds. *Screening the Male: Exploring Masculinities in the Hollywood Cinema*. New York: Routledge, 1993.

Collins, Jim, Hilary Radner, and Ava Preacher Collins, eds. *Film Theory Goes to the Movies*. New York: Routledge, 1983.

Corrigan, Timothy. *A Cinema Without Walls: Movies and Culture After Vietnam*. New Brunswick, N.J.: Rutgers University Press, 1991.

Cowie, Peter. *Coppola*. New York: Scribners, 1990.

De Laurentis, Teresa. *Alice Doesn't: Feminism, Semiotics, Cinema*. Bloomington: Indiana University Press, 1984.

Deleuze, Gilles. *Cinema 1*. Translated by Hugh Tomlinson and Barbara Habberjam. Minneapolis: University of Minnesota Press, 1986.

Diawara, Manthia. *Black American Cinema*. New York: Routledge, 1993.

Ditmar, Linda, and Gene Michaud, eds. *From Hanoi to Hollywood: The Vietnam War in American Film*. New Brunswick, N.J.: Rutgers University Press, 1990.

Dery, Mark, ed. *Flame Wars: The Discourse of Cyberculture*. Durham, N.C.: Duke University Press, 1994.

Doane, Mary Anne. *Femmes Fatales: Feminism, Film Theory, Psychoanalysis*. New York: Routledge, 1991.

Doane, Mary Anne, Patricia Mellancamp, and Linda Williams, eds. *Re-Vision:*

Essays in Feminist Film Criticism. Frederick, Md.: University Publications of America, 1984.

Doty, Alexander. *Making Things Perfectly Queer: Interpreting Mass Culture.* Minneapolis: University of Minnesota Press, 1993.

Dunne, John Gregory. *The Studio.* New York: Limelight Editions, 1968.

Dyer, Richard. *Heavenly Bodies: Film Stars and Society.* New York: St. Martin's Press, 1986.

——. *Now You See It: Studies on Lesbian and Gay Film.* New York: Routledge, 1990.

Earnest, Olen J. "*Star Wars:* A Case Study of Motion Picture Marketing." *Current Research in Film: Audiences, Economics and Law.* Ed. Bruce A. Austin. Norwood, N.J.: Ablex, 1983.

Evans, Robert. *The Kid Stays in the Picture.* New York: Hyperion, 1994.

Farber, Manny. *Negative Space.* New York: Praeger, 1971.

Finch, Christopher. *Special Effects: Creating Movie Magic.* New York: Abbeville Press, 1984.

Fiske, John. *Media Matters.* Minneapolis: University of Minnesota Press, 1994.

Friedburg, Anne. *Window Shopping: Cinema and the Postmodern.* Berkeley: University of California Press, 1993.

Friedman, David F., and Don DeNevi. *A Youth in Babylon: Confessions of a Trash-Film King.* Buffalo, N.Y.: Prometheus, 1990.

Gelmis, Joseph. *The Film Director as Superstar.* Garden City, N.Y.: Doubleday, 1970.

Gomery, Douglas. "Corporate Ownership and Control in the Contemporary U.S. Film Industry." *Screen* 4 / 5 (1984): 60–69.

——. *Shared Pleasures: A History of Movie Presentation in the United States.* Madison: University of Wisconsin Press, 1992.

Goodell, Gregory. *Independent Feature Film Production: A Complete Guide from Concept to Distribution.* New York: St. Martin's Press, 1982.

Gorbman, Claudia. *Unheard Melodies: Narrative Film Music.* London: British Film Institute, 1987.

Griffith, Nancy, and Kim Masters. *Hit and Run: How Jon Peters and Peter Guber Took Sony for a Ride.* New York: Simon and Schuster, 1996.

Guerrero, Ed. *Framing Blackness: The African American Image in Film.* Philadelphia: Temple University Press, 1993.

Harvey, Sylvia. *May Sixty-Eight and Film Culture.* New York: New York Zoetrope, 1980.

Haskell, Molly. *From Reverence to Rape: The Treatment of Women in the Movies.* Baltimore: Penguin Books, 1974.

Hellman, John. *American Myth and the Legacy of Vietnam.* New York: Columbia University Press, 1986.

Herman, Edward S., and Noam Chomsky. *Manufacturing Consent: The Political Economy of the Mass Media.* New York: Pantheon, 1986.

Hillier, Jim. *The New Hollywood.* New York: Continuum, 1994.

Hoberman, J. *Vulgar Modernisms.* Philadelphia: Temple University Press, 1991.

Horton, Andrew, ed. *Comedy / Cinema / Theory.* Berkeley: University of California Press, 1991.

Jacobs, Diane. *Hollywood Renaissance.* Cranbury, N.J.: A. S. Barnes, 1977.

James, David. *Allegories of Cinema: American Film in the Sixties.* Princeton, N.J.: Princeton University Press, 1989.

James, David, ed. *To Free the Cinema: Jonas Mekas and the New York Underground.* Princeton, N.J.: Princeton University Press, 1992.

Jameson, Fredric. *Postmodernism, or, The Cultural Logic of Late Capitalism.* Durham, N.C.: Duke University Press, 1991.

Jeffords, Susan. *The Remasculinization of America: Gender and the Vietnam War.* Bloomington: Indiana University Press, 1989.

Jowett, Garth. *Film: The Democratic Art.* Boston: Little, Brown, 1976.

Jowett, Garth, and James Linton. *Movies as Mass Communication.* Newbury Park, Calif.: Sage, 1989.

Kael, Pauline. *I Lost It at the Movies.* Boston: Little, Brown, 1965.

Kaplan, E. Ann. *Women and Film: Both Sides of the Camera.* New York: Metheun, 1983.

Kent, Nicolas. *Naked Hollywood: Money and Power in the Movies Today.* New York: St. Martin's Press, 1991.

Kerr, Paul, ed. *The Hollywood Film Industry.* London: British Film Institute, 1986.

Keyser, Les. *Martin Scorsese.* Boston: Twayne, 1986.

Kolker, Robert. *A Cinema of Loneliness: Penn, Kubrick, Coppola, Scorsese, Altman.* New York: Oxford University Press, 1988.

Koszarski, Richard, ed. *Hollywood Directors, 1941–1976.* New York: Oxford University Press, 1977.

Kuhn, Annette, ed. *Alien Zone: Cultural Theory and Contemporary Science Fiction Cinema.* New York: Verso, 1990.

Lee, David, and Spike Lee. *Five for Five: The Films of Spike Lee.* New York: Forty Acres and a Mule, 1991.

LeGrice, Malcolm. *Abstract Film and Beyond.* Cambridge, Mass.: MIT Press, 1977.

Lewis, Jon. "Disney After Disney." *Disney Discourse.* Ed. Eric Smoodin. New York: Routledge, 1993, pp. 87–105.

———. *The Road to Romance and Ruin: Teen Films and Youth Culture.* New York: Routledge, 1992.

———. "Trust and Anti-Trust in the New New Hollywood." *Michigan Quarterly Review* 35, no. 1 (1995): 85–105.

———. *Whom God Wishes to Destroy.* Durham, N.C.: Duke University Press, 1995.

MacDonald, Scott. *Avant-Garde Film: Motion Studies.* New York: Cambridge University Press, 1993.

Madsen, Axel. *The New Hollywood.* New York: Thomas Y. Crowell, 1975.

Maltby, Richard. *Harmless Entertainment: Hollywood and the Ideology of Consensus.* New York: Pantheon, 1983.

Martin, Andrew. *Reception of War: Vietnam in American Culture.* Norman: University of Oklahoma Press, 1993.

Mayne, Judith. *The Woman at the Keyhole: Feminism and Women's Cinema.* Bloomington: Indiana University Press, 1990.

McCarthy, Todd, and Charles Flynn, eds. *Kings of the B's: Working Within the Hollywood System.* New York: Dutton, 1975.

McClintick, David. *Indecent Exposure.* New York: Dell, 1983.

McGee, Mark Thomas. *Roger Corman.* Jefferson, N.C.: McFarland, 1988.

McGilligan, Patrick. *Robert Altman: Jumping Off the Cliff.* New York: St. Martin's Press, 1989.

Mekas, Jonas. *Movie Journal: The Rise of a New American Cinema, 1959–1971.* New York: Macmillan, 1972.

Mellen, Joan. *Big Bad Wolves: Masculinity in American Films.* New York: Horizon, 1978.

Miller, Don. *B Movies.* New York: Curtis Books, 1973.

Miller, Mark Crispin, ed. *Seeing Through Movies.* New York: Pantheon, 1990.

Minh-ha, Trinh T. *When the Moon Waxes Red.* New York: Routledge, 1991.

Modleski, Tania. *Feminism Without Women: Culture and Criticism in a "Postfeminist" Age.* New York: Routledge, 1991.

——, ed. *Studies in Entertainment.* Bloomington: Indiana University Press, 1986.

Monaco, James. *American Film Now: The People, the Power, the Money, the Movies.* New York: New American Library, 1979.

Mott, Donald, and Cheryl McAllister Saunders. *Steven Spielberg.* Boston: Twayne, 1986.

Mulvey, Laura. *Visual and Other Pleasures.* Bloomington: Indiana University Press, 1989.

Neale, Steve. *Cinema and Technology: Image, Sound, Color.* Bloomington: Indiana University Press, 1985.

Nichols, Bill. *Blurred Boundaries: Questions of Meaning in Contemporary Culture.* Bloomington: Indiana University Press, 1994.

——. *Representing Reality: Issues and Concepts in Documentary.* Bloomington: Indiana University Press, 1991.

Noriega, Chon. *Chicanos and Film: Representation and Resistance.* Minneapolis: University of Minnesota Press, 1992.

Obst, Lynda. *Hello, He Lied, and Other True Tales from the Hollywood Trenches.* Boston: Little, Brown, 1996.

Parenti, Michael. *Inventing Reality: The Politics of the Mass Media.* New York: St. Martin's Press, 1985.

Paul, William. *Laughing Screaming: Modern Hollywood Horror and Comedy.* New York: Columbia University Press, 1994.

Penley, Constance, and Sharon Willis, eds. *Male Trouble.* Minneapolis: University of Minnesota Press, 1993.

Pfeil, Fred. *White Guys: Studies in Postmodern Domination and Difference.* New York: Verso, 1995.

Pierson, John. *Spike, Mike, Slackers and Dykes: A Guided Tour Across a Decade of American Independent Cinema.* New York: Miramax / Hyperion, 1995.

Pollack, Dale. *Skywalking: The Life and Films of George Lucas.* Los Angeles: Samuel French, 1990.

Prince, Stephen. *Visions of Empire: Political Imagery in Contemporary American Film.* New York: Praeger, 1992.

Prindle, David F. *Risky Business: The Political Economy of Hollywood.* Boulder, Colo.: Westview Press, 1993.

Pyle, Michael, and Myles, Linda. *Movie Brats: How the Film Generation Took Over Hollywood.* New York: Holt, Rinehart and Winston, 1979.

Quart, Barbara Koenig. *Women Directors: The Emergence of a New Cinema.* Westport, Conn.: Praeger, 1988.

Ray, Robert. *A Certain Tendency in Hollywood Film: 1930–1980.* Princeton, N.J.: Princeton University Press, 1985.

Reid, Mark. *Redefining Black Film.* Berkeley: University of California Press, 1993.

Renov, Michael, ed. *Theorizing Documentary.* New York: Routledge, 1993.

Rhines, Jesse. *Black Film, White Money.* New Brunswick, N.J.: Rutgers University Press, 1996.

Rogin, Michael. *Ronald Reagan: The Movie.* Berkeley: University of California Press, 1988.

Rosen, David, and Peter Hamilton. *Off-Hollywood: The Making and Marketing of Independent Films.* New York: Grove Weidenfield, 1990.

Rosenbaum, Jonathan. *Placing Movies: The Practice of Film Criticism.* Berkeley: University of California Press, 1995.

Rowe, John Carlos, and Rick Berg. *The Vietnam War and American Culture.* New York: Columbia University Press, 1991.

Russo, John A. *How to Make Your Own Feature Movie for $10,000 or Less.* New York: Barclay / Zinn, 1994.

Russo, Vito. *The Celluloid Closet: Homosexuality in the Movies.* New York: Harper and Row, 1981.

Ryan, Michael, and Douglas Kellner. *Camera Politica: The Politics and Ideology of Contemporary American Film.* Bloomington: Indiana University Press, 1988.

Sarris, Andrew. *The American Cinema: Directors and Directions, 1929–1968.* New York: Dutton, 1968.

Salamon, Julie. *The Devil's Candy: The Bonfire of the Vanities Goes to Hollywood.* Boston: Houghton-Mifflin, 1991.

Schatz, Thomas. "The New Hollywood." *Film Theory Goes to the Movies.* Ed. Jim Collins, Hilary Radner, and Ava Preacher Collins. New York: Routledge, 1983.

Schatz, Thomas. *Old Hollywood / New Hollywood: Ritual, Art and Industry.* Ann Arbor, Mich.: UMI Research Press, 1983.

Seydor, Paul. *Peckinpah: The Western Films.* Urbana: University of Illinois Press, 1997.

Sharrett, Christopher, ed. *Crisis Cinema: The Apocalyptic Idea in Postmodern Narrative Film.* Washington, D.C.: Maisonneuve Press, 1993.

Silverman, Kaja. *Male Subjectivity at the Margins.* New York: Routledge, 1992.

Simon, Art. *Dangerous Knowledge: The JFK Assassination in Art and Film.* Philadelphia: Temple University Press, 1996.

Sitney, P. Adams, ed. *The Avant Garde Film: A Reader in Theory and Criticism.* New York: New York University Press, 1978.

Smith, Paul. *Clint Eastwood: A Cultural Production.* Minneapolis: University of Minnesota Press, 1993.

Snead, James. *White Screens, Black Images.* New York: Routledge, 1992.

Stacey, Jackie. *Star Gazing: Hollywood Cinema and Female Spectatorship.* New York: Routledge, 1994.

Stern, Lesley. *The Scorsese Connection.* London: British Film Institute, 1996.

Steven, Peter, ed. *Jump Cut: Hollywood, Politics and Counter Cinema.* New York: Praeger, 1985.

Stone, Oliver, and Zachary Sklar. *JFK: The Book of the Film.* New York: Applause Books, 1992.

Taylor, John. *Storming the Magic Kingdom.* New York: Ballantine, 1987.

Telotte, J. P. *The Cult Experience: Beyond All Reason.* Austin: University of Texas Press, 1991.

Thompson, David, and Ian Christie, eds. *Scorsese on Scorsese.* London: Faber and Faber, 1989.

Thomson, David. *Overexposures.* New York: Morrow, 1981.

Waller, Gregory, ed. *American Horrors: Essays on the Modern American Horror Film.* Champaign: Illinois University Press, 1988.

Wasko, Janet. *Hollywood in the Information Age.* Austin: University of Texas, 1994.

Weddle, David. *If They Move . . . Kill 'Em: The Life and Times of Sam Peckinpah.* New York: Grove, 1994.

Weis, Elizabeth, and John Belton, eds. *Film Sound: Theory and Practice.* New York: Columbia University Press, 1985.

Williams, Linda. *Hard Core: Power, Pleasure, and the Frenzy of the Visible.* Berkeley: University of California Press, 1989.

Willis, Sharon. "The Fathers Watch the Boys' Room: Characters in Quentin Tarantino's Films." *Camera Obscura* 32 (1993 / 94): 40–73.

Wood, Michael. *America at the Movies.* New York: Columbia University Press, 1990.

Wood, Robin. *Arthur Penn.* New York: Praeger, 1969.

——. *Hollywood from Vietnam to Reagan.* New York: Columbia University Press, 1986.

Wyatt, Justin. *High Concept: Movies and Marketing in Hollywood.* Austin: University of Texas Press, 1994.

Yule, Andrew. *Fast Fade: David Puttnam, Columbia Pictures and the Battle for Hollywood.* New York: Delacorte, 1989.

Contributors

Sabrina Barton is an Assistant Professor of English at the University of Texas at Austin. She is currently working on a book-length project on the woman's psychothriller.

Scott Bukatman is an Assistant Professor at Stanford University, where he teaches courses on cinema and visual culture. He is the author of *Terminal Identity: The Virtual Subject in Postmodern Science Fiction* (Duke University Press, 1993) and a book-length study of Ridley Scott's *Blade Runner.*

David A. Cook is a Professor of Film Studies and the Director of the Film Studies Program at Emory University, where he has taught since 1973. He is the author of *A History of Narrative Film* and is currently at work on a history of the American film industry in the 1970s.

Timothy Corrigan is a Professor of English and Film Studies at Temple University. His books include *New German Film: The Displaced Image, Writing About Film,* and *A Cinema Without Walls.*

Ed Guerrero is an Associate Professor of Film and Literature at the University of Delaware and the author of *Framing Blackness: The African American Image in Film.* He has curated film programs for the Whitney Museum and the Smithsonian Institution's National Museum of Art.

Chuck Kleinhans is coeditor of *Jump Cut: A Review of Contemporary Media* and teaches in the Radio / Television / Film Department, Northwestern University. He is currently writing a book on U.S. independent film and video.

Jon Lewis is a Professor of English at Oregon State University. He is the author of *The Road to Romance and Ruin: Teen Films and Youth Culture* and *Whom God Wishes to Destroy . . . Francis Coppola and the New Hollywood* (Duke University Press, 1995).

Ivone Margulies is an Assistant Professor of Film and Media Studies at Hunter College in New York. She is the author of *Nothing Happens: Chantal Akerman's Hyperrealist Everyday* (Duke University Press, 1996).

Tania Modleski is a Professor of English at the University of Southern California and teaches courses in film, literature, and popular culture. Her books include *Loving with a Vengeance: Mass-Produced Fantasies for Women*, *The Women Who Knew Too Much: Hitchcock and Feminist Theory*, and *Feminism Without Women: Culture and Criticism in a "Postfeminist" Age*. She is working on a book on contemporary feminist revisions of popular genres.

Fred Pfeil is a writer / teacher / activist living in Hartford, Connecticut, working in English and American Studies at Trinity College. He is the author, most recently, of *White Guys*, a work of cultural criticism; and *What They Tell You to Forget*, a collection of short fiction.

Catherine Russell is an Associate Professor of Film Studies at Concordia University in Montreal. She has written *Narrative Mortality: Closure and New Wave Cinemas* and is working on a book about experimental ethnography in film and video.

Christopher Sharrett is an Associate Professor of Communication at Seton Hall University. He is the editor of *Crisis Cinema: The Apocalyptic Idea in Postmodern Narrative Film* and is currently working on a book to be entitled *Mythologies of Violence in Postmodern Media*. In 1976 he was a consultant to the House Select Committee on Assassinations of the U.S. Congress.

Justin Wyatt is an Associate Professor of Media Arts at the University of Arizona. He is the author of *High Concept: Movies and Marketing in Hollywood* and a book-length study of Todd Haynes's *Poison*.

Index

Library of Congress Cataloging-in-Publication Data

The new American cinema / edited by Jon Lewis.
p.   cm.
Includes bibliographical references and index.
ISBN 0-8223-2087-8 (cloth : alk. paper). — ISBN 0-8223-2115-7 (paper : alk. paper)
1. Motion picture industry—Economic aspects—United States.   2. Motion pictures—
Social aspects—United States.   I. Lewis, Jon, 1955–   .
PN1993.5.U6N47   1998
384′.83′0973–dc21   97-35210